Polymer Recycling and Waste Management

Polymer Recycling and Waste Management

Saeed Ostad Movahed

Ferdowsi University of Mashhad
Mashhad, Iran

WILEY

Published by John Wiley & Sons, Inc., Hoboken, New Jersey.

For general information on our other products and services or for technical support, please contact our Customer Care Department within the United States at (800) 762-2974, outside the United States at (317) 572-3993 or fax (317) 572-4002.

Wiley also publishes its books in a variety of electronic formats. Some content that appears in print may not be available in electronic formats. For more information about Wiley products, visit our web site at www.wiley.com.

Library of Congress Cataloging-in-Publication Data has been applied for:

Print ISBN: 9781394433070
ePDF ISBN: 9781394433094
ePub ISBN: 9781394433087
oBook ISBN: 9781394433100

Cover Design: Wiley
Cover Image: BOY ANTHONY/Shutterstock

Contents

Preface

The scientific community worldwide has reached its most critical point because environmental emergencies from plastic and rubber waste continue to worsen. The time has arrived for polymer waste management to become an absolute necessity for environmental protection.

My experience as a polymer engineer spanning multiple decades has shown me that the world needs complete and advanced solutions to handle polymer waste through recycling and reclamation methods. The book "Polymer Recycling and Waste Management" represents more than **20 years of *my research and development work*** in sustainable polymer processing. The book benefits from more than 50 peer-reviewed papers, multiple industrial partnerships, and patentable discoveries which I achieved through my academic work in polymer recycling. The book functions as a technical manual which provides researchers, engineers, and decision-makers with a strategic framework to enhance circularity within the circular economy.

Scope and Structure

The book is divided into 10 chapters and correspondent sections which follow the natural sequence of polymer recycling development from basic to advanced stages. Chapter 1 establishes the worldwide polymer waste crisis and its environmental effects and explains why recycling these materials is essential. The text examines waste prevention methods alongside sophisticated separation methods to demonstrate how material separation functions as the essential first step for successful downstream processing. Chapters 2–7 dedicate their entire content to rubber recycling because this field is facing significant technical obstacles because of vulcanized rubber's crosslinked structure. Chapter 2 explains rubber structures and chemistry before moving to devulcanization methods in Chapter 6 which include sulfur-based systems, peroxide crosslinking, and microwave- and chemical-assisted decrosslinking techniques. The chapter examines different types of rubber materials including NR, SBR, BR, and EPDM. Chapter 3 deals with morphological characteristics, filler interactions, and blend behavior as crucial factors in polymer waste management, along with useful case studies. The chapter contains multiple comparative tables and SEM/TEM morphological studies and industrial examples from real-world operations. In Chapter 4, the sources and classification of waste rubbers are discussed. Chapter 5 is dedicated to available strategies in rubbers recycling which may be extended to plastics recycling as well. Chapter 6 discusses waste rubber devulcanization, fundamentals and mechanisms. Chapter 7 reviews several case studies based on published papers by the author's research team in JCR journals.

Chapter 8 focuses on plastic separation techniques which serve as the fundamental basis for plastic recycling operations. The text examines traditional separation methods and new techniques which include flotation, dissolution, and triboelectric separation. The chapter demonstrates through experimental data and process diagrams that successful separation methods are essential for enhancing plastic recyclate quality. Chapter 9 presents industrial examples of high-performance engineering plastics and blends such as PC, POM, ABS, and other common plastics while providing detailed information about flotation optimization as a novel plastics separation technique. In Chapter 10, the economy, policy, and industrial perspective of polymer recycling have been examined. The chapter terminates by introducing microplastics global challenges, the role of artificial intelligence (AI) in polymer recycling and waste management, and future research directions in circular economy.

A Researcher's Contribution to a Greener Future

The book presents a combination of theoretical models, practical information, and analytical assessments based on my published research findings. The book's sections follow a systematic order which starts with basic concepts before advancing to practical applications to provide readers with detailed information that is easily understandable. The book contains original figures, microscopic images, chemical schematics, and tables which match current recycling industry and R&D laboratory operational practices.

The book serves as a motivational call to action in addition to its scientific content, i.e., the transition of nations and industries toward green manufacturing and circular economies. It reminds that polymer recycling is an essential requirement for businesses to achieve compliance and profitability and maintain social accountability.

This book motivates people from all backgrounds to become environmental leaders and protectors through their work as researchers, engineers, policymakers, and students.

April 2026

Saeed Ostad Movahed
Ferdowsi University of Mashhad
Mashhad, Iran

Acknowledgments

The completion of this work was made possible by the backing of my academic and industrial partners together with my students and colleagues. I want to express my deepest gratitude to my former MSc and PhD students who worked on devulcanization and flotation-based plastics separation projects and to my family members, especially my spouse, Elahe, who maintained her support throughout the writing period.

Saeed Ostad Movahed
Ferdowsi University of Mashhad
Mashhad, Iran

1

Introduction to Polymer Waste and Recycling Imperatives

1.1 Introduction

The term polymers come from Greek roots, i.e., poly meaning many and meros meaning parts, to describe their structure as extended chains of repeating molecular units. The discovery of celluloid by John Wesley Hyatt in 1869 marked the start of modern polymer development, although natural polymers like cellulose and rubber had been in use for centuries. The invention of Bakelite by Leo Baekeland in 1907 marked the beginning of synthetic plastics which transformed material science and enabled industrial-scale polymer manufacturing.

Synthetic polymer and plastic manufacturing have experienced explosive growth during the last 70 years. The production of polymers started small during the mid-20th century before reaching 460 million metric tons annually in 2019, and experts predict it will exceed 800 million metric tons by 2050 based on current projections, see Figures 1.1 and 1.2 (Karali et al., 2024). The combination of polymer properties, including low cost, high durability, and versatility, has led to their widespread use in packaging, automotive, electronics, construction, and textile applications.

The fast expansion of plastic production has established a major environmental crisis because of waste accumulation. The global plastic production and approximate waste flows (2019) diagram in Figure 1.3 shows how production starts at 460 MT in 2019 before moving through consumption and waste production into four main disposal routes. The worldwide recycling capacity reaches ~11%, but mechanical recycling methods dominate this process. The energy recovery process through incineration produces toxic emissions and greenhouse gases while releasing them into the atmosphere. Most of the waste ends up in landfills, which results in extended storage of waste materials. The remaining 20% of waste ends up in rivers and oceans and spreads across terrestrial ecosystems because of poor

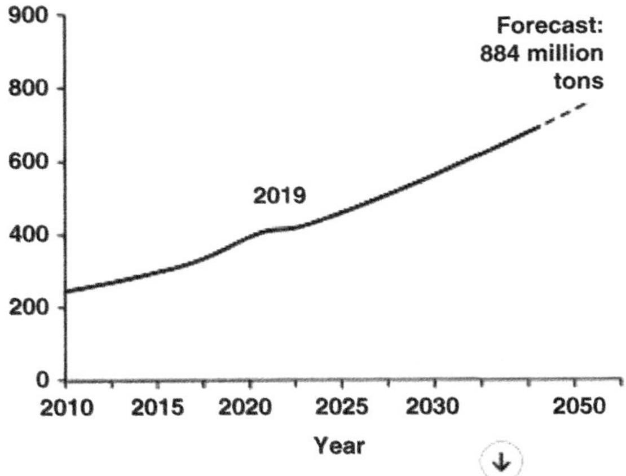

Figure 1.1 Annual production of polymers (2010–2025) and forecast for 2050. *Source*: Adapted from Karali et al. (2024).

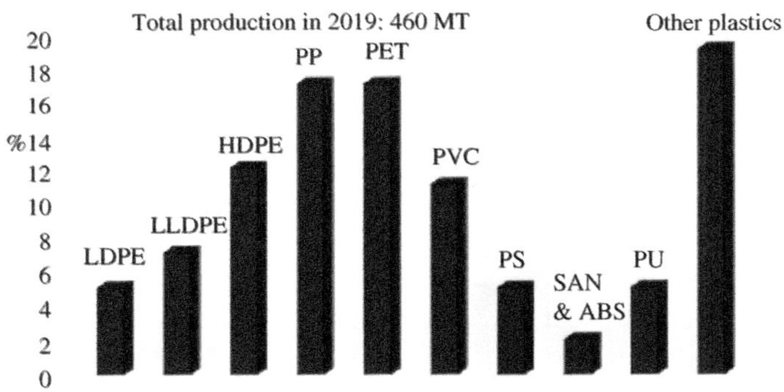

Figure 1.2 Annual production of selected polymers in 2019. *Source*: Adapted from Karali et al. (2024).

management. The ongoing poor management of plastic waste leads to microplastic pollution that affects both the environment and human health.

The long-lasting properties of polymers enable their use in different applications because they remain stable in the environment (Ghazi et al., 2022; Babakhani & Ostad Movahed, 2018). The recycling of thermosets and crosslinked elastomers faces significant obstacles because these materials show low sensitivity to recycling operations. The recycling process becomes harder to manage because mixed polymer streams contain various contaminants such as food waste, metal fragments, paper pieces, and chemical additives. Many nations in undeveloped or less developed regions lack proper polymer waste management systems because their

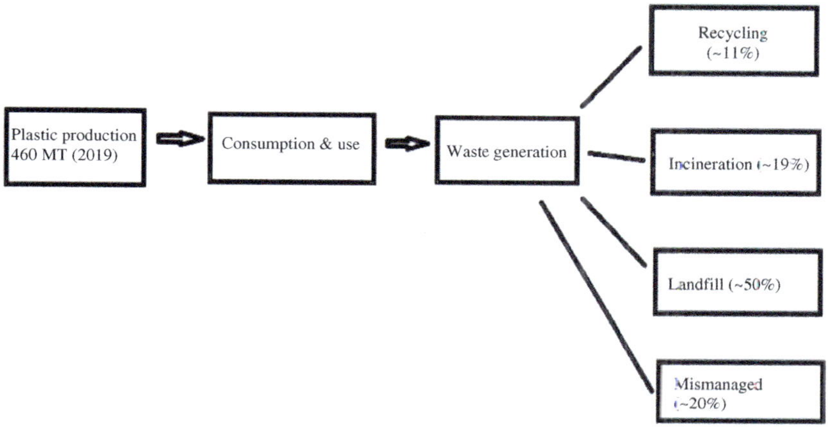

Figure 1.3 Global plastic production and waste flows (2019).

populations remain uninformed about environmental issues and their governments lack sufficient environmental regulations.

The disposal of polymers waste in developing nations occurs through three main methods which include open burning, river and ocean dumping, and unsorted landfilling. The combination of environmental damage and human health risks emerges from the practice of burning waste and dumping it without proper treatment. The World Bank indicates that 2 billion people across the globe lack sufficient waste collection services, and polymers make up a major portion of the uncontrolled waste in these areas.

The recycling performance in developed countries with modern infrastructure systems shows limited success. The United States manages to recycle about 9% of its plastic waste, but Europe reaches 30% recycling rates through Extended Producer Responsibility (EPR) schemes and deposit return systems and energy recovery from incineration. The actual recycling performance reveals fundamental problems within the system. The recycling process fails to produce usable products because waste materials in recycling bins end up in landfills or get downcycled because of contamination and sorting system limitations. The main problem emerges because designers fail to consider end-of-life strategies when creating products. Most plastic products lack built-in features for recycling purposes during their manufacturing process. The combination of multilayer packaging with pigmented polymers and incompatible resins makes recyclates less valuable and less economically feasible. The recycling process in technologically advanced systems produces high-quality results only for transparent PET (polyethylene terephthalate) bottles and a few other specific polymer streams.

The need for recycling practices became apparent when waste production reached high levels during the second half of the 20th century. The first recycling symbol emerged during the early 1970s, yet industrial-scale recycling programs started to develop during the late 1980s and 1990s when PET and HDPE (high-density polyethylene) polymers became widely recyclable. The environmental impact of unrecycled polymers continues to pose a severe threat to the environment. The combination of persistent landfill waste and ecosystem contamination and fossil fuel depletion and rising evidence of wildlife and human health damage create an immediate need for developing efficient recycling solutions. The practice of polymer recycling stands as an essential requirement for achieving sustainability. The development of improved recycling systems and innovative technologies must happen now to reduce plastic waste's impact on the environment while preserving materials for upcoming generations. Liu et al. (2024) represented the plastic solid waste (PSW) management by category mechanical/chemical recycling, incineration (not categorized as recycling), landfilling, and mismanagement expressed in million metric tons, based on the OECD ENV-Linkages model. The data presented for 2019 along with projections to 2060.

Also, Wimberger et al. (2024) showed the worldwide plastic production and recycling activities from 2017 with polymer distribution between HDPE, LDPE (low-density polyethylene), PP (polypropylene), PVC (polyvinyl chloride), PS (polystyrene), and PET. The chemical structure of these polymers determines their recyclability because green-coded polymers with carbon–carbon bonds differ from blue-coded polymers that contain heteroatoms.

1.2 Why Polymers Should Be Recycled?

The necessity of recycling polymers becomes evident through the following reasons.

- Environmental Protection
  As already stated (Figure 1.3), global plastic waste management shows that recycling reaches only up to 11% while incineration accounts for 19%, and landfills and environmental leaks make up the remaining 70%. The world has accumulated substantial plastic waste since the 1950s, which continues to stay in the environment. The discovery of microplastics in remote Antarctic areas proves that pollution spreads worldwide while causing severe damage to ecosystems, although scientists have not yet identified all potential long-term effects. The solution to this problem demands better recycling methods and circular systems which convert plastic waste into fresh material feedstocks. The main difficulty with thermoset materials stems from their

stable covalent crosslinked bonds, which make them resistant to natural degradation. These materials show resistance to environmental breakdown and remain insoluble in solvents, which leads to extended periods of environmental contamination. The manufacturing process of thermosets requires substantial fossil energy resources which generate additional greenhouse gases that contribute to climate change. The scientific field of polymer science faces an essential challenge to enhance the breakdown and recycling capabilities of thermosets. The production of plastics has experienced a 500% increase during the last several decades. The nonbiodegradable nature of petroleum-based plastics together with their disposal methods has resulted in rising oil usage and worsening environmental problems. The improper disposal of single-use items including masks and packaging materials and tissues has created substantial waste that enters both land and water ecosystems. The current environmental situation has motivated scientists to develop sustainable thermosetting polymers and bio-based materials for recycling purposes. The production of thermoplastic resins together with their applications experienced an unmatched growth during the last 70 years. Polyolefins together with polystyrenes and polyesters make up approximately 80% of the worldwide plastics market. The disposal methods for these materials at the end of their life cycle create significant environmental problems. The combination of environmental regulatory bodies and manufacturers of polymers works to develop new methods for transforming waste plastics into functional products. The implementation of sustainable practices helps decrease plastic waste while building circular economy systems and encouraging eco-friendly manufacturing technologies. The developed technologies help decrease greenhouse gas emissions while protecting limited resources and creating a sustainable environment for the future. The fast growth of additive manufacturing brings important benefits yet creates new concerns about extra plastic waste production. The negative effects of plastic pollution have become more apparent through time because of two main factors: the widespread disposal of disposable items and the expanding influence of microplastics on environmental systems and human wellness and food chain dynamics. As an illustration, the process of recycling 1 ton of PET results in more than 3 tons of CO_2 equivalent emissions reduction when compared to petrochemical feedstock production. Also, the process of devulcanizing rubber tires followed by their application in new formulations helps decrease both landfill waste and the need for synthetic rubber manufacturing.

The production methods of conventional manufacturing continue to generate substantial amounts of plastic waste which remains a major environmental concern.

- Preserving Natural Resources
  Modern lifestyle changes have led to increased energy usage, which speeds up fossil fuel depletion and worsens climate change effects while producing more air pollution. The world faces escalating environmental problems because plastic waste production has dramatically increased. The world uses 1 million plastic beverage bottles every minute and 5 million plastic bags annually. The large quantity of discarded plastics can be processed through advanced methods to produce fossil fuel fractions based on current statistics. The conversion of PSW into hydrocarbon fuels through available technologies provides an affordable solution for petroleum-based energy substitution. The conversion methods for PSW both eliminate environmental risks from waste plastics and decrease our need for crude oil resources. The circular use of polymers provides a solution to preserve limited natural resources and solve end-of-life problems, but new depolymerizable polymers need to match or surpass the functionality of present commercial plastics. The conversion of waste polyethylene into liquid aromatics at high yields stands out as a promising development. The industrial applications of these products such as surfactants and lubricants and refrigerants and insulating oils make waste plastic conversion into these substances a viable alternative to fossil fuel-based manufacturing. Plastic waste entering recycling facilities consists of two main categories which are thermoplastics and thermosets based on chemical classification. The crosslinked structure of thermosetting polymers makes them suitable for fuel and hydrocarbon feedstock production through pyrolysis or energy recovery through incineration. The two main categories of thermoplastics include addition polymers which cannot be reversed during polymerization and condensation polymers which include polyesters, polyamides, and polyurethanes (PU) that can be depolymerized through reversible condensation reactions. Plastic recycling serves as a dual-purpose solution which protects natural resources and minimizes plastic waste accumulation in the environment. The recycling process generates environmental issues because of transportation emissions and facility operations and chemical usage during treatment. The current plastic recycling rate worldwide stands at less than 14%, while most plastic waste ends up in landfills or gets incinerated which threatens both ecosystems and human health because plastic production keeps increasing. The recycling of polymers helps decrease the need for original raw materials needed in polymer production. The energy requirements for polymer recycling operations remain lower than those needed to create new polymers from petrochemical sources, thus delivering extra sustainability advantages.

- Economic Benefits
  Plastics brought functional and economic benefits to society after their introduction in the 1950s, but their manufacturing process and waste disposal create major environmental problems. The world will produce 1,050 million metric tons of plastic materials by 2060 according to projections which represent a threefold increase from the 2019 production level of 460 MT (Figure 1.2). The production of plastics continues to rise according to recent statistics, even though environmental damage from plastic use has become more widely recognized. The world's plastic production base shows that Asia leads as the largest producer since 2021 while China produces 32% of global output followed by North America at 18% and Europe at 15%. The use of postconsumer recycled (PCR) polymers in place of virgin resins provides industries with a competitive advantage through economic benefits. The use of recycled materials in production helps companies reduce their dependence on unstable crude oil markets while saving money on raw materials and building customer loyalty through sustainability initiatives. The market value of PCR plastics reached USD 7.76 billion in 2018, while experts predict a 10% annual growth rate until 2027. The North American plastic recycling industry will experience a 13.1% annual growth rate from 2020 to 2028 according to market projections. The US plastic recycling market reached USD 2.15 billion in 2022 size, and researchers predict its future development through the next 10 years.

  As stated, the plastic recycling sector grows mainly because of environmental consequences from petrochemical emissions and plastic waste problems and crude oil market price instability. The figures demonstrate that recycling and circular economy systems present substantial economic potential. The implementation of these approaches within circular economy frameworks enables industries to achieve maximum polymer value extraction while minimizing waste generation and environmental harm. The combination of cost-effectiveness with innovation and sustainability proves that recycling serves both environmental protection and industrial and economic stability for the long term.
- Legislation Frameworks
  Public health protection requires strict food safety regulations to control recycling operations that use waste materials. The laws establish specific guidelines to manage chemical substance migration into food products and microbial safety and food quality maintenance. The regulations enforce strict material tracking systems and establish recycling standards that require testing for contamination control. The certification process for recycled materials used in food contact applications follows different regulatory

approaches between regions, but all share the goal of verifying safety compliance with established standards. The European Commission operates through the European Food Safety Authority (EFSA), and the US Food and Drug Administration (FDA) maintains two prominent regulatory frameworks in Western jurisdictions. The European Union bases its long-term sustainability plan on recycling as its fundamental element. The European Union established its Plastics Strategy through the Circular Economy Action Plan (CEAP) as part of the European Green Deal to achieve climate neutrality by 2050 in 2018. The EU has established a goal to make all packaging sold in the market recyclable or reusable or compostable at affordable costs by 2030 while setting specific targets for using recycled materials in new packaging. The fast expansion of waste electrical and electronic equipment (WEEE) has emerged as a significant factor in the market. The implementation of REACH and POPs regulations has motivated both academic institutions and industrial organizations to develop safer flame-retardant alternatives and recycling methods for end-of-life plastics containing BFRs and dangerous additives. The market research shows that the halogen-free flame retardant (HFFR) market reached USD 4.129 billion in 2021, and experts predict an 8.5% annual growth rate until 2028. The regulatory frameworks for PET food packaging recycling face significant challenges because they lack clear standards despite PET being the most commonly used commodity plastic worldwide. International agencies use partial implementations of EFSA and FDA guidelines to establish minimum requirements for PET recyclability in food-contact applications because these guidelines lack complete international adoption. EU Regulation No. 10/2011 and Regulation (EC) No. 1935/2004 on food contact materials require risk assessments for nonintentionally added substances (NIAS) to protect human health. The European Union lacks specific laws that regulate thermoformed plastic packaging despite its widespread use.

The worldwide implementation of laws faces an especially difficult situation. The Asian region produces about half of all plastic materials, while the Shanghai ban of 2021 created international effects by blocking plastic exports to areas without equivalent waste management rules. The lack of proper waste management systems in sub-Saharan Africa and other developing regions makes their plastic waste problems worse. The scientific community supports waste governance participation from the public and demands a mandatory international legal system to defend human health and environmental protection from plastic waste. The existing international agreements including UNCLOS (United Nations Convention on the Law of the Sea) and MARPOL (International Convention for the Prevention of Pollution from Ships) and the London Convention on marine dumping have established pollution control requirements, but these measures have failed to stop marine plastic debris. The Basel Convention works to establish worldwide and regional systems for cleanup operations and monitoring activities. The 2019 Basel Convention amendment classified mixed and unrecyclable and contaminated plastic waste as hazardous materials which now restrict their export to developing nations and stop illegal cross-border waste transfers.

According to research from #PlasticFreeJuly and packaging retailer RAJA UK, India topped the list as biggest polluter worldwide with a staggering 126.5 million kg (about 278884430 lb.) each year that ends up in the ocean. The ocean received its highest amount of plastic pollution from China and Indonesia according to this research. The main countries which generate the most ocean plastic pollution include India (126.5 million kg), China (70.7 million kg), Indonesia (53.3 million kg), Brazil (38 million kg), Thailand (22.8 million kg), Mexico (3.5 million kg), Egypt (2.5 million kg), United States (2.4 million kg), Japan (1.84 million kg), and United Kingdom (0.703 million kg) (Source: *CIWEM*).

Extended Producer Responsibility (EPR) schemes represent an effective method to connect circular economy principles with legislative frameworks. The EPR framework transfers both financial and physical product management duties from local governments to producers and brand owners who must handle postconsumer waste. The implementation of EPR requires companies to design their products and choose materials which enable effective end-of-life recovery. The European Union implemented EPR for packaging waste management in 1994, which has become a leading policy tool for municipal waste control, according to the European waste hierarchy. The United States saw Maine become its first state to establish a packaging EPR law which created a significant example for other jurisdictions. The worldwide implementation of EPR systems shows promise to enhance waste management performance while promoting circular economy goals and environmental protection.

As a conclusion, the practice of recycling serves multiple purposes beyond its role in pollution management. The implementation of recycling leads to three main benefits: the practice of recycling helps decrease our need for extracting new fossil resources from the earth. The practice of recycling generates employment opportunities while driving forward industrial progress. The practice supports domestic economic growth while protecting people from exploitation in unregulated markets. The production of polymers generates decreased greenhouse gas emissions which stem from their manufacturing process. The successful implementation of polymer recycling demands collective effort between scientists and policymakers and manufacturers and public participation. The objective should focus on creating better designs and intelligent recovery systems and responsible usage practices rather than simply increasing recycling rates.

However, waste prevention and reduction (precycling) stands as the most effective sustainable waste management strategy because it tackles waste production directly at its origin instead of handling it after disposal. The waste hierarchy places prevention at the top position because upstream actions deliver superior environmental outcomes. The US Environmental Protection Agency (EPA) defines source reduction as the practice of using fewer materials to perform the same tasks which results in lower waste amounts and decreased treatment expenses and environmental damage. The practical application of this principle involves creating products with extended lifespans and repair capabilities which extend product use duration and decreases the need for new raw materials. Research shows that precycling actions performed by consumers at home effectively stop waste production through their use of reusable items, bulk purchases, and their reduction of packaging materials. The combination of these strategies demonstrates that waste prevention methods both support circular economy systems and directly protect natural resources while minimizing pollution and ensuring sustainability for extended periods.

1.3 What Are the General Strategies for Recycling?

The projections show that PVC and PP will continue to be the most used plastics until 2050, so these materials need to be the focus for future reduction and management plans. The wide range of industries including construction, automotive manufacturing, healthcare, packaging, and many more utilize PVC extensively. The market demand for PVC will rise 2.5 times to its current level throughout the next 30 years. The market demand for PVC keeps growing because it offers multiple benefits, including flexibility, durability, affordable prices, flame resistance, simple manufacturing, and assembly processes. The combination of strong chemical resistance, mechanical strength, and high melting point makes PP suitable for medical and laboratory equipment as well as food packaging, consumer goods, and textile applications.

The selection of recycling methods depends on the specific polymer backbone structure. The recycling process for plastics with carbon–carbon (C—C) bonds requires mechanical treatment or pyrolysis, but polymers with heteroatoms (C—X bonds in polyesters and polyamides) can be depolymerized chemically to obtain their monomers. The yearly worldwide market value of primary plastic materials exists in two main categories based on their backbone structure between C—C and C—X linkages. The classification for C—C includes polyethylene grades (high-density polyethylene (HDPE), low-density polyethylene (LDPE), and linear low-density polyethylene (LLDPE)), polyvinyl chloride (PVC), polypropylene (PP), polystyrene (PS), natural rubber (NR), and synthetic rubber (SR) with the total global annual consumption of 250.6 MMT/year. For C—X classification, polyethylene terephthalate (PET, polyurethanes (PU), nylons, unsaturated polyesters (UPE), and polycarbonate (PC), the correspondent amount was 102.6 MMT/year (Chen and Hu, 2024).

The four main strategies for polymer recycling include primary recycling which involves direct reuse, secondary recycling through mechanical reprocessing (grinding or size reduction), tertiary recycling through chemical depolymerization, thermolysis, and feedstock recovery, and quaternary recycling through incineration for energy production. The remaining plastics that do not get recycled end up in landfills or get burned in incinerators, or they remain uncontrolled in the environment.

The production of virgin-quality products through primary recycling needs scrap materials that are free from contamination. The new chemical recycling technology produces valuable monomers and fuels but requires significant amounts of energy to operate. The most common recycling method known as mechanical recycling faces two major problems which include downcycling and contamination of materials. The waste acceptance process of energy recovery takes up all types of waste but generates substantial emissions while producing minimal valuable products.

The development of new chemical recycling methods including thermolysis (pyrolysis), solvolysis, enzymatic depolymerization, and catalytic cracking has brought back interest in complete recycling systems for hard-to-recycle plastic materials including PET, PS, and multilayer films. The development of sustainable polymer management depends on three key factors: design for recycling innovations, biodegradable polymers, and advanced sorting systems that use AI and near-infrared spectroscopy.

1.4 What Is the Critical Role of Plastic Separation in Determining Recycling Strategies?

Plastic separation methods play an essential role in establishing the most suitable recycling approaches for plastic waste management. As stated, the rapid expansion of plastic manufacturing since the last several decades has brought numerous different polymer materials into the market for consumer and industrial applications. The numerous plastic types used in products enhance operational capabilities and reduce costs while extending product life yet make waste management processes more complicated. The success of recycling strategies depends heavily on effective plastic type and quality separation during the waste management process. The waste stream contains mixed industrial and postconsumer plastics which consist of various polymer types (PE, PP, PET, PS, PVC, PA, and ABS), additives (plasticizers, pigments, and UV stabilizers) together with contaminants (food waste, metals, dirt, adhesives). The current waste collection systems fail to sort plastic mixtures properly because they lack proper infrastructure, and people do not participate in source segregation programs. The recycling process becomes more complex because commercial products contain multiple plastic layers and blended polymers. The recycling process requires specific methods for each polymer stream after segregation based on material type and contamination levels and economic viability. The four-recycling strategy requires different levels of feedstock purity for each step while presenting distinct process characteristics and yield rates and environmental consequences. The recycling process of primary recycling, also known as re-extrusion or closed-loop recycling, transforms clean single-polymer waste into products that match virgin material standards. The recycling method requires pure plastic waste that includes industrial leftovers and sorted consumer packaging waste. The final product quality suffers when cross-contamination occurs because it decreases crystallinity, mechanical strength, and color quality. The secondary recycling process of recycling includes shredding followed by washing, melting, and then granulation of plastic waste to create new raw materials. The recycling process accepts various levels of contamination and polymer combinations but produces products with inferior quality. The lack of proper separation between materials results in decreased product properties which restricts their possible applications. The tertiary recycling process transforms polymers into monomers, base chemicals, and fuels through thermolysis, pyrolysis, and depolymerization methods. The process requires minimal polymer purity but needs high energy input while being vulnerable to specific contaminants. The thermal decomposition of PVC produces toxic gases which need proper treatment during pyrolysis. The quaternary recycling process includes plastic waste incineration and cement kiln coprocessing to extract energy from waste materials. The method accepts all types of plastic waste but faces criticism because it produces dioxins and heavy metals and microplastics during operation. The practice of energy recovery through incineration eliminates material value which goes against circular economy principles. Plastic separation quality determines the success rate of all recycling approaches. The implementation of sink-float separation and triboelectric methods, laser spectroscopy, and chemical tagging technologies at industrial scales enhances the efficiency of initial sorting operations. The implementation of advanced separation systems by policymakers and recyclers will create recycling systems that produce high-quality materials at affordable prices while maintaining environmental sustainability.

1.5 Economic and Environmental Context of Polymer Waste Management

The global management of polymer waste has become a major problem which links economic operations to environmental conservation and industrial regulatory systems. The production of polymers has experienced rapid growth during the previous 10 years because of increasing demand from packaging, automotive, textile, and electronic industries. The use of polymers allows for weight reduction and extended product lifespan and lower production expenses, but their disposal problems have become significantly worse. The modern circular economy requires governments, industries, and researchers to identify polymer waste management as their primary concern.

The management of polymer waste not only creates financial expenses but also generates potential benefits. The traditional linear production system which includes making and using and discarding products leads to environmental damage by sending waste outside the system while simultaneously straining both landfill space and local financial resources and supply chains. The worldwide economy experiences an annual loss of 80–120 billion USD because people discard plastics after short periods, which prevents their recycling into new products (Geyer et al., 2017). The system operates with low efficiency because most commodity plastics including PET, HDPE, PP, and PS retain their material value and heat energy after their initial use.

The recycling process becomes possible because new technologies enable us to use less natural resource-based petrochemical materials. The market value of oil-based polymers experiences regular fluctuations which makes recyclates more affordable than traditional materials when crude oil prices reach their highest point. The production of secondary raw materials through advanced recycling methods which include mechanical and chemical depolymerization creates new market opportunities. The recycling of plastics in Europe and East Asia has become a standard practice which enables these materials to enter supply chains for producing food packaging, construction materials, and automotive parts. The 2030 economic forecasts indicate that worldwide recyclate demand will increase by more than double when all current waste management laws including EPR and recycling standards and landfill bans become effective (Paloniitty and Ala-Lahti, 2024).

The cost structure of polymer recycling requires collection, sorting, separation, reprocessing, and quality assurance processes. The two most expensive operations at material recovery facilities (MRFs) are sorting and separation because they consume 40–60% of all facility expenses. The economic viability of recycling depends on efficient separation systems which include triboelectric, flotation, density separation, and advanced optical sorting. High-purity stream production enables recyclers to obtain high market values, which helps them decrease their need for unstable virgin polymer supplies. The economic case becomes stronger because of digitalization advancements and AI-based sorting systems which operate as sorting innovations. AI-powered vision systems and robotic pickers enable plants to increase their production speed while reducing defects, which leads to better financial performance. The economic models demonstrate that high-tech sorting infrastructure investments will generate sustained financial returns when supply levels stay constant, and recycling targets match regulatory requirements.

The assessment of polymer waste management systems now uses two evaluation methods which combine techno-economic analysis (TEA) with life-cycle assessment (LCA). The TEA system enables organizations to select the most cost-effective recycling approaches through its evaluation process. The LCA system allows organizations to evaluate environmental trade-offs between different options. The two tools operate together to assist decision-makers in selecting circularity-based strategies which achieve the highest circularity levels at the lowest possible environmental impact.

The TEA-LCA analysis shows that mechanical recycling provides the most cost-effective solution for recycling clean homogeneous waste materials including PET and HDPE bottles. The dissolution-based technologies demonstrate superior performance when processing contaminated waste streams which contain various polymer types. The chemical recycling process generates higher value from waste streams containing different polyolefin materials because it produces high-value monomers and fuels. The TEA-LCA framework enables complete optimization of the entire recycling system through its comprehensive optimization capabilities.

1.6 Closing Statement

The global polymer economy stands at a decisive moment according to the information presented throughout this chapter. The rapid growth of plastic usage requires immediate solutions because waste management systems remain inadequate, and plastic materials have complex structures. The current environmental and economic problems create an enormous chance for sustainable materials engineering development. The development of legislative frameworks and circular economy models requires scientists to study and enhance the basic operations of plastic waste management through precise polymer separation and efficient rubber recycling methods. The technical approaches play a vital role in creating closed-loop systems while making recycling technologies more efficient and cost-effective and larger in scale. The following chapters of this book will concentrate on two essential yet underdeveloped plastics separation (pre-recycling) and rubber recycling methods which included the author's original research about surface chemistry-based froth flotation methods that use depressants and microwave pretreatment to achieve high-purity separation of complicated polymer mixtures in waste streams. Also, the book presents the author's peer-reviewed research about devulcanizing waste rubber materials including numerous rubbers through chemical, thermos-chemical, and microwave-based methods and rheological testing and reprocessing techniques.

The following chapters present both scientific analysis and complete technical instructions which provide tested methods and evidence-based advice for experts who work with polymers including plastic and rubber waste recycling.

References

Babakhani, B., Ostad Movahed, S., Ghazi, S., and Ahmadpour, A. New formulation of polymeric separator gels for potential use in blood serum separator tubes. *Prog. Rubber Plast. Recycl. Technol.* 2018, *34*(1), 35–53.

Chen, S. and Hu, Y. H. Advancements and future directions in waste plastics recycling: from mechanical methods to innovative chemical processes. *Chem. Eng. J.* 2024, *493*, 152727.

Geyer, R., Jambeck, J. R., and Law, K. L. Production, use, and fate of all plastics ever made. *Sci. Adv.* 2017, *3*(9), e1700782.

Ghazi, S., Ahmadpour, A., Babakhani, B., and Ostad Movahed, S. Impact of a newly formulated silicone-based separator gel on clinical chemistry assays. *Iran. J. Chem. Chem. Eng.* 2022, *41*(7), 1–12.

Karali, N., Khanna, N., and Shah, N. *Climate Impact of Primary Plastic Production*: Lawrence Berkeley National Lab. Rep.; Lawrence Berkeley Natl. Lab.: Berkeley, CA, USA, 2024.

Liu, Q., Martinez-Villarreal, S., Wang, S., Tien, N. N. T., Kammoun, M., De Roover, Q., Len, C., and Richel, A. The role of plastic chemical recycling processes in a circular economy context. *Chem. Eng. J.* 2024, *498*, 155227.

Paloniitty, T. and Ala-Lahti, T. *The European Union and Plastics*. In Research Handbook on Plastics Regulation; Edward Elgar Publ.: Cheltenham, UK, 2024; pp. 187–207.

Wimberger, L., Ng, G., and Boyer, C. Light-driven polymer recycling to monomers and small molecules. *Nat. Commun.* 2024, *15*(1), 2510.

2

Fundamentals of Rubber Structure and Crosslinking Chemistry

2.1 Rubber: Structure and Types

The section will explain rubber's chemical and physical structure and describe natural and synthetic rubber types and their crosslinking properties and elastic behavior and practical applications.

2.1.1 Definition of Elastomers (Rubbers)

Rubbers and elastomers belong to a specific group of polymeric materials which exhibit superior elasticity and resilience properties. The term "elastomer" originates from "elastic polymer" because these materials demonstrate their main characteristic property of stretching under stress before returning to their original form after stress removal. The elastic properties of these materials stem from their extended polymer chains which maintain a light three-dimensional network structure that enables stretching and recoiling movements. Elastomers consist of macromolecular chains which exist as random coils when they are in their relaxed state at the molecular level. The chains extend their structure and arrange themselves in the direction of applied stress which leads to mechanical energy storage. The material returns to its original form after force removal because the chains automatically spring back to their original positions due to entropy forces. The elastic properties of elastomers differ from those of thermoplastics and crystalline materials because they derive their elasticity from entropy changes rather than enthalpic effects. The elastomer family consists of two main categories. The classical rubbers including natural rubber (NR) and synthetic rubbers (SBR [styrene–butadiene rubber], BR [butadiene rubber], and NBR [nitrile butadiene rubber]) fall under thermoset elastomers because they receive chemical crosslinks through vulcanization (sulfur crosslinking), peroxide, and resin (phenolic) processes. The crosslinking process makes these materials impossible to remelt or reshape which restricts their recyclability but provides improved thermal resistance and mechanical strength during operational performance. On the other hand and based on the reports of Madani and Ostad Movahed (2024) and Rahmani Naeim Abady and Ostad Movahed (2017), thermoplastic elastomers (TPEs) use physical crosslinks which exist as crystalline domains or block copolymer microphases (SBS, SEBS, and TPU). The combination of rubber-like elasticity with thermoplastic processing characteristics in TPEs makes them suitable for recycling operations. The mechanical performance of TPEs remains inferior to thermoset rubbers when subjected to challenging applications. Table 2.1 compares thermoplastic with thermoset elastomers. It is evident that the rubber part of TPE has crosslinking potential, although, in most applications, it is used without crosslinking (curing).

A polymer needs to fulfill these specific requirements to qualify as an elastomer:

- The material needs to have a glass transition temperature $\left(T_g\right)$ that remains below room temperature to enable chain flexibility and mobility.
- The material needs to have a suitable level of crosslinking which can be physical or chemical to stop plastic flow from occurring.
- The material needs to demonstrate high extensibility which reaches up to 500–1,000% strain in specific cases.

Table 2.1 Thermoplastic versus thermoset elastomers.

Feature	TPEs	Thermoset elastomers
Crosslink type	Physical (reversible)	Chemical (irreversible)
Recyclability	High	Low
Processing	Remeltable and moldable	Requires vulcanization
Reusability	Easily reprocessed	Not reprocessable after curing
Heat resistance	Moderate	Excellent
Mechanical properties	Moderate (depends on formulation)	Excellent (particularly fatigue and tear)
Typical applications	Soft grips, footwear, consumer goods	Tires, hoses, seals, automotive parts
Examples	SBS, SEBS, TPU, TPO, TPV	NR, SBR, NBR, EPDM, IIR
Environmental friendliness	Favorable due to easier recycling	Requires special handling or devulcanization
Cost of processing	Generally lower due to ease of molding	Higher due to curing and labor steps

- The material needs to recover its original shape completely and quickly after undergoing deformation.
- The material needs to perform well throughout multiple stress cycles without showing signs of fatigue.

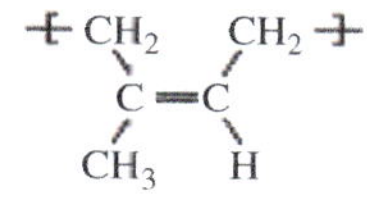

Figure 2.1 The chemical structure of a typical rubber polymer (*cis*-1,4-polyisoprene, IIR).

The properties of elastomers depend on several main factors including polymer molecular weight, branching level, unsaturation degree, crosslink density, filler content, and additive composition including plasticizers, antioxidants, and processing agents. Figure 2.1 represents the chemical structure of a typical rubber polymer (*cis*-1,4-polyisoprene, IIR).

Rubbers derive their elasticity mainly through changes in entropy rather than enthalpy variations which metals and crystalline plastics experience. The entropic elasticity theory explains this phenomenon through statistical mechanics models of polymer chain arrangements. The force required to stretch rubbery chains grows stronger when their configurational entropy decreases because this change creates the thermodynamic force which drives the material to return to its original state. The unique property of rubber distinguishes it from standard engineering materials because of this behavior. The shear modulus G of lightly crosslinked rubber networks depends directly on the density of elastically effective network chains (ν) according to the fundamental rubber elasticity equation:

$$G = \nu RT$$

The relationship between shear modulus G and temperature T and universal gas constant R serves as the foundation for various analytical models in rubber elasticity and enables rubber formulation design for specific mechanical properties.

The physical characteristics and manufacturing capabilities of rubbery polymers depend heavily on their molecular weight (M_w) and their chain entanglement properties (Jaberi Mofrad et al., 2025). The mechanical properties of elastomers including elasticity and viscosity and strength depend on these two essential molecular characteristics which affect both uncured and vulcanized states. The mass of a single polymer molecule determines its molecular weight which scientists express through number average (M_n), weight average (M_w), and z-average molecular weight (M_z). The high molecular weight of rubber polymers between 100,000 and 1,000,000 g/mol enables them to create flexible chains which stretch and store energy effectively. Material behavior depends on molecular weight according to the following pattern:

- Rubbers with low molecular weights below 100,000 g/mol show weak mechanical properties and low viscosity which makes them unsuitable for weight-bearing applications.

- Rubbers with high molecular weights show better tensile strength and elongation at break and improved abrasion resistance.

The processing behavior of materials becomes more predictable when molecular weight distribution is narrow and poly dispersion index, PDI, closes to 1 but wider distributions enhance blending and mixing operations through better flow properties.

The entanglement molecular weight (M_e) serves as a critical threshold which determines when rubber chains begin to physically link with each other. The physical network of long polymer chains creates chain entanglement through their physical connection points. The physical connections between polymer chains in un-vulcanized rubbers function like chemical crosslinks to enhance both elasticity and mechanical strength. In elastomers, the physical network of chain entanglements stops the material from flowing when it remains stationary which helps maintain its shape. The stretching process of rubber allows entangled chains to move past each other which enables elastic deformation and enhances material toughness. The density of entanglements between polymer chains depends on both chain length and flexibility because longer flexible chains produce higher entanglement densities. The noncovalent nature of chain entanglement distinguishes it from chemical crosslinking because shear stress can break these connections but they will reform after stress removal. The behavior of rubber under shear stress leads to non-Newtonian flow characteristics which result in shear thinning during processing operations. The rubbery plateau region in dynamic mechanical analysis (DMA) shows the elastic behavior of rubber because of chain entanglements before the material starts to flow. The structure of rubber molecules which includes linear or branched or star-shaped configurations determines both entanglement density and final material properties. Polyisoprene and polybutadiene exhibit high potential for entanglement because they exist as linear polymers. The processability of materials improves when using branched structures although their entanglement density decreases. Star-shaped and dendritic rubber structures enable users to control viscoelastic properties and achieve lower melt viscosity which benefits specific blending operations. Figure 2.2 illustrates the entangled polymer chains in a rubber.

Figure 2.2 The entangled polymer chains in a rubber.

The processability of rubber recycling depends heavily on molecular weight and entanglement density between polymer chains. Rubbers with high molecular weight and entangled structures become difficult to dissolve and flow which makes their reprocessing challenging. Devulcanization techniques need to break sulfur crosslinks while maintaining control over chain scission to prevent excessive molecular weight reduction in the reclaimed rubber. The evaluation of M_w and gel content during reclaiming and devulcanization operations becomes crucial because these measurements determine the final mechanical properties of reclaimed rubber.

The elastic properties of rubber materials result from their molecular structure which includes the arrangement of carbon–carbon double bonds (C=C) in the polymer backbone. The mechanical properties of elastomeric materials depend on the geometric isomerism between cis and trans configurations of the double bonds in their polymer backbone. The double bonds in diene units of polyisoprene and polybutadiene rubber create sites for unsaturation that can form either cis or trans configurations. The substituent groups on the double bond appear on the same side in cis configuration. The substituents in trans configuration are positioned on opposite sides of the double bond. The spatial arrangement of polymer chains and their crystallization potential depend on the distinction between these configurations. The kinked structure of cis isomers in NR with > 98% *cis*-1,4-polyisoprene prevents chain packing which results in high flexibility, low crystallinity at room temperature, excellent stretchability, and recovery (key for elastomers). The linear structure of trans isomers in gutta-percha and *trans*-1,4-polyisoprene allows for close packing which results in higher crystallinity, less flexibility, and more rigid behavior at ambient conditions. Rubbers with cis configuration show elastic properties while trans-configured rubbers behave as thermoplastic semicrystalline materials when exposed to the same conditions. The cis/trans ratio determines how rubber materials respond to stretching forces and their resulting strength and resistance to fatigue: The strength of NR increases through strain-induced crystallization because it contains mostly cis configurations. The high modulus and stiffness of trans-dominated rubbers result from their tendency to crystallize easily but they show poor flexibility. The distinct properties of cis-polymers and trans-polymers lead to their specific uses in engineering applications such as tire production and vibration isolation and flexible coupling systems. Trans-polymers find application in rigid insulating materials and medical devices and adhesive applications. The vulcanization process which creates finished rubber articles depends on the cis/trans ratio because it determines how the rubber reacts to sulfur or peroxide systems. Cis-polymers easily react with sulfur or peroxide systems to create flexible network structures. The formation of desirable network densities in trans-polymers requires either extreme processing conditions or specific accelerators during processing. The processing characteristics of isomer content affect compound design strategies in industry because they influence melt flow and tackiness and mixing behavior. Figure 2.3 compares the chemical structure of *cis*-1,4-polyisoprene with *trans*-1,4-polyisoprene.

The amorphous molecular structure of elastomers and rubber materials consists of randomly arranged polymer chains which differ from the crystalline structure

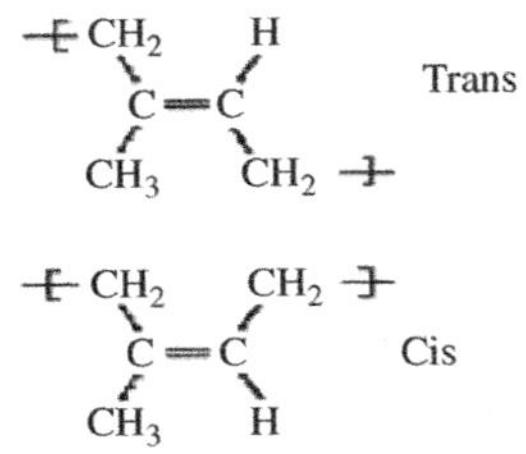

Figure 2.3 The chemical structure comparison of *cis*-1,4-polyisoprene with *trans*-1,4-polyisoprene.

found in semicrystalline polymers, plastics. The amorphous structure of rubber materials enables their flexibility and elasticity because polymer chains in this state maintain high freedom of movement when under stress. The glass transition temperature (T_g) stands as a vital thermal characteristic of amorphous polymers (Ostad Movahed et al., 2009a) which defines their transition from rigid glassy to flexible rubbery state. The glass transition temperature (T_g) marks the point where polymers shift from their rigid glassy state to their elastic rubbery state. The polymer chains remain rigid and immobile at temperatures below T_g, but they gain enough thermal energy to slide past each other at temperatures above T_g which enables rubbery material deformation. Elastomers require their glass transition temperature to stay below operational temperatures because this condition maintains rubbery behavior throughout usage. NR maintains its rubbery properties at room temperature because its glass transition temperature reaches −70 °C. Polymers with T_g values near or above room temperature (such as polystyrene at 100 °C) exhibit plastic behavior at standard conditions.

The glass transition represents a second-order transition which occurs through a gradual process of increasing molecular movement instead of a distinct phase transformation like melting. The glass transition occurs when polymer backbone segments start to move freely as a coordinated unit. The backbone flexibility, side group size, molecular weight, and crosslink density determine the glass transition temperature (T_g) of a material. The T_g value of polyisoprene remains low because its flexible backbone structure. The presence of large side groups in polystyrene leads to a higher T_g value. The degree of crosslinking determines chain mobility which directly affects the glass transition temperature (T_g).

The process of rubber formulation heavily depends on T_g control because it determines the final product properties. The selection of appropriate elastomers with suitable T_g values or polymer blending enables formulators to create materials with specific mechanical properties for different temperature ranges of temperature.

The following examples demonstrate how different rubber materials perform in various applications:

- The low-temperature flexibility of butyl rubber (IIR) stems from its T_g value which reaches −60 °C.
- The T_g value of SBR exists between −50 and −60 °C which enables it to provide both durability and elasticity for tire tread applications.

- The addition of plasticizers and fillers together with crosslinking agents enables the adjustment of T_g for particular uses.

2.1.2 Sulfur Vulcanization (Conventional, EV, and SEV Systems)

The chemical transformation of elastomeric materials into durable rubber through polymer chain crosslinking forms the basis of vulcanization (Abgorg and Ostad Movahed, 2024). The process of sulfur vulcanization which Charles Goodyear discovered in 1839 continues to be the leading method for crosslinking diene-based rubbers including NR, SBR, BR, and NBR. The vulcanization process creates rubber products with enhanced elastic recovery, mechanical strength, chemical resistance, and thermal stability, which makes them suitable for demanding applications including tires, industrial goods, and seals.

Sulfur vulcanization follows a specific chemical reaction mechanism. The process requires heat and accelerators to activate sulfur species which then react with diene rubber unsaturated sites containing carbon–carbon double bonds. The sulfur crosslink network consists of different bond lengths and types which include monosulfidic, disulfidic, and polysulfidic bonds (Ansarifar et al., 2013).

Sulfur vulcanization consists of three main stages which start with initiation followed by propagation and then crosslinking. The activation of sulfur through accelerators (thiurams and sulfenamides) produces reactive sulfurating agents which start the process. The reactive species in the process target both allylic hydrogen atoms and double bonds that exist within the rubber backbone structure. The sulfur bridges link different polymer chains together to create an enduring network structure. Figure 2.4 represents the proposed vulcanization (sulfur curing or crosslinking) mechanism of unsaturated, i.e., polyisoprene rubbers.

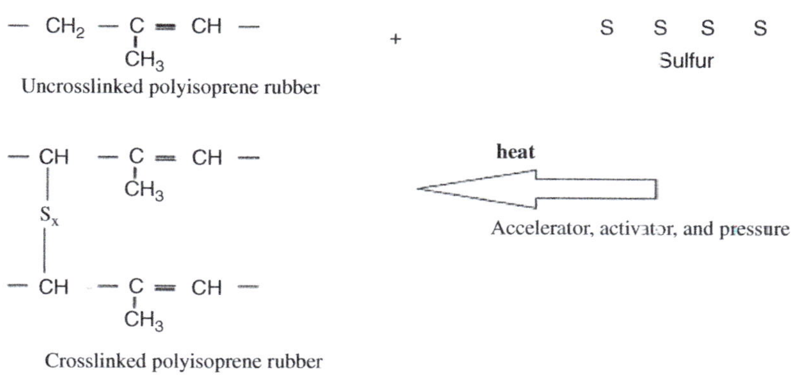

Figure 2.4 The proposed vulcanization (sulfur curing or crosslinking) mechanism of unsaturated, i.e., polyisoprene rubbers.

The selection of sulfur-accelerator ratios enables users to create different vulcanization systems which produce distinct crosslink densities and types. Types of sulfur vulcanization systems include (Ostad Movahed et al., 2015):

- Conventional vulcanization (CV): the sulfur to accelerator ratio in this system exceeds 2.5. The crosslink structure consists mainly of polysulfidic bonds. The material exhibits excellent flexibility together with strong resistance to dynamic fatigue. The applications for this material include tire sidewalls, conveyor belts, and rubber mounts. The heat stability of polysulfidic bonds remains poor because they break down when exposed to heat.
- Effective vulcanization (EV): the sulfur to accelerator ratio in this system reaches below 0.4. The crosslink structure contains mostly monosulfidic bonds. The material exhibits excellent heat and oxidative stability and maintains high modulus values. The applications for this material include heat-resistant products, industrial rubber goods, and gaskets. The material shows reduced resistance to fatigue and lower resilience in its properties.
- Semi-effective vulcanization (SEV): the sulfur to accelerator ratio in this system reaches approximately 1. The crosslink structure contains a mix of monosulfidic, disulfidic, and polysulfidic bonds at balanced levels. The material shows intermediate properties which combine heat resistance with flexibility. The material serves applications in tire tread production, seal manufacturing, and dynamic system operation. The material shows minimal compromise between its dynamic performance and its aging characteristics.

Table 2.2 compares the sulfur curing systems, CV, EV, and SEV in brief.

Table 2.2 Comparison of EV, SEV, and CV vulcanization systems.

Feature	CV	SEV	EV
Sulfur/accelerator ratio	High (>2.5)	Intermediate (~1)	Low (<0.4)
Crosslink type	Predominantly polysulfidic	Mix of poly- and mono-/disulfidic	Mainly monosulfidic
Cure time	Short	Moderate	Long
Heat aging resistance	Low	Moderate	High
Reversion resistance	Low	Moderate	High
Mechanical properties	High tensile and tear strength	Balanced performance	High modulus and hardness
Application areas	General-purpose products, tires	Dynamic products	High-temperature applications

2.1.3 Peroxide and Radiation Crosslinking

Peroxide (Imanifar et al., 2018) and radiation crosslinking serve as two essential crosslinking methods which work best for rubbers that do not react with sulfur systems (such as EPDM and silicone rubbers and fluoro-elastomers) and need heat and oxidative stability. The elastomer matrix obtains superior thermal resistance and oxidative durability through C—C crosslinks which these methods generate.

Organic peroxides including dicumyl peroxide (DCP) and benzoyl peroxide (BPO) decompose thermally to produce highly reactive free radicals. The free radicals produced from decomposition initiate crosslinking reactions by removing hydrogen atoms from polymer chains to form C—C bonds as shown in the following sequence:

Initiation: RO—OR → 2 · RO. The peroxide fragment exists as RO in this reaction. Propagation: RO· + R—H → ROH + R· and R· + R· → R—R. The polymer chain segment R—H serves as the starting point for this reaction.

Peroxide crosslinking produces C—C bonds, which maintain their stability at temperatures up to 200 °C, thus enabling peroxide-cured rubbers to function in hot and oxidative conditions. The process eliminates the requirement for sulfur-based additives which reduces the formation of unpleasant odors and prevents corrosion and blooming. The process produces unwanted by-products which can cause color and odor changes (such as acetophenone from DCP). The process fails to work properly crosslink rubbers that contain easily oxidizable components like NR. The cure rate depends heavily on both the decomposition speed of peroxides and their uniform distribution throughout the material. The process applies to EPDM, EVM (ethylene vinyl acetate rubber), HNBR, and silicone rubber materials.

The process finds application in automotive radiator hoses, wire insulation and high-voltage cable production.

The process of radiation crosslinking uses ionizing radiation from Co-60 gamma rays or electron beams (EB) to create free radicals which then form permanent crosslinks without needing chemical initiators. The process involves two main steps which include chain scission and radical recombination. The process forms C—C bonds at points where crosslinks occur. The process occurs in N_2 environments to stop oxidative breakdown from happening. The process operates without catalysts and produces no by-products. Also, the process allows users to achieve exact crosslinking levels through radiation dose adjustments. However, the process requires protection from radiation through specialized equipment and facilities. As another disadvantage, the process requires specialized equipment for radiation protection and acceleration systems. The process faces challenges because EBs have restricted penetration ability when working with thick materials. The process shows different effectiveness on various rubber materials because radiation can cause chain scission in specific polymers. The process finds application in polyethylene-insulated wire production as well as heat-shrink tubing, medical device manufacturing, and sterilizable rubber component creation.

2.1.4 Resin (Phenolic) Crosslinking

The thermally activated chemical curing process of alkyl-phenolic resins with catalysts creates C—C crosslinks between polymer chains at elevated temperatures. The vulcanization process through phenolic crosslinking serves as a crucial method for halogenated rubbers including chlorobutyl rubber (CIIR) and bromobutyl rubber (BIIR) because these materials do not respond well to standard sulfur-based curing systems.

The methylol (—CH_2OH) groups in phenolic resins enable them to undergo electrophilic substitution reactions with halogenated elastomers during crosslinking. The rubber chain allylic halogen sites on rubber molecules serve as the main reaction points for phenolic resin to create ether bonds and carbon–carbon crosslinks.

The crosslinking process uses Novolac resin as its alkylated phenol resin component. The crosslinking process requires metal oxides (ZnO, MgO) or Lewis acids (stannous chloride) as activators. The condensation reaction needs temperatures between 150 and 180 °C to initiate the process. The reaction between R—CH_2—Br and phenolic resin produces a crosslinked network through heat treatment.

Typical applications:

- The application of tubeless tire inner liners represents a primary use case for this material.
- The material serves as a component for hoses and gaskets and diaphragms which operate in automotive and chemical processing settings.
- The material finds application in high-performance seals and bladders which need to withstand oil exposure and heat and gas exposure.

Table 2.3 compares sulfur (vulcanization), peroxide, and phenolic resins curing (crosslinking).

2.2 Classification of Rubber Types

2.2.1 Natural Rubber

NR exists as a high molecular weight elastomer which scientists extract from *Hevea brasiliensis* tree latex (Figure 2.5) that originates from South America but now grows throughout tropical regions including Southeast Asia

Table 2.3 Sulfur (vulcanization), peroxide, and phenolic resins curing (crosslinking) comparison.

Feature	Sulfur system	Peroxide system	Phenolic system
Crosslink type	Sulfur bridges	C—C bonds	Ether/C—C bonds via phenol
Suitable rubbers	NR, SBR, BR	EPDM, HNBR, silicone	CIIR, BIIR, IIR
Thermal stability	Moderate	High	High
Cure speed	Fast	Moderate	Slow
Color/staining	Can stain	May yellow	Clean, nonstaining
Postprocessing smell	Possible (due to sulfur)	Possible (peroxide by-products)	Negligible
Key applications	Tires, general rubber goods	Wires, automotive hoses	Tire inner liners, diaphragms

(Thailand, Indonesia, and Malaysia), Africa, and South American territories (Ostad Movahed et al., 2008).

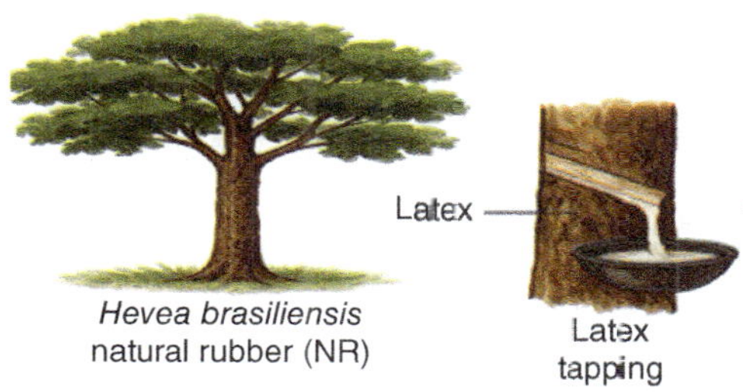

Figure 2.5 *Hevea brasiliensis* tree latex.

The latex contains 25–35% rubber resins and water in a milky colloidal suspension. The rubber extraction process starts with tree bark hydrocarbon (*cis*-1,4-polyisoprene) along with proteins and carbohydrates and incisions for latex collection through a method known as tapping. The latex undergoes acid coagulation with formic or acetic acid to create rubber precipitation. The extraction process of coagulated rubber results in sheet or block forms which become ready for industrial use. Key characteristics of NR are as follows:

- Molecular structure: the high elasticity and tensile strength of NR stem from its composition of *cis*-1,4-polyisoprene which makes up nearly 100% of its molecular structure. The amorphous state of the polymer chains enables their high flexibility because of their predominantly cis configuration.
- Physical properties: the mechanical properties of NR include superior tear, abrasion resistance, elasticity, and dynamic fatigue performance. The rubber exhibits better low-temperature flexibility than most synthetic rubber materials.

- Biodegradability: the natural composition of NR together with its chemical makeup enables it to undergo partial biodegradation which sets it apart from synthetic elastomers.
- Applications: the main applications of NR include tire production for trucks and aircraft as well as conveyor belts, vibration dampers, hoses, and adhesives. The material finds its best use in applications which need both high mechanical strength and elastic properties.
- Limitations: NR shows several weaknesses because it fails to resist ozone, UV light, oils, and oxidative aging processes. The material requires stabilizers and antioxidants to maintain its performance when exposed to high heat or chemical aggression.

The main component of NR consists of *cis*-1,4-polyisoprene which demonstrates outstanding green tack properties through its ability to stick to itself without any external pressure or adhesive (Ostad Movahed et al., 2009a). The manufacturing sector heavily relies on this tacky property because it enables the assembly of multilayer structures before vulcanization in tire building and rubber lamination operations. The specific property of raw or lightly compounded materials which exhibits this characteristic is known as high tack. The green tack properties of NR reach their peak among all elastomers which makes it essential for multi-ply assembly operations. The molecular structure and physical properties of NR create its high tackiness through the following factors:

- The cis configuration of NR molecules creates flexible chains that enable better molecular contact between them.
- The high molecular weight of NR combined with its narrow molecular weight distribution leads to extensive chain entanglements and cohesive interactions.
- The surface adhesion properties of raw NR become more effective because *H. brasiliensis* latex contains proteins and lipids.
- The uncured state of NR maintains its amorphous structure which enables rubber chains to move freely and create intermolecular bonds between adjacent rubber sheets.

The measurement of tack exists in distinct forms. The first type of tack measurement occurs before curing and is known as green tack. The measurement of tack uses three different testing methods which include rolling ball tests, probe tack tests, and peel adhesion tests. Quality control of raw rubber batches depends on tack testing because it determines their suitability for tire carcass construction and conveyor belt layering applications. The tire industry requires high tack properties for its operations. The process of rubber sheet lamination requires high tack properties. The internal sheet adhesion of carcass and tread layers depends on NR blends which provide strong bonding properties. The high tack properties of NR enable manufacturers to operate at increased speeds while reducing bonding failures which results in better product quality and higher production rates.

2.2.2 Synthetic Rubbers

2.2.2.1 Styrene–Butadiene Rubber

SBR stands as the leading synthetic rubber worldwide because it dominates tire production for passenger cars and light trucks as well as other rubber applications. The synthetic rubber compound consists of styrene and butadiene (BD) polymers which duplicate NR properties while delivering superior durability, aging performance, and manufacturing flexibility. The chemical composition of SBR contains between 23% and 30% styrene and between 70% and 77% BD which creates a flexible material with elastic properties. The ratio between styrene and BD determines how hard the rubber will be and how it will perform under heat and wear conditions. The rubber material serves multiple uses which include tire tread and sidewall production through blending with NR and BR as well as conveyor belt, gasket applications, footwear sole manufacturing, seal, and vibration dampening component production. The research by Ostad Movahed et al. (2009d) demonstrates that SBR blends serve three essential functions by lowering production expenses and enhancing wet traction and managing tire compound heat generation. The research by Ostad Movahed et al. in "silica-filled SBR/BR blends for tires" also demonstrates SBR's ability to work with advanced filler materials such as silica which results in better rolling resistance and fuel efficiency. The production of SBR rubber occurs through two main methods which include emulsion polymerization (E-SBR) and solution polymerization (S-SBR). The two production methods result in different end products which exhibit unique structural characteristics, molecular weight patterns, and functional properties. Table 2.4 compares emulsion versus solution polymerization of SBR.

The production of E-SBR remains prevalent because it offers affordable to manufacture and easy to process while serving applications in tire carcasses, sidewalls and budget tire production. S-SBR replaces E-SBR in premium tire tread production because it delivers superior wet grip and lower rolling resistance performance which meets EU tire labeling system requirements. The research by Ostad Movahed et al. (2009e) investigates silica-reinforced SBR systems and SBR/BR blends to enhance dynamic properties for tire tread applications. The research shows that the solution SBR exhibits superior silica compatibility because of its functional end groups. The performance of abrasion and hysteresis in blends depends on the precise control of morphology and crosslink density. The use of SEV and EV curing systems with S-SBR results in improved network structure and extended aging performance. The synthetic

Table 2.4 Emulsion versus solution polymerization of SBR.

Feature	E-SBR	S-SBR
Initiation	Free-radical emulsion	Anionic polymerization
Microstructure	Random copolymer with ~25% vinyl	Controlled microstructure (styrene blocky or tapered)
Molecular weight distribution	Broad	Narrow (tailorable)
Branching	Present (chain transfer to polymer)	Mostly linear
Styrene control	Limited	Precise (for grip/wear tuning)
Glass transition (T_g)	Typically ~−50 °C	Adjustable with vinyl content
Abrasion resistance	Good	Excellent
Wet traction	Moderate	High (better dynamic properties)
Rolling resistance	Moderate	Lower (better fuel economy)
Cost	Lower	Higher
Common use	General-purpose tires, mechanical goods	High-performance tires, eco tires, treads

rubber industry depends on SBR as its foundation while emulsion and solution polymerization methods provide different cost–performance–processing characteristics. The increasing demand for S-SBR in environmentally friendly and high-performance tires stems from stricter emission regulations and better fuel efficiency requirements. The process of polymerization requires understanding because it enables the creation of compounds which serve various needs in automotive and industrial sectors.

2.2.2.2 Butadiene Rubber

The synthetic elastomer BR serves as a primary material for industrial applications because it demonstrates exceptional elasticity, resilience and maintains excellent performance at low temperatures (Ostad Movahed et al., 2009e). The production of BR occurs through 1,3-BD polymerization which results in different structural forms based on the polymerization techniques and catalyst systems used. The polymer structure of BR consists of linear chains which contain *cis*-1,4, *trans*-1,4, and 1,2-vinyl configurations. The high *cis*-1,4 content in BR structures provides excellent elasticity and flexibility to the material. The presence of *trans*-1,4 units in the structure leads to increased crystallinity which results in decreased cold flexibility. The 1,2-vinyl units in the structure create side-chain branches which influence both glass transition behavior and manufacturing characteristics.

The polymerization technique determines the final structure of the material. The polymerization of 1,3-BD using neodymium (Nd) or cobalt (Co) catalysts in solution produces polymers with *cis*-1,4 content reaching up to 97%. The polymerization of lithium-based anionic systems produces polymers with higher 1,2-vinyl content. The T_g (glass transition temperature) of BR is less than −100 °C. The material exhibits exceptional elasticity through its high rebound and flexible nature. Excellent abrasion resistance exists in this material. The material shows better cold resistance than all other synthetic rubbers available in the market. The material exhibits average tear strength which becomes better when blended with other materials or filled with additives. The material serves as a fundamental component in high impact polystyrene (HIPS) and acrylonitrile–butadiene–styrene (ABS) plastic formulations. Table 2.5 compares BR, NR, and SBR.

Table 2.5 Comparison of BR, NR, and SBR.

Feature	BR	NR	SBR
Glass transition temperature (T_g)	~−100 °C	~−70 °C	~−50 °C
Abrasion resistance	Excellent	Good	Moderate
Rolling resistance	Low (beneficial for fuel economy)	Moderate	Moderate to high
Cold flexibility	Superior	Moderate	Lower
Fatigue resistance	Excellent	Good	Fair
Heat build-up	Low	High	Moderate
Compression set	Low	High	Moderate
Tear resistance	Moderate	Excellent	Moderate
Oil resistance	Poor	Poor	Moderate (depending on formulation)
Rebound resilience	High	Very High	Moderate
Tensile strength	~18–25 MPa (raw)	~22–28 MPa (raw)	~20–25 MPa (raw)
Elongation at break	High (~500–600%)	Very high (~700–800%)	Moderate (~400%)
Processing ease	Good (solution or emulsion)	Moderate (coagulation)	Good (emulsion preferred)
Main applications	Tires, belts, hoses	Tires, adhesives, rubber goods	Tires, seals, gaskets
Weathering resistance	Fair	Poor	Fair to good (can be improved with additives)

2.2.2.3 Nitrile Butadiene Rubber

The unsaturated copolymer family of NBR or acrylonitrile butadiene rubber results from polymerizing acrylonitrile (ACN) and BD monomers. The NBR grade determines the ACN content which ranges between 18% and 50%. The chemical structure of NBR contains the following sequence: $-(CH2-CH{=}CH-CH_2)-x-(CH_2)-CH(CN)-y-$. The chemical structure of NBR contains BD and ACN segments. NBR shows superior resistance to mineral oils, fuels and hydraulic fluids which makes it suitable for automotive and aerospace applications. The material demonstrates strong resistance to nonpolar solvents. The operating temperature range of NBR extends from −30 to +120 °C. The material exhibits average elasticity which decreases when the ACN concentration reaches higher levels. The material shows good resistance to tears and abrasion but performs worse than NR and BR. The material shows poor resistance to weather and ozone exposure which requires additional protection through antioxidants and other protective agents. The material exhibits lower gas permeability than NR and BR because of its polar nitrile groups. The material demonstrates strong resistance to compression at temperatures below 120 °C. The material costs fall within a moderate range based on the amount of ACN present and the manufacturing process used. NBR finds extensive use in various applications because it demonstrates superior resistance to oils, fuels, and numerous chemicals. The automotive sector employs NBR for manufacturing fuel hoses, gaskets, and seals and O-rings. Industrial applications of NBR include oil-resistant gloves, conveyor belts, and diaphragms. The aerospace industry uses NBR for fuel system seal applications. Medical and household applications use disposable protective gloves made from nitrile rubber materials which dominate the nitrile glove market.

The performance characteristics of NBR change according to the amount of ACN present in the material. ACN content affects properties. The material exhibits better flexibility and lower oil resistance and improved low-temperature properties when the ACN content ranges from 18% to 25%. The material shows better chemical and oil resistance with medium (26–35%) ACN content but its flexibility and processing characteristics deteriorate when ACN content reaches 36–50%.

The material requires protective additives to resist ozone and UV light exposure because it lacks natural protection. The material experiences limited usage in outdoor settings and weather-resistant applications. The material shows poor performance when exposed to strong polar solvents including ketones and esters.

2.2.2.4 Ethylene–Propylene–Diene Monomer

EPDM (ethylene–propylene–diene monomer) exists as a terpolymer structure which contains ethylene and propylene monomers together with a small amount of nonconjugated diene (ENB or DCPD) (Rahmani Naeim Abady and Ostad Movahed, 2017).

The diene monomer within the saturated backbone creates unsaturation points which allow crosslinking through sulfur or peroxide systems.

EPDM exhibits several essential properties which make it suitable for various applications. EPDM shows outstanding weather and ozone resistance which makes it one of the most durable elastomers for outdoor use. The material demonstrates superior UV resistance because it maintains its elasticity and color when exposed to prolonged sunlight. The material shows strong resistance to acids and alkalis and polar solvents but fails to resist hydrocarbons and oils. The material operates within a temperature range of −50 to +150 °C, but peroxide curing extends this range. EPDM shows exceptional resistance to water and steam exposure. The rubber exhibits good to excellent elastic properties. The material functions as an excellent electrical insulator which makes it suitable for wire and cable applications. It has average flame retardancy but needs flame-resistant additives for protection. The material shows weak resistance to oil exposure; thus, it should avoid petroleum and oil environments. It is a cost-moderate rubber.

The chemical structure of EPDM features a saturated C—C backbone with pendant unsaturation from the diene which enables crosslinking without compromising thermal-oxidative stability. The ethylene-to-propylene ratio determines the material's crystallinity because higher ethylene content produces a more crystalline and rigid material. The diene content between 3% and 9% enables sulfur vulcanization which enhances material processability and makes it suitable for dynamic applications.

EPDM finds its main applications in outdoor weather-resistant and thermally stable and electrically insulating products including, the automotive industry which uses EPDM for door seals, window seals, radiator hoses, and belts. The construction industry employs EPDM for roofing membranes, weatherstripping, and expansion joints. The electrical applications include high-voltage cable insulation, grommets, and connectors.

The industrial sector uses EPDM for steam hoses, pump diaphragms, and HVAC gaskets. The material shows poor compatibility with petroleum-based oils and fuels which prevents its use in various automotive engine applications. The material needs particular compounding approaches to achieve proper mechanical properties and crosslink density levels. The material shows weak bonding properties with rubbers and metals unless surface treatments or adhesives are applied. Table 2.6 compares the main features of studied (common) rubber types.

2.2.2.5 Specialty Rubbers for High-performance Applications

The industrial sector depends on general-purpose rubbers including NR, SBR, and BR for most applications but synthetic elastomers exist for specialized uses in harsh environments. Here, it examines four key specialty rubber types which include IIR, CR, VMQ, and FKM.

Table 2.6 The main features of studied (common) rubber types.

Rubber type	Key properties	Main applications	Resistance	Processability	Cost
NR	High tensile strength, elasticity, tackiness	Tires, adhesives, vibration isolators	Poor weather and ozone resistance	Excellent	Moderate
SBR	Good abrasion resistance, low cost	Passenger car tires, shoe soles	Poor weather and ozone resistance	Good	Low
BR	High resilience, cold flexibility	Tire treads, golf balls	Moderate chemical/weather resistance	Good	Low
NBR	Excellent oil and fuel resistance	Oil seals, hoses, gaskets	Excellent oil/solvent resistance	Moderate	Moderate
EPDM	Outstanding weather, ozone, and heat resistance	Automotive seals, roofing membranes	Excellent	Moderate to good	Moderate

Source: Reproduced with permission from Estaji et al. (2016)/SAGE Publications.

The production of IIR involves combining isobutylene with isoprene at levels between 1% and 3% to create a copolymer. The rubber exhibits low permeability to gases and moisture while providing excellent damping, shock absorption capabilities, showing good resistance to ozone, oxidation, heat exposure, and moderate chemical resistance. The product applications range includes inner tubes for tires, pharmaceutical stoppers, seals, gaskets, and protective clothing. IIR shows weak performance against petroleum oils and solvents while its resilience level remains lower than NR and BR.

The chloroprene rubber (CR) or Neoprene structure consists of 2-chloro-1, 3-butadiene molecules. The material demonstrates a perfect combination of mechanical strength, flexibility, abrasion resistance, and shows outstanding weather, ozone, and flame protection. The rubber shows average performance when exposed to oils and greases. The product range includes industrial belts and hoses together with wetsuits, diving equipment, electrical insulation materials and adhesives. The material fails to perform well when exposed to strong acids or esters while its elasticity levels remain lower than NR and SBR.

The structure of silicone rubber (VMQ) consists of a siloxane backbone with methyl and vinyl side groups. The material provides outstanding thermal stability across the temperature range of −100 to +250 °C. The material shows complete chemical stability and nonreactive properties with bodily fluids while maintaining biocompatibility. The material demonstrates superior dielectric properties and shows excellent resistance to ultraviolet light. The product range includes medical and food-grade applications such as tubing and implants as well as aerospace oven seals, gaskets, and electrical insulation for harsh environments. The material shows weak tear resistance while failing to protect against aromatic hydrocarbons and petroleum oils.

The production of fluoro-elastomers (FKM; Viton® and related) involves combining vinylidene fluoride with hexafluoropropylene and additional fluoro-monomers through copolymerization. The material demonstrates superior heat, chemical, solvent, and oil resistance while performing well in harsh conditions. The operating temperature range of this rubber extends from −20 to +250 °C. The applications include aerospace fuel systems, automotive seals, gaskets, and chemical processing facilities. The high cost of this material stands out as its main disadvantage while its use with ketones and low-temperature flexibility remains restricted.

2.3 Rubber Blends and Alloys

Rubber blends and rubber–plastic alloys function as hybrid elastomeric materials that unite different rubber compounds and rubber–plastic combinations to produce materials with improved mechanical strength, thermal stability, chemical resistance, and processing capabilities (Ostad Movahed et al., 2008). The commercial polymer blend market sector has demonstrated continuous growth. A successful blend requires robust interfacial connections between all components of the polymers. The blend industry faces a traditional challenge because of weak interphase connections. These blends serve multiple applications throughout tire production, automotive components, sealing systems and cable insulation manufacturing. Materials scientists use rubber blending as a strategic method to develop elastomers with tailored properties for industrial and engineering applications. The combination of NR with synthetic rubbers including SBR, BR, and NBR enables manufacturers to achieve enhanced mechanical, thermal, and chemical properties through synergistic effects. The main reason for rubber blending exists to achieve cost reduction. NR provides excellent elasticity and tensile strength but shows deficiencies in oil and ozone resistance. The combination of NR with synthetic rubbers NBR or EPDM enables manufacturers to create products with tailored performance characteristics for different environmental settings. The combination of SBR with BR in tire tread production enhances both abrasion resistance and rolling resistance because of BR's exceptional low-temperature properties and resilience. Rubber blends provide manufacturers with the capability to create products with customized curing properties. The vulcanization kinetics of different rubber materials vary independently, yet blending them enables formulators to modify cure profiles and enhance scorch safety and processing characteristics. The combination of NR with EPDM in rubber compounds provides superior weather and ozone protection than pure NR compounds while preserving suitable elasticity levels. The use of rubber blends enables producers to decrease their reliance on expensive or scarce raw materials. The combination of different materials in automotive tires, footwear, sealing systems, and conveyor belts enables producers to fulfill demanding performance standards and regulatory needs while maintaining affordable raw material expenses. The final properties of rubber blends depend heavily on the way their phases distribute and interact with each other. The use of block copolymers and reactive blending agents serves as compatibilization methods to enhance interfacial adhesion and achieve better homogeneity in rubber blends. The combination of dynamic vulcanization with reactive extrusion and solution blending allows for precise control of phase structure and resulting material properties.

2.3.1 Common Rubber Blends

The combination of different rubber materials leads to better performance while reducing production expenses. The combination of NR with SBR creates tires that have better abrasion resistance. The combination of NR with BR produces materials that demonstrate superior dynamic characteristics. The market features of several important rubber blends were shown in Table 2.7.

Table 2.7 The market features of several important rubber blends.

Blend	Constituent rubbers	Key properties	Applications
NR/SBR	Natural rubber/styrene–butadiene rubber	High abrasion resistance, good wet grip	Tires, industrial rollers
NR/BR	Natural rubber/butadiene rubber	Excellent resilience and cut resistance	Heavy-duty tires, conveyor belts
NBR/PVC	Nitrile rubber/polyvinyl chloride	Enhanced oil resistance and flame retardancy	Fuel hoses, gaskets
EPDM/NR	Ethylene–propylene–diene monomer/natural rubber	Weather resistance with high tack	Weather seals, roofing membranes
IIR/NR	Butyl rubber/natural rubber	Improved impermeability and flexibility	Inner tubes, seals

Table 2.8 Common TPVs and their features.

TPV system	Rubber phase	Thermoplastic matrix	Key applications
EPDM/PP	EPDM	PP	Automotive seals, tubing
NR/PS	Natural rubber	Polystyrene	Soft-touch grips, shoe soles
NBR/PA	Nitrile rubber	Polyamide (Nylon)	Fuel lines, high-performance gaskets

2.3.2 Thermoplastic Vulcanizates

The rubber–plastic blend known as thermoplastic vulcanizates (TPVs) contains EPDM rubber which undergoes dynamic vulcanization inside a thermoplastic polypropylene (PP) matrix. The dynamic vulcanization technique allows TPVs to be processed like thermoplastics, yet they retain rubber-like elasticity. The combination of recyclability and high performance makes TPVs suitable for automotive weather seals, soft-touch grips, and medical tubing applications.

Table 2.8 represents the common TPVs and their features.

2.3.3 Importance of Rubber Compatibility and Phase Morphology in Blends

The study of rubber compatibility together with phase morphology structure plays a vital role in blend development (Estaji et al., 2016a). The thermodynamic interaction between two or more elastomers at molecular levels defines compatibility in rubber blends. The resulting single-phase structure emerges from fully compatible blends but incompatible blends separate into multiple distinct phases. The following elements determine how well different rubber phases will mix together:

- The rubber phases show different polarity levels between them (e.g., NR and NBR). The mechanical properties of rubber blends depend on the combination of molecular weight and glass transition temperature (T_g) values. The presence of interfacial agents which react with each other during processing determines the level of compatibility between rubber phases. The final properties of rubber blends depend on the specific conditions used during processing including shear forces, temperature levels and duration of mixing.
- Influence of phase morphology: the mechanical properties, fatigue resistance, and dynamic behavior of blends depend on their phase structure. The performance of partially compatible systems depends on achieving optimal domain dispersion levels. The mechanical strength and elongation properties

of rubber materials improve when their phases are fine and well distributed. The combination of properties becomes more effectively occurs when materials have cocontinuous structures. The processability of materials becomes affected when phase inversion occurs at elevated concentration ratios.

Table 2.9 shows the compatibility in selected rubber blends.

As a conclusion, rubber blends serve as a flexible and cost-effective method to create customized elastomeric materials which meet various modern application requirements. The field of rubber blending advances through modern characterization tools and reactive processing technologies which enable sustainable high-performance elastomer design.

In following subsections, author illustrates and reviews several important rubber blends with full detail for EPDM/SBR.

2.3.4 NR/SBR/BR in Tire Treads

Tire tread production depends on the combination of NR/SBR/BR materials for achieving optimal tire performance and product durability (Ostad Movahed, 2009).

The tire manufacturing process requires tread formulation to achieve performance goals and safety standards. BR serves as the standard material for passenger car and truck tire tread production because it provides optimal mechanical and dynamic properties. The combination of high tensile strength, excellent tear resistance, and superior resilience makes NR an essential component in tire production. The heavy-duty and off-road tire applications benefit from NR because it provides superior grip and extended durability. The heat build-up resistance of NR remains high in NR which enables tread performance under

Table 2.9 Compatibility in selected rubber blends.

Rubber blend	Compatibility level	Morphology observed	Remarks
NR/SBR	Partial	Cocontinuous	T_g shift observed in DMA
NR/BR	Good	Dispersed phase	Improved elasticity, lower T_g
NR/EPDM	Low	Phase-separated	Use of compatibilizer required
SBR/NBR	Moderate	Fine domain structure	Enhanced oil resistance
NR/NBR	Poor	Large dispersed domains	Brittleness at high NBR content

demanding high-load and high-speed operating conditions. The combination of SBR provides better abrasion protection and lower rolling resistance to vehicles. The manufacturing process of SBR through emulsion or solution polymerization enables performance adjustments while maintaining consistent traction in wet and dry conditions. The use of SBR in passenger car treads provides excellent wear performance together with improved fuel efficiency. The combination of BR with NR and SBR provides improved low-temperature flexibility and abrasion resistance while reducing rolling resistance and wet or ice grip. The low glass transition temperature (T_g) of BR makes it suitable for improving traction in cold weather and extending tread wear life. The three rubber components are combined in specific ratios to create a formulation that delivers optimal wet grip, rolling resistance, and treadwear and durability performance. The combined effect of NR/SBR/BR blends produces high-performance characteristics which work well across different road.

2.3.5 EPDM/NR in Automotive Applications

Another important rubber blend is the blend of NR with EPDM. The blend of EPDM and NR materials functions as door seals and trunk seals and window channels because it provides both elasticity and durability. The combination of NR and EPDM materials functions as radiator and heater hoses because they withstand coolants and heat and mechanical stress. The combination of NR provides dynamic damping properties while EPDM shows superior aging resistance. The combination of materials finds application in engine mounts and bushings. The materials serve to protect tires from thermal changes and ozone damage in their sidewalls and inner liners. The general crosslinking (curing) methods used in EPDM/NR blends are sulfur and peroxide curing. The use of sulfur crosslinking produces materials with enhanced fatigue properties. The heat aging performance of peroxide systems exceeds that of other crosslinking systems. The SEV sulfur system serves as a common choice for achieving balanced cure characteristics. The process of blending EPDM/NR faces multiple technical obstacles. The different polarities between the materials lead to phase separation in the blend. The curing process of these materials faces challenges because their kinetic properties and compatibility levels differ. The distribution of fillers throughout the system experiences unevenness. The addition of plasticizers or compatibilizers becomes necessary to achieve proper material blending.

2.3.6 EPDM/SBR, a Novel Technique for Compatibility

Estaji et al. (2016b) developed a novel chemical technique for compatibility of EPDM rubber and SBR in their blends.

As stated, the main objective of polymeric material development focuses on creating high-quality products through affordable production methods. The high expense of creating new polymeric materials leads manufacturers to blend different polymers which results in the desired combination of properties needed for product success. The commercial polymer blend market has demonstrated continuous growth since the 1980s, and it had already exceeded to 2.6 billion pounds by 2018. The optimal blend requires strong connections between all sections of its polymer components. The blend industry faces a fundamental challenge because different blend components lack strong interphases due to their chemical differences and varying surface energies. The creation of any heterogeneous system leads to interface formation which becomes known as an interphase or boundary layer that controls essential system properties particularly those of polymers (Ostad Movahed et al., 2009d and 2008). The process of thermal diffusion between two polymers that can mix partially or fully leads to the development of an interfacial phase. The interface thickness grows with time while the concentration distribution transforms in most cases. The interfacial layer or interphase contains enough atoms or molecules to develop its own properties including modulus and strength, heat capacity, density, and others.

The interphase exists as a transitional area between two contacting phases which exhibits changing composition, structure, and properties that differ from both individual phases. The formation of interphase occurs through molecular diffusion between two pure phases. Ostad Movahed et al. (2008) developed a practical method to measure interphase formation between different polymers using modulated differential scanning calorimetry.

The aging resistance of EPDM/SBR blends is excellent while their compression set properties remain good. The rubber materials show limited miscibility because their solubility parameters (δ) amount to 17.6 and 15.9 $MPa^{1/2}$ for SBR and EPDM, respectively. The solubility parameters can be divided into three parts:

$$\delta^2 = \delta_d^2 + \delta_p^2 + \delta_h^2$$

where the subparameter δ_d is the component from the dispersive forces, δ_p is the component from polar forces, and δ_h is from hydrogen bonding. The mutual solubility between two rubbers 1 and 2 depends on the value of $\Delta\delta$ which is calculated as:

$$\Delta\delta = \left[(\delta_{d1} - \delta_{d2})^2 + (\delta_{p1} - \delta_{p1})^2 + (\delta_{h1} - \delta_{h2})^2\right]^{1/2}$$

The rubbers show complete immiscibility when $\Delta\delta$ exceeds 5 $MPa^{1/2}$. The literature presents multiple approaches to achieve compatibility between incompatible polymer blends. The methods to compatible immiscible polymer blends include cocrystallization of two phases, in-situ immobilization of one phase through dynamic vulcanization, the use of a third material as a compatibilizer agent through separate addition of a compatibilizer agent, copolymer incorporation into the immiscible polymers, and generic processes and specific types of reactions to form copolymer in a reactive compatibilization process. The reactive compatibilization process requires both general copolymerization mechanisms and particular reaction types that create copolymers. The last strategy consists of two subcategories which include direct reaction between reactive functionalities on each polymer and the addition of a third reactive species to the system as described for this study to compatible EPDM and SBR in their blends. In this study, the reaction includes two polymers EPDM and SBR, and the third component is the Lewis acid aluminum chloride $AlCl_3$. The Friedel–Crafts alkylation process was used to attach EPDM to the benzene ring of SBR as shown in Figure 2.6.

The rubber compounds and blends received preparation through a two-roll mill with 14 cm diameter and 32 cm length rolls. The two-roll mill operated with a 3–4 mm gap setting. The two-roll mill operated with a built-in cooling system. The mixture of EPDM and SBR with carbon black and oil received five minutes of processing before the simultaneous addition of $AlCl_3$ and filled EPDM, and SBR for blend preparation during three minutes. The rubber compounds A–J together with reference compound (RC) from Table 2.10 underwent compression molding at 180 °C under 160 atm pressure using ODR cure curve determined optimum cure times. The mechanical properties of cured rubbers were evaluated using 15 cm × 15 cm sheets with 2.3 mm thickness. The cure properties of all compounds and blends are represented in Table 2.11.

The FTIR operates by sending infrared radiation (energy) into samples to measure the absorption and transmission of IR radiation that chemical bonds between atoms in molecules produce. The FTIR spectra in Figure 2.7 demonstrate the

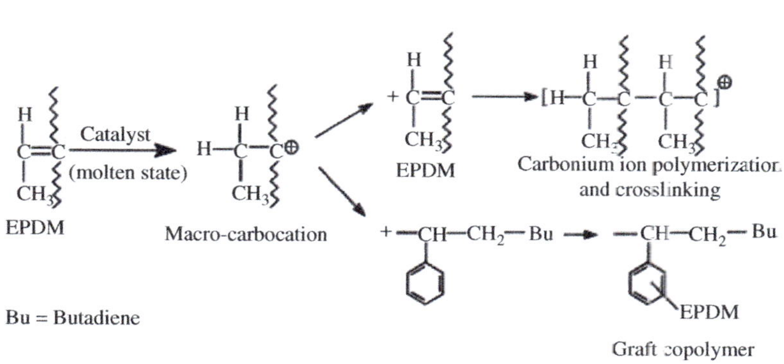

Figure 2.6 The proposed mechanism of the crosslinking of EPDM and the formation of EPDM-*g*-SBR copolymer. *Source*: Reproduced with permission from Estaji et al. (2016b). SAGE Publications.

Table 2.10 Mooney viscosities and formulation of rubber compounds and blends (all values in phr).

Compound	EPDM	SBR	Carbon black (N550)	Paraffinic oil	$AlCl_3$	Mooney viscosity (MU)
A	100		110	44	1	60
B		100	110	44	1	76
C	40	60	110	44	0.06	63
D	40	60	110	44	0.13	63
E	40	60	110	44	0.25	63
F	40	60	110	44	0.40	63
G	40	60	110	44	0.50	63
H	40	60	110	44	0.80	63
I	40	60	110	44	1	63
J	40	60	110	44	2	63
RC	40	60	110	44	0	63

Source: Reproduced with permission from Estaji et al. (2016b).

Table 2.11 The cure properties of rubber blends and compounds, cured by $AlCl_3$.

Compound	RC	A	B	C	D	E	F	G	H	I	J
t_{s2} (min: s)	—	—	0:46	—	—	—	—	2:01	0:51	0:24	0:10
t_{90} (min: s)	—	—	4:41	4:18	4:36	4:49	4:38	4:34	4:52	4:30	1:22
ΔTorque (dN · m)	—	—	9.92	1.2	0.7	1.2	1.6	5.7	9.7	18.0	26.5
CRI (1/s)	—	—	0.43	—	—	—	—	0.50	0.40	0.40	1.40
CLD (mol/m^3)	—	—	166	84	82	84	111	138	145	288	871

Source: Reproduced with permission from Estaji et al. (2016b).

comparison between N550-filled EPDM and SBR (compounds A and B) and compound I (Table 2.11). The FTIR spectra show the cured extracted gels (rubbers). The FTIR spectra show peaks at 3,424, 3,422, 3,441, 1,629, 1,630, and 1,628 1/cm which confirm the presence of carbon black (N550) in the compound formulations (gels). The EPDM spectrum shows a distinct peak at 2,918 1/cm but SBR does not display this peak. However, compound I showed the same peak position. The presence of EPDM in compound I gel structure proves the

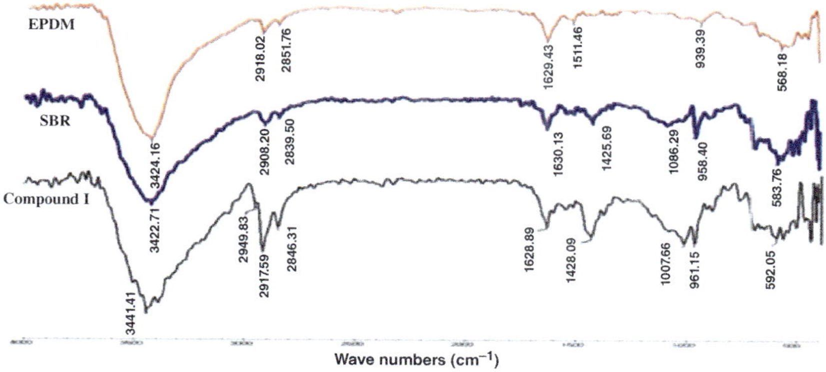

Figure 2.7 FTIR spectrum for filled EPDM and SBR with 110 phr N550 (compounds A and B, Table 2.10) and compound I after $AlCl_3$ curing. *Source*: Reproduced with permission from Estaji et al. (2016b).

formation of SBR-g-EPDM copolymer formation. The EPDM spectrum indicates macro-carbocation formation which leads to EPDM chain growth through carbonium ion polymerization and crosslinking of the EPDM as shown in Figure 2.7. The SBR and compound I spectra show two peaks at 1,426 and 1,428 1/cm which correspond to benzene ring signals. The EPDM spectrum lacks this peak which proves that the benzene ring exists in the gel structure of compound I and Friedel–Crafts reactions take place.

The TGA curves showed a comparison between the thermal stability of RC and compound I. The degradation process started at 100 °C for paraffinic oil degradation and ended at 610 °C for RC and 675 °C for compound I. The weight loss percentages for RC and compound I reached 98% and 94.5%, respectively, from their original weights. The thermal stability of compound I exceeded that of RC because the EPDM-g-SBR copolymer formed in compound I.

The compounds C–J and RC (Table 2.10) contain 40 phr EPDM, 60 phr SBR, 110 phr carbon black, and 44 phr paraffinic oil with various amounts of $AlCl_3$, i.e., 0–2 phr. The formulation of RC, lacking $AlCl_3$, ΔTorque and CLD values (Table 2.11) demonstrated a steady rise when $AlCl_3$ΔTorque values ranged from 1.2 to 26.5 dN·m, while CLD values ranged from 84 to 871 mol/m^3 when adding 0.06–2 phr of $AlCl_3$ to the blends. The trend becomes more pronounced when $AlCl_3$ exceeds 0.5 phr. The cure times for compounds J and H and CRIs for the same compounds range between 1:22 to 4:52 min:s and 0.4–1.4 1/s, respectively. The compound J exhibited ΔTorque and CLD values reached their highest points among all tested compounds. The addition of 2 phr $AlCl_3$ proved sufficient to create effective EPDM–SBR and SBR–SBR crosslinks.

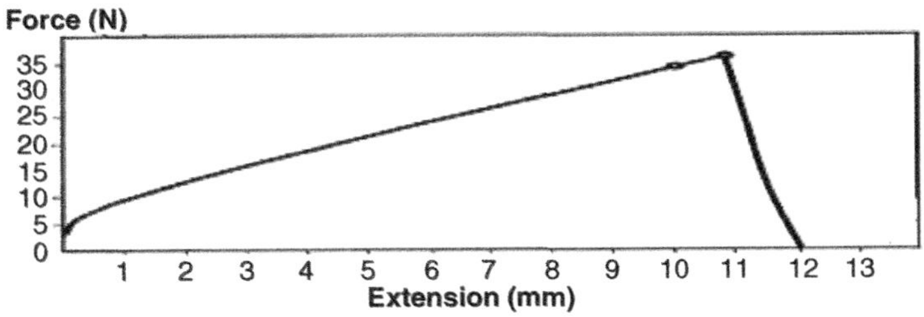

Figure 2.8 Tensile curve for compound G in Table 2.10. *Source*: Reproduced with permission from Estaji et al. (2016b).

Table 2.12 Mechanical properties of selected cured rubbers.

Compound no.	Hardness (Shore A)	Tear strength (kJ/m^2)	Tensile strength (MPa)	Elongation at break (%)	Modulus at 50% elongation (MPa)	Modulus at 100% elongation (MPa)
A	None[a]	None[a]	None[a]	239	None[a]	None[a]
B	72	11	7	86	4.4	—
E	79	11	4	50	4.0	—
G	77	9	4	59	3.6	—
I	92	—[b]	5	10	—	—

[a] This compound was not cured, and it was soft.
[b] This compound was very brittle, and it's tear strength could not be measured.
Source: Reproduced with permission from Estaji et al. (2016b).

The tensile behavior of compound G with 0.5 phr $AlCl_3$ is shown in Figure 2.8. The extension level of this compound reaches 59% according to Table 2.12 before it breaks. The mechanical properties of cured compounds A, B, E, G, and I with 1, 1, 0.25, 0.5, and 1 phr of $AlCl_3$ are compared in Table 2.12. The mechanical properties of compound A with 100% EPDM remain low because the compound failed to cure properly with $AlCl_3$. The mechanical properties of compound B with 100% SBR in its formulation reached acceptable levels at 72, 11 kJ/m^2, 7 MPa, 86%, and 4.4 MPa for hardness, tear strength, tensile strength, elongation at the break, and modulus at 50% elongation, respectively. The mechanical properties of compound I with 1 phr of $AlCl_3$ (Table 2.10) show the worst performance among all tested compounds. The tensile strength decreased to 5 MPa and elongation at break reached only 10%. The compound became brittle, and its tear strength measurement failed. The addition of $AlCl_3$ to Compounds E, G, and I resulted in improved hardness and tensile strength values.

The research results show that SBR–EPDM and SBR–SBR macromolecules formed covalent bonds through Friedel–Crafts reactions when Lewis acid $AlCl_3$ was used as a catalyst. The cure parameters including ΔTorque values in cure traces confirmed that Friedel–Crafts reactions took place. The DSC curve shows an exothermic heat flow peak occurred at 111.28 °C which indicates the formation of carbocations during Friedel–Crafts reactions. The FTIR spectra of SBR and SBR–EPDM blends show two peaks at 1,425.7 and 1,428.09 1/cm which represent benzene ring absorption. The absence of this peak in EPDM spectrum confirms that SBR-g-EPDM copolymer forms during SBR–EPDM blend processing. The thermal stability of SBR-g-EPDM exceeded that of SBR–EPDM blends which lacked chemical bonds between their polymer chains. The addition of 2 phr $AlCl_3$ proved to be an optimal amount for creating links between EPDM–SBR and SBR–SBR molecules. The tensile strength of cured blends showed lower values than the corresponding values of the SBR-based prepared compound. The SEM micrograph of the cured blend with 1 phr $AlCl_3$ showed excellent polymer compatibility and strong interphases between the two materials.

2.4 Comparison of Rubber Types

The physical and chemical performance of rubbers depends heavily on their molecular structure and degree of unsaturation and the specific crosslinking system used during vulcanization. The prediction of recyclability and application potential for mixed rubber streams requires knowledge about the fundamental differences between various rubber types. The following section evaluates NR, SBR, BR, NBR, EPDM, and IIR based on their mechanical properties, thermal resistance, chemical stability, and economic value.

NR maintains its position as the best material for ultimate tensile and fatigue resistance because it develops crystalline structures when subjected to stress. The synthetic tire elastomers SBR and BR demonstrate average performance between processing characteristics and wear resistance. The polar nitrile groups in NBR create its stiffness which enhances oil resistance yet reduces its flexibility. The softness of EPDM and IIR materials makes them suitable for weather resistance and gas barrier applications. The recycling process of NR and SBR and BR tire materials produces good mechanical properties after devulcanization because their molecules inter diffuse well. The addition of NBR or EPDM to blends requires compatibilization to prevent phase separation during recycling. Table 2.13 compares and summarizes the important mechanical properties of the mentioned rubbers.

The lowest T_g values in BD-based rubbers result in their exceptional low-temperature flexibility. The elastic properties of NR and IIR remain intact

Table 2.13 The important mechanical properties of the common rubbers.

Rubber type	Tensile strength (MPa)	Elongation at break (%)	Modulus (100% strain, MPa)	Remarks
NR	20–25	700–800	1–1.5	High resilience, strain-induced crystallization gives excellent strength and tear resistance
SBR	15–20	400–500	1.5–2	Good abrasion resistance but poorer tear and fatigue resistance than NR
BR	18–22	500–600	0.9–1.2	High elasticity and resilience; low heat build-up
NBR	12–18	300–450	2–3	Tough but less flexible; mechanical properties decrease with higher ACN content
EPDM	10–15	350–600	1–1.5	Excellent elastic recovery; limited reinforcement due to saturated backbone
IIR	9–12	400–600	1–1.3	Low rebound resilience; superior air impermeability

until temperatures reach −60 °C. The thermal and oil resistance of NBR exceeds its cold temperature performance because of its elevated T_g value. The saturated backbone structure of EPDM provides excellent resistance to oxidation and heat which makes it suitable for automotive seal applications and roofing membranes. The recycling process causes EPDM and IIR to maintain their chain structure (Khani and Ostad Movahed, 2024), but NR/SBR/BR systems experience oxidative breakdown so stabilizers need to be present during the reclaiming process. Table 2.14 compares thermal behavior of the same rubbers.

The fast degradation of NR and BR rubbers under ozone and UV radiation exposure leads to crack formation which damages their structural strength. The styrene units in SBR provide slightly improved resistance against degradation.

The saturated and partially saturated backbones of EPDM and IIR materials lead to exceptional resistance against ozone degradation and aging processes. The polar nitrile groups in NBR rubber provide excellent resistance to hydrocarbons and oils. The end-of-life tire materials NR/SBR/BR need additional protection through antioxidant and UV stabilizer additives but EPDM waste can be used directly for outdoor applications. Table 2.15 compares the oils, weathering, and solvents resistance of common rubbers.

NR stands as the leading elastomer in worldwide production because it provides the most affordable solution but its market value fluctuates based on plantation

Table 2.14 Thermal behavior of the selected rubbers.

Rubber type	Glass transition (T_g, °C)	Thermal decomposition (°C)	Usable service range (°C)
NR	≈ −70	> 300	−50–+80
SBR	−50	> 300	−40–+100
BR	−120–−90	> 320	−80–+80
NBR	−25–−10	> 300	−20–+120
EPDM	−60–−50	> 350	−50–+150
IIR	−70–−60	> 350	−60–+120

Table 2.15 The oils, weathering, and solvents resistance of common rubbers.

Rubber type	Oil resistance	Ozone and UV weathering	Solvent resistance	Water/air impermeability
NR	Poor	Poor	Poor	Moderate
SBR	Poor	Moderate	Poor	Moderate
BR	Poor	Poor	Poor	Moderate
NBR	Excellent	Good	Good	Moderate
EPDM	Good	Excellent	Moderate	Good
IIR	Good	Excellent	Good	Excellent

output and environmental conditions which drives manufacturers to develop synthetic rubber alternatives. The production of SBR for tire tread compounds remains widespread because of its affordable price and simple manufacturing process while BR enhances tire durability and reduces rolling friction. The market (Table 2.16) segment for NBR, EPDM, and IIR products exists because these materials provide specific properties including oil resistance, weather resistance, and gas barrier properties. The market value structure determines recycling profitability because tire recycling of NR/SBR/BR materials remains profitable but industrial recycling of NBR and IIR materials faces restrictions. The development of efficient recycling and devulcanization methods depends on complete knowledge of how rubbers react mechanically, thermally, and chemically when compared to each other. The three rubber types, NR, SBR, and BR, show excellent recyclability because they are compatible and widely available, but NBR, EPDM, and IIR need customized recycling methods to become reusable. The economic value of different rubber types determines which circular economy investments will receive funding.

Table 2.16 Cost and market share of common rubbers.

Rubber type	Approx. price (USD/kg, 2024 avg.)	Global market share (%)	Key applications
NR	1.6–2.2	≈ 42	Tires, conveyor belts, vibration isolators
SBR	1.4–1.8	≈ 25	Tires, footwear, adhesives
BR	1.8–2.5	≈ 12	High-performance tires, golf balls
NBR	2.2–2.8	≈ 8	Fuel hoses, seals, gloves
EPDM	2.5–3.2	≈ 7	Automotive seals, roofing membranes
IIR	3–3.8	≈ 6	Inner tubes, pharmaceutical stoppers

References

Abgorg, H. K. and Ostad Movahed, S. Effect of nano-zinc oxide on the mechanical properties of cured ethylene–propylene–diene monomer rubber with a semi-efficient vulcanisation system. *Plast. Rubber Compos.* 2024, *53*(5–7), 168–177.

Ansarifar, A., Saeed, F., Ostad Movahed, S., Wang, L., Yasin, K. A., and Hameed, S. Using a sulfur-bearing silane to improve rubber formulations for industrial applications. *J. Adhes. Sci. Technol.* 2013, *27*(2), 371–384.

Estaji, S., Ostad Movahed, S., Yazdanbakhsh, S., and Karim Nezhad, M. Study of the Friedel–Crafts co-curing of ethylene–propylene–diene rubber and styrene–butadiene rubber. *Rubber Chem. Technol.* 2016a, *89*(4), 683–698.

Estaji, S., Ostad Movahed, S., Yazdanbakhsh, S., and Karim Nezhad, M. A novel chemical technique for compatibility of ethylene–propylene–diene monomer rubber and styrene–butadiene rubber in their blends. *J. Elastomers Plast.* 2016b, *49*(1), 298–314.

Imanifar, M., Ostad Movahed, S., and Ahmadpour, A. Effects of peroxide and phenolic cure systems on characteristics of filled ethylene–propylene–diene monomer rubber (EPDM). *J. Appl. Polym. Sci.* 2018, *135*(3), 45732.

Jaberi Mofrad, F., Ahmadpour, A., and Ostad Movahed, S. Effect of silane-modified carbon black and nano-silica, individually and in combination, on the performance of ethylene–propylene–diene monomer rubber. *Polym. Compos.* 2025, *46*(1), 547–566.

Khani, Z. and Ostad Movahed, S. Impact of industrial methyl diethanolamine solution on the aging of cured ethylene–propylene–diene monomer and butyl rubber compounds and associated blends for gas refinery sealant applications. *J. Appl. Polym. Sci.* 2024, *141*(37), e55939.

Madani, S. M. and Ostad Movahed, S. Effect of polyvinyl chloride and polyethylene terephthalate on the mechanical properties of butyl rubber (IIR)-based thermoplastic elastomers using a semi-efficient vulcanisation system and nano zinc oxide as activator. *Plast. Rubber Compos.* 2024, *53*(5–7), 143–153.

Ostad Movahed, S. Assessing the effects of different processing parameters on interphases between dissimilar rubbers using modulated-temperature differential scanning calorimetry. *J. Rubber Res.* 2008, *11*(1), 13–28.

Ostad Movahed, S. *Development of a New Tyre Tread Using Silanized Pre-Treated Silica Nanofiller*. Ph.D. Thesis, Loughborough University, Loughborough, UK, 2009.

Ostad Movahed, S., Ansarifar, A., and Mirzaie Mahmouie, F. Effect of various efficient vulcanization cure systems on the compression set of a nitrile rubber filled with different fillers. *J. Appl. Polym. Sci.* 2015, *132*(1), 41266.

Ostad Movahed, S., Ansarifar, A., and Song, M. Comparing the dynamic behaviour of several rubbers filled with silanized silica nanofiller. *Polym. Int.* 2009a, *58*(2), 209–217.

Ostad Movahed, S., Ansarifar, A., and Song, M. Effect of silanized silica nanofiller on tack and green strength of selected filled rubbers. *Polym. Int.* 2009b, *58*(4), 424–429.

Ostad Movahed, S., Ansarifar, A., and Song, M. Measuring dynamic behaviour of several rubbers filled with silanized silica nanofiller. *Rubber World* 2009c, *239*(3), 23–30.

Ostad Movahed, S., Ansarifar, A., and Song, M. Using modulated-temperature differential scanning calorimetry to study interphases in blends of styrene-butadiene and polybutadiene rubbers filled with silanized silica nanofiller. *J. Appl. Polym. Sci.* 2009d, *111*(2), 1644–1652.

Ostad Movahed, S., Ansarifar, A., and Song, M. Effects of different interphases on mechanical properties of cured silanized silica-filled styrene–butadiene/polybutadiene rubber blends for use in passenger car tire treads. *J. Appl. Polym. Sci.* 2009e, *113*(8), 1868–1878.

Ostad Movahed, S., Ansarifar, A., Song, M., and Hameed, S. Comparing the effects of silanized silica nanofiller on crosslinking and mechanical properties of natural rubber and synthetic polyisoprene. *J. Appl. Polym. Sci.* 2008, *109*(7), 869–881.

Rahmani Naeim Abady, A. and Ostad Movahed, S. Effect of curing system on the properties of thermoplastic elastomer foams of ethylene–vinyl acetate copolymer and styrene–butadiene and ethylene–propylene–diene monomer rubbers. *J. Appl. Polym. Sci.* 2017, *134*(10), 45357.

3

Morphology and Filler Interactions in Polymeric Composites, a Crucial Factor in Polymers Waste Management

The selection of suitable recycling methods for polymeric composites depends on main factors which include filler type, concentration, and their distribution pattern inside the polymer matrix. The recycling process for neat thermoplastics depends on polymer chemistry and molecular weight degradation, but filled and reinforced polymeric systems present additional challenges because of filler–matrix interface interactions and structural filler effects and their influence on thermal, rheological, and mechanical properties during reprocessing. The filler-related parameters determine how waste materials will be managed through different recovery methods, including reusing, mechanical processing, chemical treatment, and energy recovery.

The main factor that determines filler performance is the filler type (Ostad Movahed, 2014). The thermal stability and chemical inertness of inert mineral fillers, including calcium carbonate, talc, and kaolin, make them suitable for use. The mechanical properties of these fillers remain mostly unchanged when they experience melting extrusion followed by remolding after recycling. The recycling process becomes difficult because carbon-based fillers including carbon black (CB), graphite, graphene, and carbon fibers add electrical conductivity, UV protection, and strengthening capabilities. The thermal and microwave energy absorption properties of CB result in uneven heating that causes specific areas to deteriorate during reprocessing or devulcanization. The processing of glass and carbon fibers becomes challenging because their fibers tend to break apart, and their interfaces lose strength while their reinforcing properties deteriorate significantly during multiple secondary processing steps. The recycling methods for fiber-reinforced composites need to change from mechanical processing to chemical depolymerization, solvolysis, and thermal recovery because of their poor recyclability.

The amount of filler material in the waste stream determines how well materials can be recovered. The recycling process for mechanical applications remains possible when filler content does not exceed 10 wt.% because it preserves the material's tensile strength, impact resistance, and processing characteristics. The polymer melt develops higher viscosity and non-Newtonian flow characteristics and thermal instability when filler content exceeds 30 wt.%. The addition of high filler amounts to materials leads to increased melt viscosity and non-Newtonian flow behavior and faster tool degradation, and more difficult extrusion filtration operations. The economic and technical feasibility of mechanical recycling becomes less viable when filler content reaches more than 30 wt.%, so alternative recycling methods become more appropriate. The high filler content in flame-retardant systems containing metal hydroxides and phosphorus minerals requires waste streams to go straight to energy recovery or cement-kiln coprocessing because the inorganic materials can integrate into clinker mineral structures.

The filler's morphological status requires equal attention because it depends on how well the filler disperses, its agglomeration state, its ability to bond with the polymer matrix, and its position within the matrix. The materials undergo significant changes in their crystallization process, viscoelastic properties, and thermal stability when nanosilica, nanoclays, and carbon nanotubes form percolated networks through their well-dispersed state. The nanostructured filler shapes in reprocessing serve two purposes by defending polymer chains from degradation and by speeding up degradation through stress points and oxidative chemical reactions. The properties of materials deteriorate significantly after multiple recycling cycles because poorly dispersed or agglomerated fillers create defect sites that spread cracks and decrease fatigue resistance.

The strength of filler–polymer interface bonds determine how a material will react to recycling processes. The initial mechanical properties of materials become stronger because of covalent bonds and silane-mediated bonds, yet these bonds prevent materials from flowing during remelting and decrease their recyclability. The bonding strength between fillers and polymers determines how well a material can be recycled because weak bonds enable easy melting but result in quick property deterioration during multiple processing cycles. The relationship between these factors determines whether a composite material can undergo multiple mechanical recycling cycles or needs to move to chemical recycling after reaching its service lifespan.

The waste management process depends on these connected elements to determine which recycling method should start first. The recycling process for thermoplastics containing inert fillers at low levels needs to follow a closed-loop mechanical recycling system. The recyclability of systems with functional fillers at moderate filler levels requires additional processing steps which include compatibilization, stabilization, and partial depolymerization. The technical boundaries of mechanical recycling exceed the capabilities of highly filled composites which contain fibers and flame retardants, so these materials need to move toward chemical recycling or energy recovery.

The selection of fillers for polymeric composites determines their service life performance and their final disposal destination. The development of sustainable composite materials requires designers to create material structures from the beginning which prioritize recyclability through proper filler choices and loading amounts and morphological designs.

The distribution pattern of fillers within rubber matrices determines how rubber compounds perform mechanically and dynamically. The rubber compounding industry faces this current challenge because fillers exist as minerals while rubber materials are organic substances.

Rubber compounding requires reinforcing fillers which serve to boost the mechanical properties, durability, and operational performance of vulcanized rubbers. The two leading fillers in rubber compounding remain CB and silica because they demonstrate effective elastomeric matrix reinforcement through physical and chemical bonding. The amorphous carbon material known as CB possesses a high surface area and develops strong bonds with rubber materials. The addition of CB enhances rubber materials by increasing their tensile strength, tear resistance, and abrasion resistance and by improving hysteresis behavior which directly affects tire performance. The selection of CB grades including N220, N330, and N550 depends on achieving optimal reinforcement levels with suitable processing characteristics.

The combination of silica with silane coupling agents in tire compounds produces silica-based materials that deliver reduced rolling resistance and enhanced wet traction performance (Ansarifar and Ostad Movahed, 2009). The rubber interaction with silica requires surface treatment because silica lacks van der Waals forces for bonding with rubber. The combination of silica with silane coupling agents makes rubber compounds suitable for modern "green tires" because they provide both energy efficiency and enhanced safety performance. Specialty applications and cost-effective formulations use reinforcing fillers including clays, talc, mica, nanoclays, and calcium carbonate. Nanotechnology has brought forth carbon nanotubes, graphene oxide, and layered silicates which provide outstanding mechanical strength and barrier properties at minimal filler content because of their extensive surface areas and high aspect ratios.

3.1 Carbon Black

The rubber industry depends on CB as its primary filler material. The rubber industry consumes 90% of all CB production that occurs worldwide. The manufacturing process of CB occurs through the incomplete combustion or thermal decomposition of hydrocarbon materials. The production of industrial CB depends on crude oil products together with by-products obtained from coal processing, natural gas, and acetylene operations. The carbon content in commercial CB materials exceeds 97% while hydrogen, oxygen, and sulfur make up the remaining percentage.

CB exists as a fine black powder which contains nearly pure elemental carbon. The rubber compounding industry uses CB as its most popular reinforcing filler material because it serves the tire industry above all else. The rubber industry depends on CB to enhance tensile strength, tear resistance, abrasion resistance, and durability in natural rubber (NR), styrene–butadiene rubber (SBR), and butadiene rubber (BR) blends.

The properties of CB depend on its particle size and structure as well as its surface area and surface activity. The reinforcing potential and processability of rubber compounds depend directly on these specific factors. Table 3.1 shows selected physical properties of CB.

The term structure in CB science describes the intricate arrangement of CB aggregates while aggregate size refers to the total size of the CB particles. The level of structure in CB determines its reinforcing power, but it also makes the material more viscous during manufacturing processes.

The furnace black process stands as the main production method for CB because it enables precise control over particle dimensions and structural formation. The production of CB through the channel black process, acetylene black process, and thermal black process serves specific market requirements. The dominant method for producing furnace black uses heavy aromatic oils derived from residual petroleum fractions as its feedstock. The reactor operates with partial combustion through a controlled air-to-fuel ratio in a sealed environment. The

Table 3.1 Important physical properties of CB.

Property	Typical range	Effect on rubber
Primary particle size	10–100 nm	Smaller particles yield higher surface area and reinforcement
Surface area (BET)	30–150 m^2/g	Higher surface area increases filler–polymer interaction
DBP absorption	50–150 $cm^3/100$ g	Indicates structure/aggregates; affects viscosity and elasticity
Color	Deep black	Used in black rubber compounds for appearance and UV protection
Density	~1.8 g/cm^3	Influences loading and dispersion
Moisture content	<1%	Affects mixing behavior

process includes a quick cooling step which prevents additional particle enlargement. The powder collection process uses cyclone separators together with bag filters to extract the material.

CB significantly improves tensile, tear strength, abrasion resistance, cut and chip resistance, UV and ozone resistance, and electrical conductivity (in conductive grades) of the rubber compound. These effects originate from three main factors:

- Physical interlocking between filler aggregates and rubber matrix
- Surface interactions at polymer–filler interfaces
- Crosslink density (CLD) enhancement during vulcanization

The ASTM D1765 system uses the names N220 and N330 to classify CB grades. The "N" in the prefix indicates normal cure rate, and the following number shows the relative surface area and structure. Table 3.2 compares common CB grades.

The N110 grade with its larger surface area provides additional reinforcement but results in higher compound viscosity while N660 grade delivers better processability and resilience for dynamic application.

The production of CB from fossil resources continues to receive attention. However, scientists investigate sustainable alternatives, i.e., to produce bio-based CB through the processing of biomass materials and algae, as well as to produce recovered carbon black (rCB) through pyrolysis of discarded tires.

The spherical shape of CB primary particles leads to agglomeration because of the van der Waals forces that exist between the nanoparticles. Hence, the surface chemistry and energy of CB require modification to enhance its dispersion within the polymer matrix which leads to better mechanical, electrical, thermal, and optical properties (Mofrad and Ostad Movahed, 2024c).

Researchers have developed multiple methods to modify CB surfaces through grafting functional groups by using oxidation and surfactants, and polymer

Table 3.2 Common CB grades.

Grade	Typical use	Properties
N110	High-performance tires, conveyor belts	Very high reinforcement, high surface area
N220	Truck tire treads	High wear resistance, good processability
N330	Passenger tires, general-purpose use	Balanced properties, standard for many applications
N550	Inner liners, mechanical goods	Lower hysteresis, good dispersion
N660	Sidewalls, vibration mounts	Lower surface area, improved elasticity

grafting, halogenation, thiolation, sulfonation, plasma treatment, thermal treatment, and silane coupling agents. The surface oxidation techniques used to create oxygen-containing groups (hydroxyl, ketone, phenol, ether, and carboxylic acid) on CB surfaces for subsequent covalent, electrostatic, hydrogen bonding modification, and better polar media dispersion. The most frequently used oxidizing agents include nitric acid, sulfuric acid–nitric acid mixtures, hydrogen peroxide, citric acid, maleic acid, and persulfates. The silanization reaction mechanism indicates that silane coupling agents need to bond with hydroxide-rich surfaces to enhance filler dispersion throughout the polymer (rubber) matrix. The research has employed three silane coupling agents including 3-[methoxy (polyethyleneoxy)6-9 propyl]trimethoxysilane (SiPEG), γ-methacryloxypropyl-trimethoxysilane (MEMO), and 3-mercaptopropyltrimethoxysilane (3-MPTMS) to react with surface functional groups of multiple CB grades. The literature contains numerous studies about CB functionalization. A study presents a sustainable method to functionalize CB through dry ball milling with tetraphenyl phosphonium bromide. The modification process improves CB thermal stability and electrical conductivity, which makes it appropriate for developing advanced rubber composites. The researchers also attempted to enhance CB dispersibility and aqueous compatibility through functionalization with natural substances including humic acid and tannic acid. These researches enable scientists to create electrochemical sensors based on CB for environmental monitoring applications. The research also examined how acid-treated CB affects the performance characteristics of tyrosinase biosensors. The modification of CB through functionalization enables better enzyme immobilization and bio-electrocatalytic performance, which leads to biosensing applications in food and medical industries. The study demonstrates the functionalization of CB through diamino-bipyridine and polydopamine to create a platform for Ru complexation in oleic acid catalytic conversion. The research shows that functionalized CB can be used for sustainable catalytic operations.

The research showed that the durability of carboxylated CB nanoparticles when exposed to different solution conditions. The research demonstrated that humic acid creates substantial stability improvements, which proves the need to study how functionalized CB interacts with its environment. A research team developed phenolic formaldehyde resin-treated silica CB which will be used to evaluate its impact on mechanical properties of NR/BR composites containing fillers. The modification process resulted in improved CLD and enhanced mechanical properties of the resulting composite materials. Another research demonstrated that CB nanoparticles show promise for rubber reinforcement because they possess large surface areas and adjustable characteristics. The functionalization of these nanoparticles enables their use in multiple rubber-based applications. As a conclusion, CB modification by different techniques enables better rubber matrix compatibility while addressing its natural drawbacks.

As an illustration, the research involved surface modification of commercial N660 grade CB through different oxidation procedures conducted by Jaberi and Ostad Movahed (2024c) will be described here in detail. In this research, the modified CBs underwent additional treatment with a standard coupling agent which led to results that were evaluated through multiple characterization methods. The treated CB material showed promise to replace traditional rubber compound fillers. The research used four different oxidation techniques to boost the surface functional groups of commercial CB through hydroxyl (—OH) and carboxyl (—COOH) group enhancement. The research details all oxidation procedures in Table 3.3 while specifying the sample designations for each method.

- The first method involved heating CB in concentrated HNO_3 at 90 °C through reflux.
- The second method used a 3:1 solution of concentrated HNO_3 and concentrated H_2SO_4 at 65 °C for reflux treatment.
- In third method, the researchers applied ultrasonic vibration to a mixture of citric acid monohydrate and distilled water for treatment (Sono Swiss, Switzerland, 280 W, 50/60 Hz).
- In fourth method, the researchers treated the material with maleic acid solution and distilled water while using ultrasonic vibration equipment (Sono Swiss, Switzerland, 280 W, 50/60 Hz).

The third oxidation method produced a citric acid monohydrate (CB-Oxi) which became the chosen premodification method for CB surface treatment based on obtained results. The choice of this method resulted from evaluating factors that included technical practicability alongside cost efficiency, safety standards, environmental sustainability, and large-scale manufacturing potential.

The researchers used bis-(triethoxysilylpropyl) tetrasulfide (TESPT) silane coupling agents to treat CB-Oxi samples in this research. The researchers applied the

Table 3.3 The used oxidation methods for CB surface modification along with their corresponding operational parameters.

Method	Used oxidant	Temperature (°C)	Time (min)	Remarks
1	Nitric acid	90	180	Reflux
2	Nitric acid and sulfuric acid	65	120	Reflux
3	CB-Oxi	25	15	Ultrasonic vibration
4	Maleic acid	25	15	Ultrasonic vibration

same silane treatment to CB samples through an identical process for comparison purposes. The process of CB surface modification through oxidation and silane coupling agent functionalization of CB-Oxi follows the sequence shown in Figure 3.1.

The preparation of citric acid monohydrate solution in distilled water required magnetic stirring followed by 15 minutes of ultrasonic treatment (Sono Swiss, Switzerland, 280 W, 50/60 Hz). The CB-Oxi underwent drying in an oven after the oxidation process. The ultrasonic bath treated CB-Oxi with ethanol for 30 minutes to achieve dispersion. The solution received the silane coupling agent after which it underwent reflux treatment at 65 °C for three hours. The solution reached room temperature after the reaction finished so the researchers used a centrifuge (HS 18500, Farzaneh Arman Co., Iran, 230 V, and 50 Hz) to extract the product before performing three ethanol washing cycles. The TESPT-functionalized CB-Oxi underwent drying at 50 °C for 12 hours before receiving the designation "Modified-CB-Oxi" for upcoming examinations. The researchers performed the same steps on CB to obtain "Modified-CB" as their final product. The diagram in Figure 3.2 shows the theoretical reaction pathways and chemical bonding events between CB-Oxi and the silane coupling agent.

The FTIR spectrum of CB appears in Figure 3.3 together with pretreated CBs that underwent different oxidation procedures. The FTIR spectra showed hydroxyl (O—H) and carbonyl (C=O) group signals of carboxylic acid at, 3,300–3,500 and 1,500–1,700 1/cm which confirmed successful surface modification of CB through all applied methods. The spectral intensities reached their highest point when using method 3 compared to all other methods. The carbonyl functional group stretching in carboxylic acid appeared at 1,711, 1,699, and

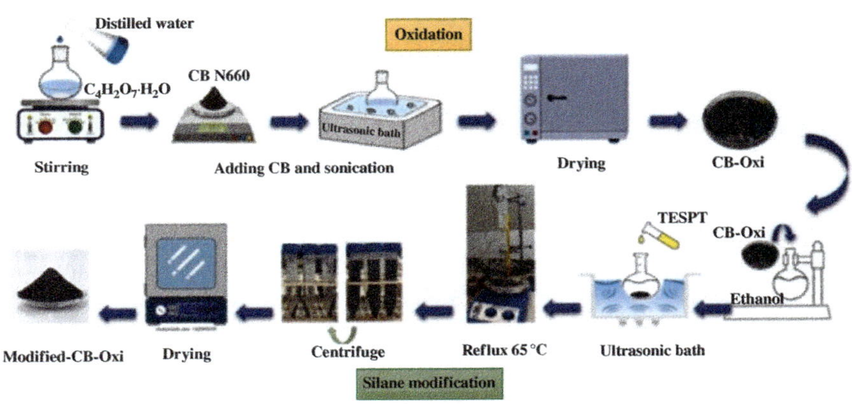

Figure 3.1 The process starts with sequential oxidation of CB followed by functionalization with silane coupling agents.

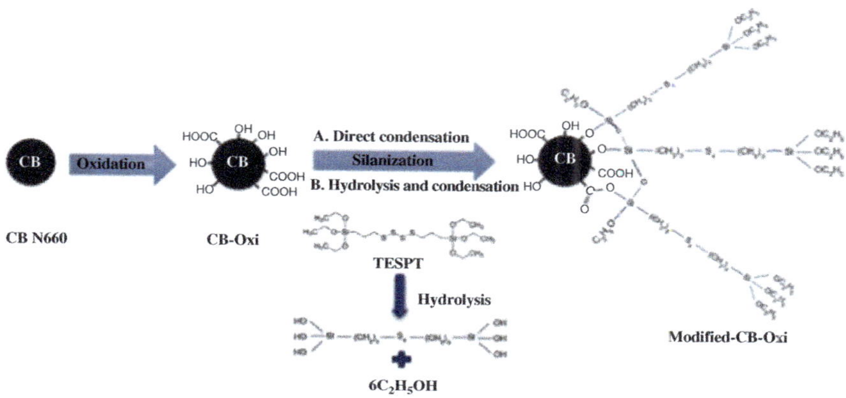

Figure 3.2 The proposed mechanism describes the chemical interactions between CB and oxidant and TESPT.

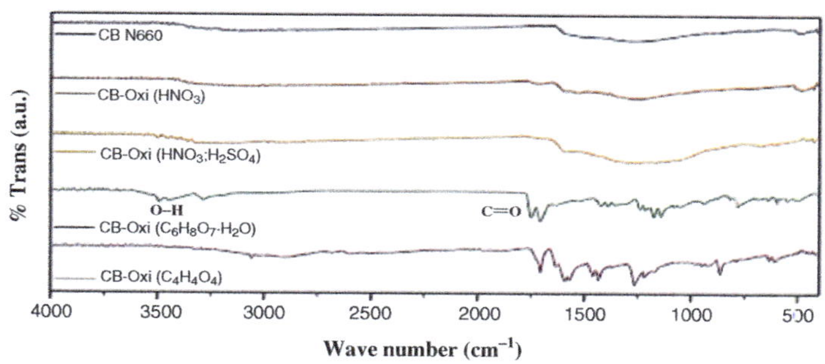

Figure 3.3 The comparison of FTIR spectra of CB without premodification with those premodified with different oxidation methods as described in Table 3.3.

1,705 1/cm for oxidation methods 1, 2, and 4, respectively. The peaks from method 3 showed two distinct carbonyl functional group signals at 1,748 and 1,706 1/cm with enhanced intensity (Figure 3.4). The peaks at 3,493 and 3,447 1/cm correspond to the stretching vibrations of the —OH group. The hydroxyl group intensities from all methods showed the highest values in method 3 according to Figures 3.3 and 3.4. The surface functionalization of commercial CB succeeded through all applied techniques for hydroxyl and carboxylic groups. The results showed better clarity through method 3. The treatment process of method three eliminates the need for extensive heating and multiple washing and filtering steps that are required in other methods.

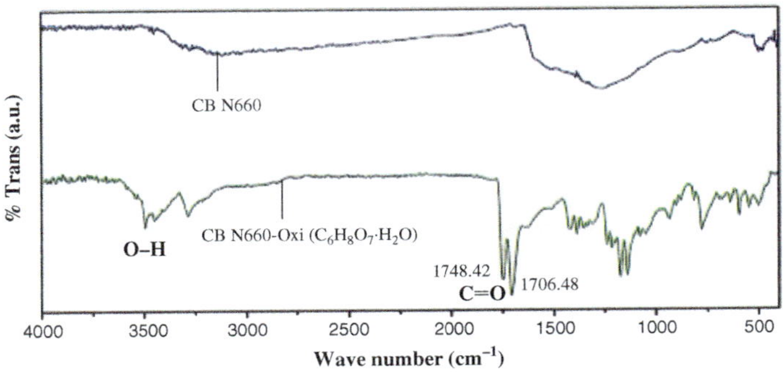

Figure 3.4 The comparison of FTIR spectra of CB without premodification with those premodified with method 3 as described by Table 3.3.

The authors directed readers to observe how these methods produce different results through their analysis. The main points of evaluation become clear through the following statements:

- The use of citric acid proves superior to HNO_3 as an environment-friendly acid because it provides a sustainable alternative.
- The citric acid method demonstrates both cost-effectiveness and operational simplicity when compared to other available methods.
- The method shows promising potential for large-scale production because it maintains its effectiveness during scale-up operations.
- The reaction conditions of citric acid processing prove safer than other available methods. The method operates at room temperature while requiring only 15 minutes of reaction time which differs from other methods that need elevated temperatures.
- The FTIR spectrum analysis showed that citric acid treatment makes CB more reactive for subsequent functionalization procedures.

The authors selected method three for CB-Oxi silane modification because of the above provided information and results.

The traditional silane coupling agent TESPT served to modify the surfaces of CB and CB-Oxi. The FTIR spectra of TESPT, CB-Oxi, and Modified-CB-Oxi appear in Figure 3.5. The FTIR spectrum of TESPT shows three distinct bands at 2,973, 2,926, and 2,885 1/cm which correspond to CH_2 and CH_3 symmetric and asymmetric stretching vibrations. The FTIR spectrum shows Si—O stretching vibrations from methoxy (Si—O—CH_3) and siloxane (Si—O—Si) groups between 925 and 1,195 1/cm. The C—O stretching vibrations between 1,084 and 1,290 1/cm overlap with the Si—O stretching vibrations of the silane group. The two peaks at 1,058 and 1,111 1/cm in Modified-CB-Oxi correspond to Si—O and/or C—O stretching vibrations.

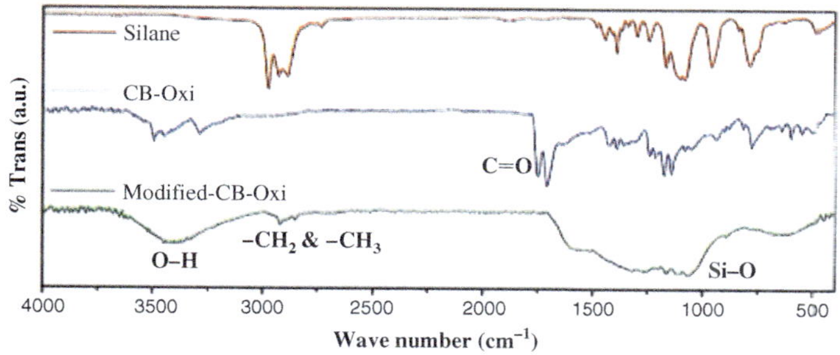

Figure 3.5 Comparison of FTIR spectra of silane coupling agent (TESPT), CB-Oxi, and Modified-CB-Oxi with silane groups.

Table 3.4 Elemental analysis data of Modified-CB and Modified-CB-Oxi derived by CHNS analysis.

	CHNS analysis result	
Sample	**Element**	**Content (%)**
Modified-CB	C	87
	H	0.2
	N	0.1
	S	2.2
	Rest (O & Si)	10.5
Modified-CB-Oxi	C	57.2
	H	0.3
	N	0.1
	S	6.3
	Rest (O & Si)	36.1

The Modified-CB-Oxi spectrum displays two peaks at 2,912 and 2,847 1/cm which stem from TESPT's CH_2 and CH_3 symmetric and asymmetric stretching vibrations. The FTIR spectra of TESPT-treated CB and Modified-CB show no noticeable differences between them. The FTIR technique fails to establish the exact bonding state of TESPT on Modified-CB and Modified-CB-Oxi surfaces. The samples underwent CHNS and EDS (energy-dispersive X-ray spectroscopy) elemental analysis to determine the state of silane (TESPT) bonding on their surfaces.

The CHNS elemental analysis results in Table 3.4 show the weight distribution of C, H, N, and S elements between Modified-CB and Modified-CB-Oxi samples.

Table 3.5 The elements wt.% for CB, Modified-CB, and Modified-CB-Oxi obtained by EDS.

Sample	C (wt.%)	S (wt.%)	Si (wt.%)	O (wt.%)
CB	98.4	1.6	—	—
Modified-CB	98.3	1.7	—	—
Modified-CB-Oxi	89.1	2.1	0.9	7.9

The weight percentage of elements beyond C, H, N, and S in the samples appears in Table 3.4 as the cumulative total of O and Si elements. The weight percentages of sulfur and silica and oxygen in Modified-CB-Oxi exceed those of Modified-CB by three times according to the data in Table 3.5. The weight percentages of Modified-CB-Oxi contain 42.4%, while Modified-CB contains 12.7%. The surface modification of CB through oxidation (method 3) created CB-Oxi which developed additional surface hydroxyl and carboxylic functional groups. The increased number of functional groups on the CB surface enables better TESPT bonding according to Figure 3.2. The oxidation modification of CB through method 3 produces increased hydroxyl and carboxylic functional groups on the surface, which enhances performance beyond untreated CB. The CHNS analysis method fails to detect the presence of silicon elements in the sample. The researchers performed EDS analysis to determine the elemental composition of silicon.

The weight distribution of elements in CB, Modified-CB, and Modified-CB-Oxi samples appears in Table 3.5 through EDS analysis. The untreated CB contains 98.4% carbon and 1.6% sulfur according to its elemental composition. The Modified-CB material consists of 98.3% carbon and 1.7% sulfur content. The chemical composition of Modified-CB-Oxi shows 89.9 wt.% carbon, 2.1 wt.% sulfur, 0.9 wt.% silica, and 7.9 wt.% oxygen. The EDS spectra in Figure 3.6 show that CB and Modified-CB lack the silicon (Si) peak at 1.8 keV but Modified-CB-Oxi displays this peak. The successful attachment of silane to CB surface material is confirmed by these results. The absence of Si in CB and Modified-CB samples together with its presence in Modified-CB-Oxi proves that silane functional groups successfully bonded to CB particles. The FTIR spectrum of Modified-CB shows C—O stretching vibrations because the CHNS and EDS results indicate no bond formation between CB and silane.

The FESEM images of CB (a), Modified-CB (b), CB-Oxi (c), and Modified-CB-Oxi (d) appear in Figures 3.7 and 3.8 at two different magnifications of 2,000 nm (2 μm) and 200 nm. The FESEM images of CB (Figures 3.7a and 3.8a) show spherical particles that measure less than 200 nm in diameter. The oxidation treatment of CB (method 3) resulted in observable particle accumulation and

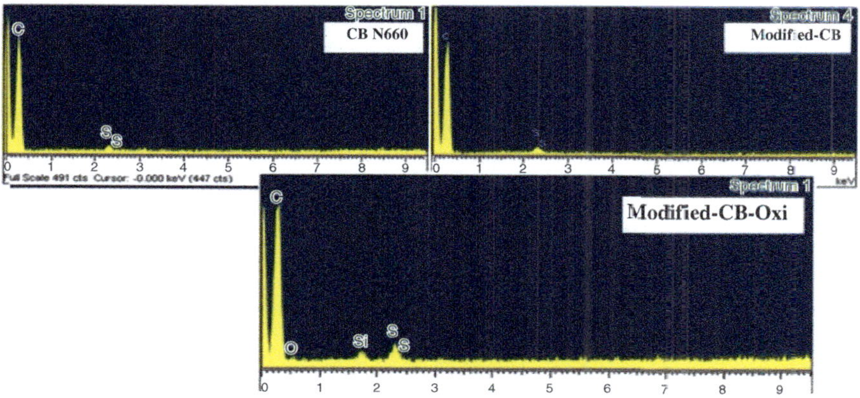

Figure 3.6 EDS spectra CB, Modified-CB, and Modified-CB-Oxi.

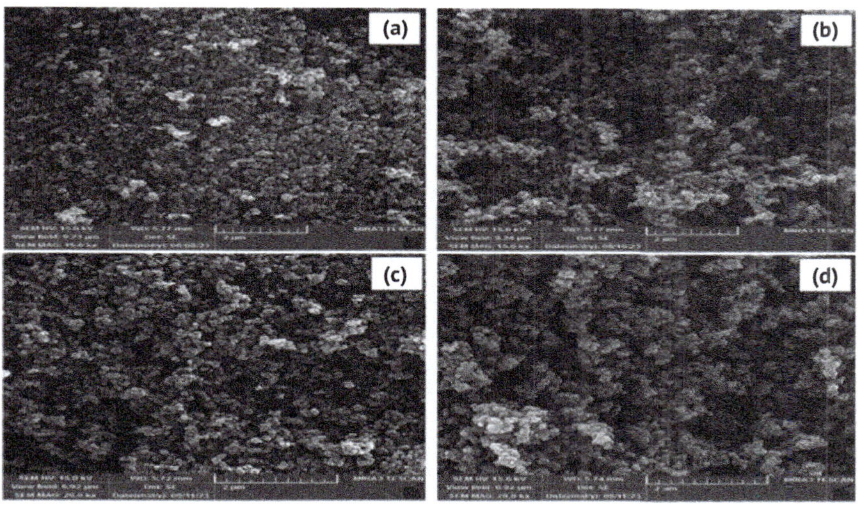

Figure 3.7 The SEM micrographs of (a) CB, (b) Modified-CB, (c) CB-Oxi, and (d) Modified-CB-Oxi at a magnitude of 2,000 nm (2 μm).

agglomeration as shown in Figures 3.7c and 3.8c. The surface functional groups including hydroxyl and carbonyl and carboxyl groups on CB particles enhance their intramolecular forces through hydrogen bonding. The surface treatment of CB with method 3 oxidation resulted in observable agglomeration as shown by comparing Figures 3.7c and 3.8c with Figures 3.7a and 3.8a. The surface examination of CB and Modified-CB particles through Figures 3.7a, 3.8a, 3.7b, and

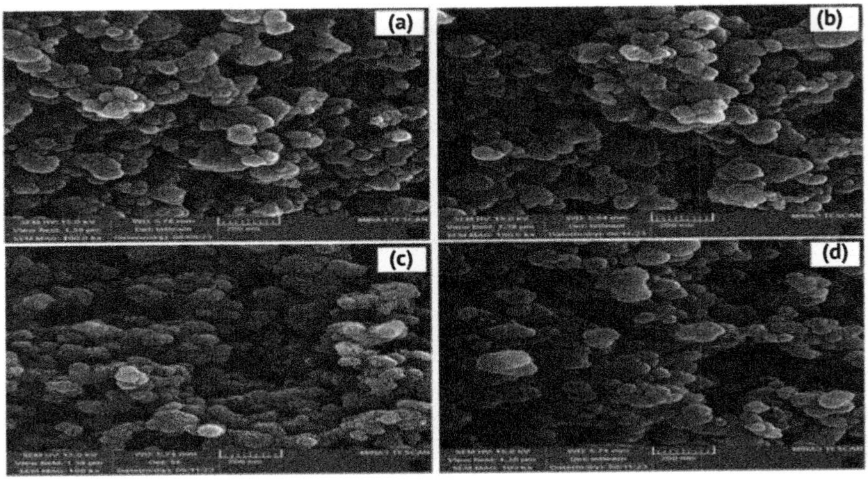

Figure 3.8 The SEM graphs of (a) CB, (b) Modified-CB, (c) CB-Oxi, and (d) Modified-CB-Oxi at the magnitude of 200 nm.

3.8b reveals no significant differences between them. The results from FTIR, CHNS, and EDS analysis support these findings. The TESPT molecules present on Modified-CB-Oxi particles create new particle separation and reduce particle agglomeration as shown in Figures 3.7c and 3.8c compared to Figures 3.7d and 3.8d. As already stated, the silane molecules successfully bonded to the CB surface.

As an important conclusion, the surface modification of CB through functionalization resulted in enhanced hydrophobic characteristics of the material. The resulting effect of this modification should lead to better CB dispersion.

The Raman spectra of carbonaceous materials contain two dominant peaks which appear at 1,600 1/cm (G-band "graphite") for sp^2 hybridized carbons and 1,350 1/cm (D-band "defect or disorder") for sp^3 hybridized carbons. The $I(D)/I(G)$ ratio serves as an indicator to determine the microstructure of carbonaceous materials. The $I(D)/I(G)$ ratio showed a decrease from 0.982 in CB to 0.972 in Modified-CB-Oxi and 0.918 in CB-Oxi. It was concluded that the addition of —OH and —COOH functional groups together with silane groups minimized disruptions to the graphite structure of CB.

The XRD patterns of CB, Modified-CB, and Modified-CB-Oxi are shown in Figure 3.9. The diffraction peak at $2\theta = 25.1°$ from the irregular graphite plane (002) of CB indicates its coke-like structure with disordered carbon interlayers. The (100) peak at 43° shows low intensity because CB contains poorly crystalline carbon structures. The (002) diffraction peak of CB remained unchanged in Modified-CB but shifted to $2\theta = 24.85°$ (smaller angles) in Modified-CB-Oxi. The silane groups in Modified-CB-Oxi caused an increase in the interlayer distance

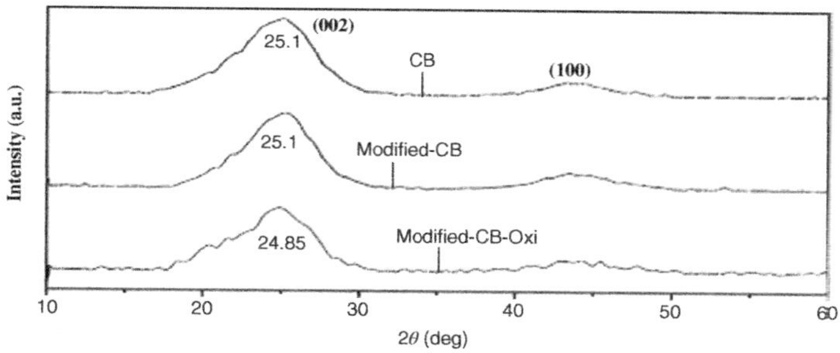

Figure 3.9 The XRD patterns of CB, Modified-CB, and Modified-CB-Oxi.

(002) which resulted in this peak shift. The peak intensity at (100) decreased in the Modified-CB-Oxi sample according to Figure 3.9.

The findings from the above sections of the study create a connected relationship between them. The Modified-CB-Oxi shows increased oxygen and silicon content according to CHNS and EDS analysis because oxidation creates hydroxyl and carboxyl functional groups which enable silane attachment. The Raman and XRD results show that Modified-CB-Oxi has a lower $I(D)/I(G)$ ratio and its (002) peak position moves to the left which indicates minimal structural damage and expanded interlayer distance that enables effective silane penetration and bonding. The oxidation process generates functional groups which enable silane molecules to bind to the material. The surface characteristics of CB undergo changes through silane bonding, which also alters its chemical makeup. The combined effects from these modifications lead to better interaction between Modified-CB-Oxi and nonpolar environments.

The contact angle measurement stands as the most widely used technique for surface energy comparison between different solids. The contact angle measurement of water droplets placed on solid surfaces allows researchers to determine their surface energy properties. The contact angle indicates how surface energy between the solid and liquid interacts with their interfacial energy. The contact angle with water indicates solid surface wettability because high contact angles with water (73 mN/m surface tension) show low wettability which corresponds to low surface energy. The contact angle measurement of CB reached 95°. The contact angle measurements showed that Modified-CB maintained its original value, but CB-Oxi decreased to 80° and Modified-CB-Oxi increased to 112° after surface treatment which resulted in surface energy changes.

From this research, it was concluded:

- The FTIR analysis results demonstrate a successful method to modify commercial-grade CB surfaces through citric acid oxidation.

- The researchers chose this method because it provides a simple approach that is affordable, scalable, and environmentally friendly.
- The FTIR spectra shows that CB surface modification succeeded through the addition of hydroxyl and carboxyl functional groups.
- The researchers successfully applied a commercial silane coupling agent to modify the surface of CB that had undergone previous treatment.
- The grafting process achieved high efficiency according to FTIR, FESEM, CHNS elemental analysis, XRD, and EDS results.
- The complete evaluation confirmed that the chosen coupling agent successfully transformed the surface properties of the commercial CB material.
- The modified CB shows promising characteristics that make it suitable for use as a rubber compounding filler. The proposed functionalization approach proves effective for improving commercial CB performance when used in rubber-based materials.

3.2 Silica

The rubber industry uses silica (silicon dioxide, SiO_2) as its second most popular reinforcing filler after CB. The material provides multiple performance advantages which make it suitable for tire and nontire applications that need low rolling resistance and high wet traction and aging resistance. Table 3.6 shows important physical properties of used silica in the rubber industry.

The production of precipitated silica occurs through controlled acid–base reactions between sodium silicate and sulfuric acid:

$$Na_2SiO_3(aq) + H_2SO_4(aq) \rightarrow SiO_2(s) + Na_2SO_4(aq) + H_2O(l)$$

The reaction between sodium silicate solution and sulfuric acid solution produces SiO_2 solid particles, Na_2SO_4 solution, and liquid water. The final product

Table 3.6 Physical characteristics of silica.

Property	Typical range/value
Surface area (BET)	120–250 m^2/g
Particle size (aggregate)	10–40 nm
Pore volume	1–2.5 cm^3/g
pH (5% in water)	6–7.5
Bulk density	2.20–2.65 g/cm^3
Moisture content	≤7%

Table 3.7 Advantages of silica in rubber compounds.

Property enhanced	Mechanism
Rolling resistance (tires)	Reduced hysteresis loss
Wet traction	Increased filler–polymer interfacial friction
Heat aging resistance	Improved oxidative stability
Electrical insulation	Inherent dielectric property of SiO_2

Table 3.8 Comparison of CB with silica in rubber compounding.

Feature	Silica	Carbon black
Reinforcing efficiency	High with silane	Naturally high
Abrasion resistance	Moderate	Excellent
Electrical conductivity	Insulating	Conductive
Compatibility	Needs coupling agents	Good with nonpolar rubbers
Rolling resistance	Lower hysteresis (better)	Higher hysteresis

properties emerge from precise management of temperature, pH levels, agitation speed, and aging duration. Table 3.7 summarizes the advantages of silica as a filler in rubber compounds.

The tire industry utilizes silica–silane systems to create environmentally friendly tires which reduce fuel usage while providing better traction in wet conditions. The use of silica in truck and bus radial tires (TBRs) has become more widespread because it enhances both tire durability and extends their operational life (Ostad Movahed, 2014). Table 3.8 compares CB and silica in brief.

However, the processing of rubber matrices with silica faces special challenges because of its surface characteristics and structural features. The hydrophilic nature of silica emerges from its extensive silanol group presence on its surface, which creates poor interactions with hydrophobic rubber materials. The compound viscosity increases, and mechanical performance decreases when filler–filler interactions occur because of poor dispersion unless proper coupling agents and mixing optimization are applied. The BET surface area measurement of silica reaches between 120 and 250 m^2/g. The numerous silanol groups on the surface of silica enable hydrogen bonding to occur. The rubber compounds experience a significant Payne effect because of silica presence.

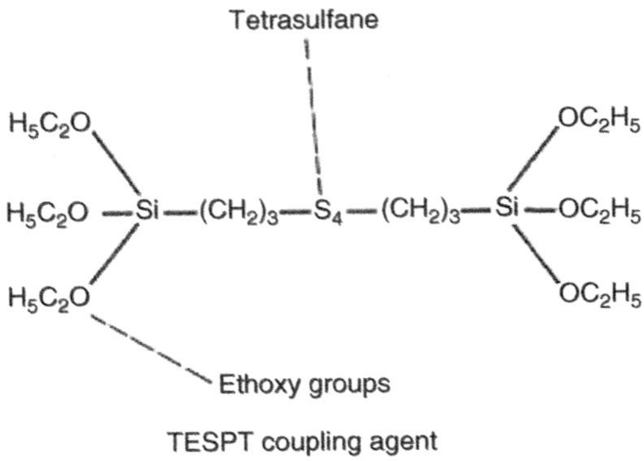

Figure 3.10 Coupling agent TESPT.

To improve silica dispersion, several techniques were proposed by researchers. Silane coupling agents serve as the primary method to enhance silica dispersion. The coupling agent TESPT (see Figure 3.10) creates chemical bonds between silica surfaces and rubber molecular chains. The combination of silane coupling agents with rubber chains leads to improved filler dispersion and reduced Payne effect and rolling resistance. Another technique may be the addition of silica through multiple stages in stepwise mixing, which produces better dispersion results. Internal mixers with high shear capabilities break down silica agglomerates into smaller particles. The activation of silane requires elevated temperatures between 140 and 160 °C but operators need to monitor the process duration to prevent premature chemical reactions. Also, the production of modern precipitated silica results in silica particles with more uniform dimensions. The controlled surface treatment and porosity structure of these materials enable better dispersibility.

The fine powder nature of silica creates problems during dusting operations and handling procedures. The mixing process creates viscosity increases because of hydrogen bonding due to the existence of silica (Payne effect). The risk of premature silane activation during processing creates scorch safety problems. The production of consistent batches with silica-filled rubber compounds proves more difficult than working with CB-based systems.

As an important conclusion, the implementation of silica into rubber matrices for improved rolling resistance and wet traction requires proper compounding methods, effective dispersion aids, mixing techniques, and suitable processing conditions.

3.3 Payne Effect

The Payne effect represents a well-documented viscoelastic effect which occurs in filled elastomers that contain reinforcing materials such as CB or silica. The storage modulus (G') shows a nonlinear decrease when strain amplitude increases during dynamic mechanical analysis (DMA) tests at low strain values below 10%. The Payne effect gets its name from A. R. Payne because it shows how rubber matrix weak physical bonds between filler aggregates break down when subjected to strain which causes the material's storage modulus to decrease. The Payne effect becomes most noticeable in rubber compounds with high filler content because their initial strong filler–filler and filler–polymer network structures deteriorate when subjected to rising deformation levels.

The Payne effect emerges at the molecular and microstructural level because of the following:

- Filler–filler interactions: the filler particles create a network structure through van der Waals forces and hydrogen bonding at rest and under minimal strain conditions especially in silica-based systems. The initial modulus receives substantial contributions from this network structure.
- Breakdown of filler network: the weak physical bonds between components start to break down when stress increases, which results in a decrease of modulus values. The storage modulus reaches a plateau after the filler network breaks down because the matrix takes over as the dominant material.
- Reformation hysteresis: the filler network shows partial recovery after strain removal, but it usually experiences hysteresis which results in energy loss that generates heat during dynamic operations.

The Payne effect becomes visible through storage modulus (G') measurements against strain amplitude (γ) at fixed frequency (10 Hz) and temperature conditions (room temperature or 60 °C for tire analysis) by DMA. The Payne effect measurement involves calculating the Payne amplitude through the comparison of G' values at 0.1% strain and 10% strain. The strength of filler–filler interactions can be quantitatively assessed through the measurement of $\Delta G'$.

The Payne effect requires thorough analysis because it directly affects tire and dynamic rubber product design through the following reasons:

- The energy loss from high $\Delta G'$ values results in higher tire rolling resistance.
- The dispersion quality of fillers particularly silica with silane coupling agents determines the magnitude of $\Delta G'$.
- The Payne effect reduction leads to better fuel economy and enhanced dynamic response in tires.

3.4 Mullins Effect

The Mullins effect received its name from L. Mullins who discovered this stress-softening behavior in filled elastomers during his work at the MRPRA (Malaysia Rubber Producers' Research Association) in the 1940s. The phenomenon occurs in CB or silica-filled rubbers when these materials experience cyclic loading and unloading processes. The material shows reduced stress response when it undergoes reloading after its first loading cycle at the same strain level. The first stretch of rubber makes it more pliable for subsequent stretching during the second deformation cycle when the strain levels remain constant.

The Mullins effect appears most strongly in elastomeric materials containing CB or precipitated silica fillers because these materials experience several main processes. Fillers in the network experience disruption during the first loading phase because their physical bonds between aggregates and between aggregates and matrix chains get destroyed. The material loses its stiffness when it undergoes additional loading cycles at the same strain levels.

Also, because of chain uncoiling and slippage, the initial resistance of long polymer chains against deformation leads to their partial alignment and disentanglement. The material shows reduced resistance to deformation, which results in lower stress during reloading.

Besides, the formation of voids occurs near filler particles because of debonding when the matrix-filler bond strength remains low. The microstructural damage from this process becomes permanent after each loading cycle and continues to grow with subsequent cycles.

The Mullins effect shows its characteristics through cyclic stress–strain curves. The first loading cycle produces higher stress values than strain levels. The stress values decrease in subsequent cycles while creating a hysteresis loop between the curves. The stress difference between the first and second loading cycle serves as the standard method to quantify the Mullins effect. The Ogden–Roxburgh model along with other constitutive models describes softening behavior through damage parameters which track strain history evolution.

The Payne effect appears in DMA measurements at low amplitude ranges but the Mullins effect becomes dominant in quasi-static and low-frequency cyclic loading conditions. The two effects stem from filler–filler and filler–matrix interactions, yet they operate at different time periods and strain levels.

The following methods help decrease the severity of the Mullins effect:

- Improving filler dispersion.
- The use of coupling agents such as silanes for silica represents one possible solution.
- The rubber matrix requires specific modifications to its polarity levels.
- The rubber component needs prestraining before application in specific uses.

3.5 Network Structure and Bound Rubber Concept in Rubber Compounds

Rubber compound performance depends heavily on two main factors which include the crosslinking process that creates molecular networks and the way reinforcing fillers interact with the rubber material. The mechanical strength, elasticity, and durability of filled rubber systems depend on two fundamental factors as follows:

- Network structure which refers to the three-dimensional pattern of polymer chains that link together through chemical or physical crosslinking bonds.
- The polymer segment which becomes trapped at the filler interface through physical or chemical bonding processes is known as bound rubber.

The knowledge of these concepts enables the creation of rubber formulations which achieve high performance through optimized viscoelastic and mechanical properties.

The vulcanization process creates three-dimensional networks through covalent bonding between polymer chains. The properties of vulcanized rubber depend on both the density and type of crosslink bonds which include mono-, di-, and polysulfidic in sulfur-based and carbon–carbon in peroxide-based crosslinks. The properties of vulcanized rubber include elastic recovery, modulus, swelling resistance, and heat aging performance. The rubber material forms a pseudo-network structure through physical junctions and entanglements which exist before chemical curing and crystallites in NR. The physical network elements contribute to three main properties of rubber materials. The precured green strength, initial stiffness, and orientation effects during processing depend on these physical network elements. However, the network structure of rubber compounds consists of three main components which include permanent covalent crosslinks, temporary physical crosslinks, and filler-induced restrictions.

The tightly adsorbed or chemically grafted polymer fraction remains nonextractable by common solvents before curing constitutes bound rubber. The filler particles receive a core–shell structure from bound rubber which functions as a vital component for filler–matrix interaction, reinforcement efficiency, stress transfer, and dynamic behavior.

The formation of bound rubber occurs through:

- The combination of van der Waals forces with CB leads to the formation of bound rubber.
- The combination of silica with NBR or SBR rubber through hydrogen bonding results in bound rubber formation.
- The process of covalent bonding occurs when silane-treated fillers interact with rubber. The polymer chains become trapped inside the filler aggregates, which results in the formation of bound rubber.

For measurement of bound rubber, a common method involves combining rubber that has not been cured with filler materials. Subsequently, the free polymer can be extracted from the mixture by using toluene as a solvent and finally, the insoluble residue is dried before obtaining weight.

Bound rubber content is calculated as:

$$\text{Bound rubber (\%)} = \frac{W_{\text{residue}} - W_{\text{filler}}}{W_{\text{polymer}}} \times 100$$

where

- W_{residue} is the weight of insoluble residue after extraction
- W_{filler} is the weight of filler originally added
- W_{polymer} is the weight of total polymer (rubber)

The mechanical reinforcement of filled rubbers receives support from bound rubber through the interphase as a stress transfer, which connects the matrix to the filler. Also, bound rubber restricts the movement of polymer chains, resulting in higher material stiffness. The material absorbs more energy and consequently dissipates it when subjected to dynamic loading with increasing bound rubber. The presence of bound rubber decreases the tendency of silica particles to form floccules.

It was concluded that the fundamental principles of rubber technology depend on understanding network structure and bound rubber concepts which explain how rubber properties relate to its structure. Rubber technologists use filler selection, surface modification, and formulation design to manage these features which enable them to create elastomers with specific properties for multiple uses.

3.6 SEM/AFM/TEM Analysis of Polymer–Filler Interfaces

Rubber compound performance and morphology depend heavily on the interface quality between rubber matrix and reinforcing filler particles including CB and silica. The direct observation of filler dispersion and interfacial bonding and phase morphology requires microscopic tools, namely, scanning electron microscopy (SEM), transmission electron microscopy (TEM), and atomic force microscopy (AFM) to provide microscopic details beyond what mechanical testing and swelling studies can show. Figure 3.11 illustrates a typical SEM image.

The SEM image shows both the surface shape and the broken material structure. The method assesses how well fillers are distributed and how they form clusters and separate from the matrix material. The method detects voids and cracks, and filler pull-outs which indicate weak adhesive bonds. Figure 3.12 shows the filler (silica) dispersion in the IR rubber matrix. The process of cryofracturing samples in liquid nitrogen as a presample preparation for SEM technique reveals the interactions between fillers and matrix materials. The application of gold or

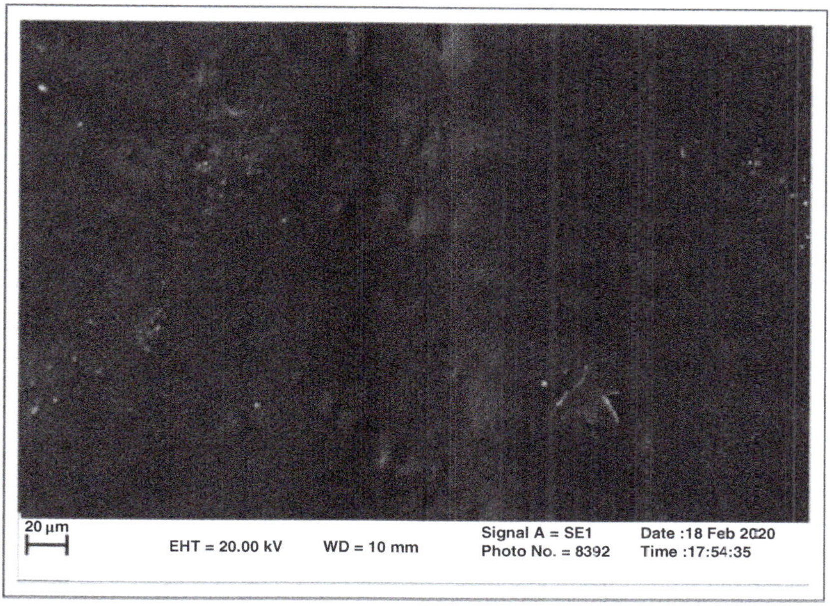

Figure 3.11 The interphase between EPDM and SBR. *Source*: Jourabchi and Ostad Movahed (2024) / with permission of Sage Publications.

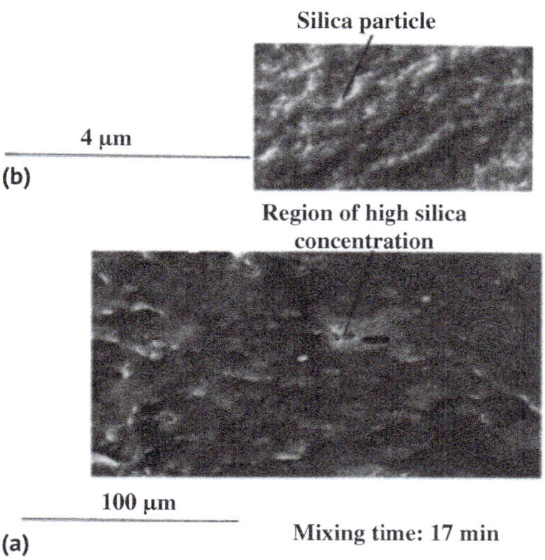

Figure 3.12 SEM images represent the silica dispersion in the IR rubber matrix.

platinum coatings on samples serves to decrease charging effects because rubbers do not conduct electricity.

The TEM technique provides detailed nano-scale images which show how fillers arrange themselves and how they interact with each other. The technique provides excellent results for studying nanocomposites containing silica and nanoclay in rubber materials. The technique shows how rubber layers bond to each other and how different phases penetrate each other and how core–shell structures appear. In this technique, the preparation of rubber compounds requires ultramicrotomy to produce thin sections which range from 50 to 100 nm in thickness. The samples need to be embedded in resin or undergo cryosectioning procedures. The use of osmium tetroxide as a staining agent improves the visibility between different material phases. Figure 3.12 illustrates the SEM images representing the silica dispersion in IR rubber matrix.

The AFM serves as a powerful nanoscale imaging and measurement tool which scientists use throughout materials science to study rubber technology and composite systems through surface analysis, mechanical property evaluation, and interfacial interaction assessment. The AFM provides researchers with unique capabilities to study the surface interactions between rubber matrices and their reinforcing fillers (CB, silica, and nanoclays) through detailed high-resolution imaging. Figure 3.13 shows the AFM image for interphase measurement in a rubber blend.

The AFM operates with a small tip of a few nanometers radius which is attached to a cantilever. The scanning process of the tip across the sample surface generates deflection measurements of the cantilever through laser beam and photodiode detection during surface force interactions. The AFM enables surface analysis through three different interaction modes, which include contact mode, tapping mode, and noncontact mode. The instrument generates three types of maps which include height maps, mechanical property maps, and phase contrast or friction maps. The AFM technique enables researchers to detect both filler particles within the matrix and bound rubber layers surrounding the fillers and

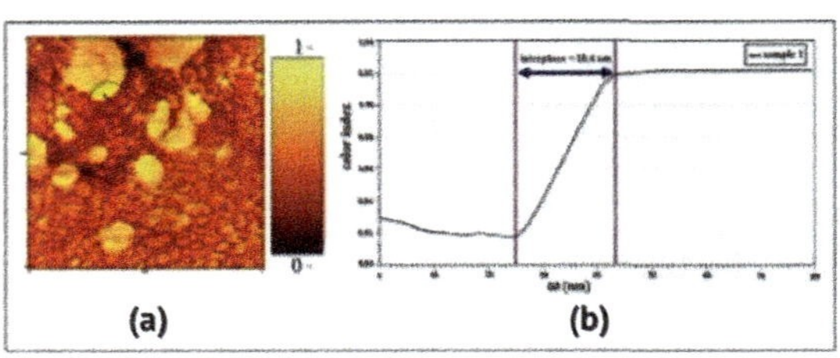

Figure 3.13 The application of AFM image for measurement of interphase in a rubber blend. *Source:* Jourabchi and Ostad Movahed (2024) / with permission of Sage Publications.

various morphological defects such as voids and debonding and microcracks. The technique enables researchers to detect both filler particles inside the matrix material and bound rubber layers surrounding the fillers and various morphological defects including voids and debonding and microcracks. The technique provides nanometer-scale measurements of spatial modulus and stiffness variations across the material surface.

The tip maintains contact with the surface during contact mode measurements, but this method causes damage to soft materials including rubber. The intermittent contact method of tapping mode allows the tip to oscillate and touch the surface lightly, which makes it suitable for soft polymer systems because it decreases both lateral forces and sample damage. The peak-force or fast force mapping mode generates force–distance curves at each point to create maps of indentation modulus and adhesion force and deformation which are essential for polymer composite analysis. The phase imaging technique adjusts tip response phase lag to identify different viscoelastic properties between filler and rubber matrix and bound rubber regions.

The AFM insights following main approaches on rubber–filler systems:

- Bound rubber and interphase layers: the thin bound rubber layer surrounding filler aggregates becomes detectable through AFM because its local modulus exists between the filler particle modulus and the bulk matrix modulus. The modulus mapping technique reveals a smooth transition between filler and shell and matrix phases which proves the core–shell structure of the material.
- Filler dispersion and agglomeration: the topography images from AFM reveal the distribution of filler clusters together with their dimensions and the distances between them in the matrix material. The quality of dispersion becomes evident through small uniform particles that maintain a tight size range while poor dispersion results in big clusters which create wide gaps between them and leads to decreased macroscopic performance.
- Mechanical modulus contrast force mapping modes enable researchers to measure the stiffness levels across different areas of the material. The distribution of filler particles and bound rubber domains becomes visible through mapping allows researchers to measure how filler content, surface modifications, and mixing techniques affect material properties.
- Debonding, micro-voids and damage sites: the rubber–filler interface becomes detectable through AFM height and phase imaging when microscopic voids or debonding zones appear which can trigger mechanical failure under dynamic loading conditions. The identification of these specific areas enables researchers to study how materials develop fatigue and where cracks start to form.

However, there are several practical considerations and challenges in AFM for rubber composites which should be considered. The preparation of rubber-filled compounds for AFM imaging requires microsectioning through cryo-ultramicrotomy or polishing to achieve flat surfaces with roughness below 1 μm. The interaction between the AFM tip and sample surface requires gentle forces because soft rubber materials break easily when subjected to excessive pressure. The measurement of success depends heavily on selecting appropriate cantilever stiffness and controlling oscillation amplitude and scan speed. The large difference in modulus between fillers and matrices creates difficulties for measurement interpretation, which requires proper calibration procedures. The small area scanned by AFM (μm^2) requires multiple scans and statistical averaging to achieve representative results.

The results from AFM studies enable direct formulation choices for product development. The results from AFM studies enable formulators to make changes to coupling agent chemistry and loading levels for better rubber layer adhesion. The observation of large agglomerates through AFM analysis requires formulators to modify their mixing procedures and vacuum venting protocols. The presence of microvoids and debonding zones in imaging data enables researchers to predict material failure through fatigue and dynamic modulus loss measurements. The evaluation of surface treatments becomes possible through nanoscale analysis which measures the improvement of filler dispersion and interphase quality between treated and untreated fillers.

In summary, the study of rubber technology depends heavily on SEM, TEM, and AFM for understanding the microscopic and nanoscopic arrangement of fillers throughout the material (rubber). They also assess the surface treatment methods and compatibilizer performance and the presence of bound rubber, and the strength of interfacial connections. The findings from these techniques enable scientists to create advanced rubber compounds which enhance tire performance, seal durability, and mechanical system stability.

3.7 DMA of Filled Polymers

The analysis of filled rubber materials through DMA serves as a key testing method. The analysis of rubber compounds through DMA enables researchers to study their viscoelastic properties when filled with CB, silica, and hybrid nanofillers. The testing method reveals information about polymer chain movement, filler–polymer bonding, and polymer composite structural stability when subjected to different thermal and mechanical conditions.

DMA tests rubber specimens by applying sinusoidal stress or strain to measure their response. The test reveals three essential properties of materials through its results. The material stiffness appears through its stored energy which the storage modulus (E' or G') measures. The viscous energy dissipation properties of materials appear through their loss modulus measurements which are denoted as E'' or G''. The ratio of loss to storage modulus in tan delta ($\tan\delta = E''/E'$) shows both energy dissipation and glass transition temperature (T_g). DMA tests rubbers by running temperature-controlled tests from −100 to 100 °C to study material transitions and filler responses. Figure 3.14 represents the tanδ versus temperature for several filled rubbers, SBR, BR, NR, and IR with silanized silica nanofiller.

DMA allows to observe the glass transition temperature in filled rubbers. The addition of fillers to rubber materials results in changing the T_g of the gum (uncured without filler) rubber. The T_g of filled rubbers shows a wider peak that shifts upward temperatures because polymer chains near the filler surface become restricted (bound rubber formation). The peak tanδ values decrease based on how well the fillers are dispersed throughout the material and the extent of network formation. Also, the rubbery region of filled rubbers shows a substantially increased storage modulus than unfilled rubber materials because of reinforcement effects. The level of damping in hybrid or semireinforced systems depends on the combination of filler materials and their dispersion quality. The material exhibits high modulus at low strain values because the filler–filler network remains unbroken. The Payne effect causes the modulus to decrease when the material experiences increasing strain because the network structure breaks down. The speed and degree of modulus reduction help scientists understand how fillers interact with each other and their strength of interaction.

DMA provides superior performance compared to traditional tensile testing methods through its ability to detect small changes in material structure and its

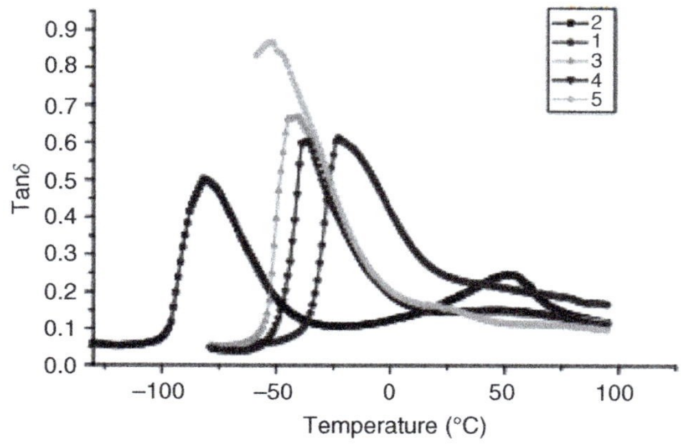

Figure 3.14 The tanδ versus temperature for several filled rubbers, SBR, BR, NR, and IR with silanized silica nanofiller at 20 Hz oscillation amplitude.

dual measurement of elastic and viscous properties and also, its simultaneous assessment of time and temperature effects.

The data from DMA tests enable engineers and scientists to achieve the following goals:

- The process of developing tire compounds, mounts, and gaskets requires optimization through DMA data analysis.
- The system uses predictive models to forecast how materials will behave when subjected to moving loads such as tire rolling resistance and wet surface traction.
- The system tracks the curing process and aging behavior of vulcanized rubber materials.

The analysis of material properties enables scientists to create better designs for applications that need damping or precise operations.

3.8 Differential Scanning Calorimetry and Modulated Temperature DSC (MTDSC) in Filled Polymers Analysis

The characterization of rubber materials requires a thermal analysis to detect their glass transition temperature (T_g), melting points, crystallization points, and curing behavior. The analysis of rubber compounds depends on two essential tools which are differential scanning calorimetry (DSC) and modulated temperature DSC (MTDSC) (Ostad Movahed et al., 2009b). The following text explains the application of these methods for rubber compound analysis.

The DSC technique measures the thermal transition heat flow of samples while tracking temperature changes or time progression. The technique finds application in three main areas, the measurement of T_g (glass transition temperature), the detection of melting and crystallization peaks and the calculation of crystallinity levels in semicrystalline rubber materials, and the evaluation of filler, plasticizer, and crosslinking system effects. The glass transition temperature (T_g) of rubber formulations determines their flexibility and segmental movement and blend compatibility between different materials such as NR/SBR and ethylene–propylene–diene monomer (EPDM)/NBR.

The thermal analysis of rubber compounds faces difficulties because their signals show multiple overlapping transitions and weak broad signals. The heat flow responses of rubber compounds become affected by the presence of CB and silica fillers. The combination of these factors creates difficulties for the precise identification of glass transition, an event which remains challenging. Also, the detection of secondary transitions becomes difficult when analyzing multiphase systems. In addition, the measurement of small changes in specific heat (ΔC_p) proves to be challenging.

The technique of MTDSC solves multiple above-mentioned analysis problems through its method of adding sinusoidal temperature variations to standard linear heating protocols. The method enables researchers to distinguish between thermal events that reverse direction and those that do not. The main benefits of MTDSC included the method enables exact T_g measurement in rubber materials containing fillers and blends, the technique provides enhanced detection capabilities for detecting weak thermal transitions, and the technique enables researchers to distinguish between T_g and enthalpic relaxation events when they occur simultaneously. Also, the technique enables researchers to detect phase separation and partial miscibility within rubber blends. Figure 3.15 illustrates typical MTDSC curves.

Among the applications of DSC and MTDSC in rubber compounds may include the identification of the miscibility status of rubber blends which becomes evident through MTDSC measurements which show miscibility or partial miscibility or phase separation. Also, the evaluation of devulcanized rubber requires T_g and ΔC_p measurements between devulcanized rubber and virgin rubber to determine chain scission or renetworking effects. In addition, the addition of high amounts of CB to rubber materials results in T_g broadening and shifting which MTDSC can detect. The thermal history of rubber compounds becomes detectable through DSC/MTDSC measurements which also track ageing-related transformations in rubber compounds.

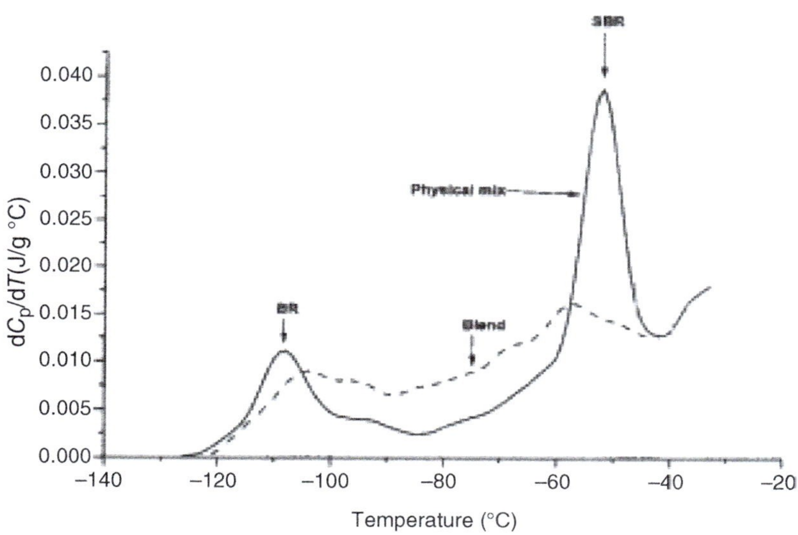

Figure 3.15 Typical MTDSC curves for SBR and BR individuals and in blend.

Typical DSC/MTDSC experimental setup included as the heating rate for DSC measurements reaches 10 °C/min, but MTDSC operates at 2 °C/min base rate with 0.5 °C amplitude at 60-second period. The sample amount used for analysis consists of 5–10 mg of material sealed inside aluminum pans. The system operates under nitrogen gas flow at 50 mL/min to stop oxidation reactions.

3.9 Thermogravimetric Analysis

The thermal analysis method TGA (thermogravimetric analysis) tracks sample mass changes while temperature rises or stays constant in a controlled environment with specific gas conditions. TGA analysis of rubber compounds enables researchers to measure the amounts of volatile substances including processing oils and plasticizers present in the material. It provides data about the temperature at which polymers start to break down and their degradation rates. It also determines the amount of filler materials present in the sample along with the remaining carbon-based residue. The process of char formation and the stability of oxidative reactions will be characterized by TGA. The thermal breakdown patterns of devulcanized rubber materials and recycled rubber materials become visible through this method.

A standard TGA thermogram of rubber compound material will display the following pattern:

- The thermogram shows two distinct weight loss areas which start at low temperatures because of moisture and volatile substances.
- The main weight loss section of the thermogram represents the decomposition process of rubber polymers when they break down either as NR or synthetic elastomers.
- The thermogram shows a flat section or a slow weight loss area which indicates the presence of fillers, char, and inorganic materials.
- The thermogram shows additional weight loss of CB and carbonaceous char when the atmosphere switches to oxidative conditions.

The technical note demonstrates an EPDM-based compound which lost its polymer structure under nitrogen until 600 °C before CB started burning in air at higher temperatures. The onset temperature of decomposition (T_d) serves as a stability indicator for thermostability. The peak rate temperature which emerges from derivative of TG curve (DTG) analysis indicates the highest point of decomposition speed. The amount of filler and char remains constant at high temperatures.

TGA analysis will be applied in filled rubber compounds in various ways. The thermal decomposition analysis of filled elastomers enables scientists to determine the amounts of fillers present and identify the chemical composition of remaining materials. The amount of residual mass increases when polymer decomposition and carbon burn-off steps occur because of high CB content in the material. The change from inert to oxidative atmosphere during analysis enables scientists to measure carbon-based fillers independently from other inorganic fillers. The thermal stability of materials becomes measurable through TGA when researchers compare virgin rubber to devulcanized rubber and rubber composites with various fillers and additives. The research used TGA to study rubber composite decomposition kinetics which demonstrated how different filler materials and their concentrations affected material stability. The thermal analysis method TGA enables researchers to detect hidden substances within their compounds. A research investigation used TGA to detect a specific weight loss peak at 280 °C which proved that PVC contaminants or additives existed in the NBR rubber O-ring sample. In addition, the TGA method enables researchers to study how fillers interact with bound rubber through weight loss measurements of polymer layers that attach to filler surfaces. The study of bound rubber formation kinetics on modified silica surfaces used TGA as its analytical method. Figure 3.16 illustrates a typical TGA analysis.

The choice between inert nitrogen gas and oxidative air gas for testing purposes affects the results because it produces either polymer pyrolysis or filler oxidation. The rate of heating affects when degradation starts because fast heating causes the process to begin at higher temperatures while high-resolution TGA measurements help distinguish between multiple occurring events.

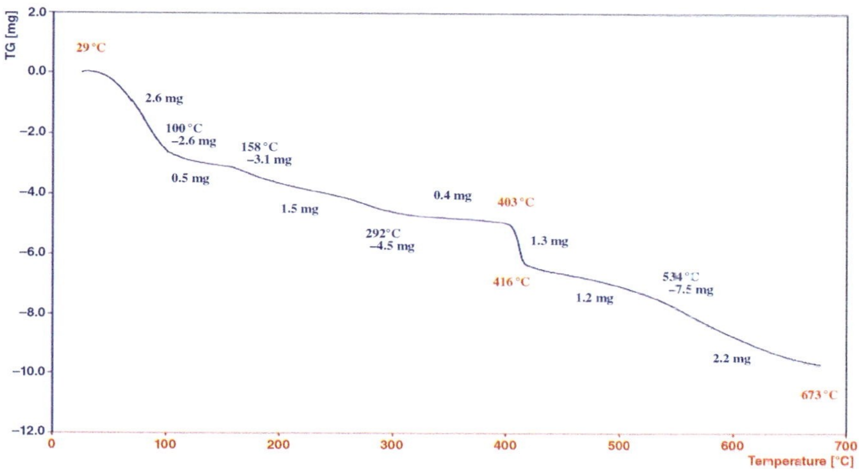

Figure 3.16 A typical TGA curve.

The accuracy of heat/mass transfer measurements decreases when using bigger samples. The high filler content in filled rubber causes the residue to become the main component which distorts the baseline. The high amount of filler materials including CB and silica in the mixture functions as heat absorbers while simultaneously altering the decomposition process through char production and catalytic reactions. The process of decomposition does not necessarily mean failure because plasticizer volatilization and low-mass fraction components can cause weight loss events. The complex degradation process of rubber requires knowledge about how its polymer chain scission and crosslink breakdown and char oxidation events interact with each other.

The TGA test shows rubber compound decomposition patterns and final residue composition, so it serves multiple essential purposes. The test helps users determine if rubber materials will withstand high temperatures when used in hose and belt applications. The devulcanized rubber evaluation through TGA shows lower decomposition temperatures when rubber chains experience damage or when chemical agents remain present. The TGA test enables researchers to determine when rubber waste will produce volatile substances and char during pyrolysis and energy recovery operations. The TGA test enables quality control teams.

In brief, the TGA technique serves as a flexible analytical method which provides essential information about rubber compound composition. The technique delivers weight-loss and residue data, but users can obtain essential thermal decomposition information and filler content and stability data content in filled rubber systems, blended, and recycled rubber materials. The technique delivers essential information about thermal stability, decomposition patterns, and filler.

3.10 Case Studies

In the subsequent three case studies, the researchers employed various fillers, including modified types whose surfaces were altered using the techniques outlined in this text (Mofrad et al., 2024c).

3.10.1 Case Study I

The influence of silane-treated CB and nanosilica, both separately and together, on the properties of EPDM rubber compounds is discussed.

This case study has been written based on author's published paper "The effect of silane-modified carbon black and nano-silica, individually and in combination, on the performance of ethylene–propylene–diene monomer rubber" in *Polymer Composites* (2025, Wiley).

The research assesses the operational efficiency of surface-modified CB when used alone and in combination with nanosilica as a hybrid filler in EPDM rubber compounds. The silane coupling agent treatment during alteration procedures resulted in enhanced curing properties of the rubber compounds.

The synthetic rubber material EPDM emerges from the combination of ethylene and propylene with a diene comonomer. The material shows excellent resistance to heat, ozone, steam, and weathering conditions, which makes it suitable for multiple applications. The material maintains its structure at temperatures up to 150 °C which allows it to function outdoors for extended periods without showing any signs of deterioration. The selection of EPDM grades depends on the specific requirements of the final product because different grades contain different proportions of their components. The material suits well for automotive, construction, and electrical applications because of its advantageous physical properties. The material faces two main limitations which affect its ability to resist solvents and its adhesive properties.

Scientific research uses EPDM rubber with different fillers to improve its performance while decreasing manufacturing costs. The rubber industry uses CB and silica as their most common filler materials. The rubber industry uses CB as its main filler because it accounts for 90% of worldwide production which serves the tire industry. The material offers affordable processing and excellent abrasion resistance while it improves multiple mechanical, electrical, and thermal properties of EPDM compounds. The material becomes challenging to work with because of its poor distribution throughout the rubber matrix. The rubber compound's dynamic properties improve significantly when silica replaces traditional fillers because it enhances both wet and icy surface traction and reduces rolling resistance. The high cost of silica stands as a major drawback because it fails to provide satisfactory wear performance due to its poor dispersion in rubber blends. The use of nanosilica as a filler material has become popular because it demonstrates outstanding properties.

The development of hybrid fillers represents a major advancement in rubber technology because these materials unite with different types of fillers. The combination of different fillers in hybrid systems produces synergistic effects which lead to better rubber–filler interactions and superior reinforcement and dispersion compared to using individual fillers. The use of hybrid fillers enables manufacturers to achieve desired properties at lower filler concentrations, which results in cost savings and environmental benefits. The combination of CB with nanosilica in a hybrid filler system presents an attractive method to leverage both fillers' advantages while addressing their individual limitations. Research studies have investigated CB–silica blends with various rubber materials including IR, NR, SBR, NR/BR, NR/NBR, NR/SBR, SBR/BR, EPDM, and multiple other rubber types.

The scientific community has not documented any research about silane-modified CB that forms covalent bonds with silane coupling agents in EPDM compounds. The authors conducted previous research to modify commercial CB surfaces through silane coupling agent treatment (Mofrad et al., 2024c). The authors conducted multiple analytical tests to evaluate the properties of silane-modified CB (Modified-CB-Oxi). The research findings indicated that Modified-CB-Oxi demonstrates outstanding potential as a reinforcement filler for rubber materials. The addition of silane-modified CB to rubber compounds resulted in better filler distribution and rubber compatibility and simultaneously improved curing properties, mechanical strength, dynamic performance, thermal stability, solvent resistance and thermodynamic properties in rubber compounds.

The research investigates how commercial CB, modified CBs, and nanosilica affect the curing process, mechanical, thermal, and dynamic properties of EPDM rubber compounds when used individually and in combination. The research will study both the filler distribution patterns and the morphological characteristics of rubber compounds containing EPDM rubber. The research will determine if EPDM rubber compounds suit the requirements for automotive rubber component formulations.

The research used EPDM rubber KEP270 from KUMHO in South Korea as its main raw material. The rubber contains 57.5% ethylene by weight and 4.3% diene monomer ethylidene norbornene (ENB) with a Mooney viscosity of ML(1 + 4) at 125 °C that measures 71. The research used CB (N660) from Pars Carbon Ltd. in Saveh, Iran as its reinforcing material. The product functions as the primary filler because it represents a typical furnace CB with 99.9% purity. The research used ULTRASIL® nanosilica from Evonik Industries AG in Germany as an alternative filler material which shared 99.9% purity with the other materials. The rubber formulation required base rubber and fillers and additional curing chemicals for completion. The research used TBBS (*N-tert*-butyl-2-benzothiazyl sulfenamide) from Flexsys in Belgium as an accelerator and nano-zinc oxide (99.8%) from Robin Arman Nano Technology Co as an activator.

The research methodology from previous description in this text guided the initial process which involved CB oxidation through citric acid monohydrate solution with distilled water under ultrasonic vibration (Sono Swiss, Switzerland, 280 W, 50/60 Hz). The research team used this method to produce CB-Oxi through chemical surface modification of commercial CB (N660) by adding hydroxyl (—OH) and carboxyl (—COOH) functional groups. The researchers expected these surfaces modifications would help the silane coupling agent react with the material. The sonication process of CB-Oxi in ethanol lasted for 30 minutes before TESPT (Figure 3.10) was incorporated into the solution. The mixture underwent reflux at 65 °C for three hours. The solution reached room temperature after the grafting reaction was completed. The modified CB-Oxi required centrifugation to separate it while undergoing three ethanol washing steps. The TESPT-functionalized CB-Oxi underwent dehydration at 50 °C for 12 hours to produce "Modified-CB-Oxi." The researchers applied the same procedure to commercial CB to develop "Modified-CB" which served as a control sample (Figure 3.1).

The detailed information about EPDM rubber compound development appears in Table 3.9. The two-roll mill operated with 25 cm diameter and 75 cm length rolls to blend materials. The mill operated with a cooling system while maintaining a 3–4 mm distance throughout the process. Figure 3.17 shows the step-by-step process of rubber compounding. The rubber compounding process started with EPDM grinding using a two-roll mill. The following step involved a 10–15 minutes mixing process with fillers. The curing agents including stearic acid, nano-zinc oxide, accelerators, and sulfur were added to the mixture until it reached a uniform state. The complete mixing process required 30 minutes to complete. The rubber materials underwent compression molding in a hydraulic hot press (SPH-500, Santam, Iran) at 160 °C and 150 bar pressure after determining the optimal cure times from ODR cure curves. The hot press produced a 15×15 cm^2 sheet mold with a 2 mm thickness to evaluate the mechanical properties of cured rubber. The hot press produced cylindrical specimens with 12.5 ± 0.5 mm thickness and 29 ± 0.5 mm diameter for hardness testing. The experiments were conducted three times with each run producing a median result. Figure 3.18 shows the proposed mechanism which explains the chemical interactions between Modified-CB-Oxi and EPDM rubber.

The prepared EPDM compounds (Table 3.9) receive their detailed vulcanization properties from Table 3.10. The total filler content in each compound formulation reaches 20 parts per 100 rubber (phr). The compounds E1–E5 contain specific fillers including nanosilica, CB, Modified-CB, Modified-CB-Oxi, and CB-Oxi. The compounds E6–E17 contain nanosilica at different concentrations (15, 10, and 5 phr) with CB, Modified-CB, Modified-CB-Oxi, and CB-Oxi fillers to achieve 20 phr total filler content. The compounds contain the same amounts of curing agents TBBS, nano-ZnO, stearic acid, and sulfur at 4, 6, 2, and 3 phr, respectively. The minimum torque of uncured rubber materials shows their viscosity and flow properties. The lowest value of 1.15 dN·m occurred in compound E5 which used 20 phr of CB-Oxi while compound E4 with Modified-CB-Oxi achieved the highest value of 1.42 dN·m. The physical barriers between EPDM macromolecule diene segments and TESPT grafts on Modified-CB-Oxi (Figure 3.17) restricted rubber macromolecule movement which increased viscosity. The compounds from E6 to E17 showed similar viscosity patterns when nanosilica replaced CB and Modified-CBs in their formulations.

The maximum torque serves as an indicator to measure the highest CLD which rubber achieves during curing because it determines the stiffness and hardness

Table 3.9 Formulating EPDM compounds with various fillers (phr).

		Fillers								
Compound no.	**EPDM**	**Nanosilica**	**CB**	**Modified-CB**	**Modified-CB-Oxi**	**CB-Oxi**	**TBBS**	**NanoZnO**	**Stearic acid**	**Sulfur**
E1	100	20	—	—	—	—	4	6	2	3
E2	100	—	20	—	—	—	4	6	2	3
E3	100	—	—	20	—	—	4	6	2	3
E4	100	—	—	—	20	—	4	6	2	3
E5	100	—	—	—	—	20	4	6	2	3
E6	100	15	5	—	—	—	4	6	2	3
E7	100	10	10	—	—	—	4	6	2	3
E8	100	5	15	—	—	—	4	6	2	3
E9	100	15	—	5	—	—	4	6	2	3
E10	100	10	—	10	—	—	4	6	2	3
E11	100	5	—	15	—	—	4	6	2	3
E12	100	15	—	—	5	—	4	6	2	3
E13	100	10	—	—	10	—	4	6	2	3
E14	100	5	—	—	15	—	4	6	2	3
E15	100	15	—	—	—	5	4	6	2	3
E16	100	10	—	—	—	10	4	6	2	3
E17	100	5	—	—	—	15	4	6	2	3

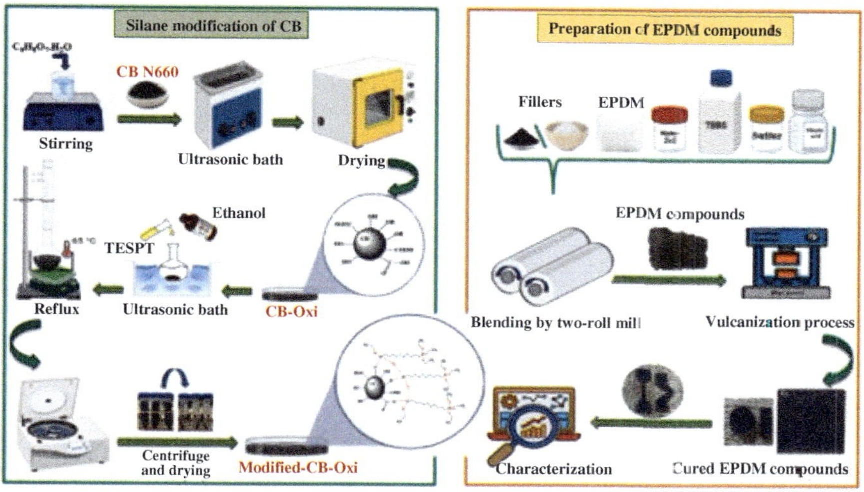

Figure 3.17 The method for the preparation of Modified-CB-Oxi/EPDM formulations.

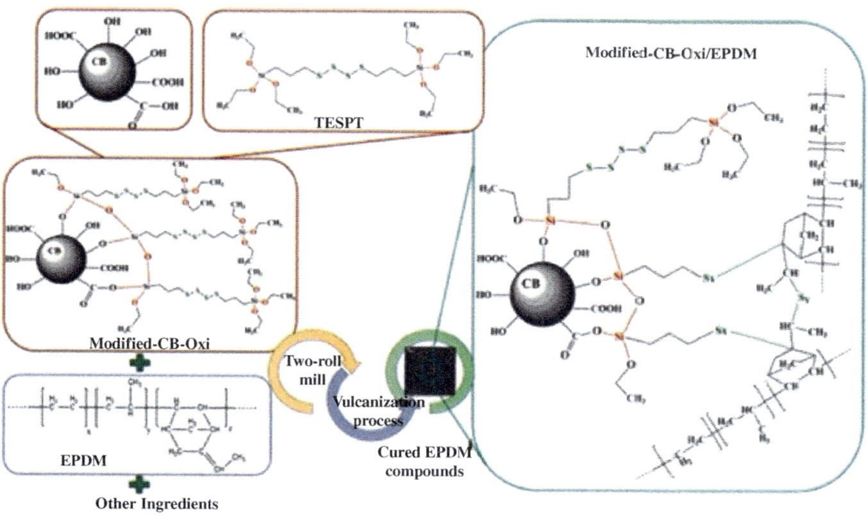

Figure 3.18 The suggested mechanism for interactions between Modified-CB-Oxi and EPDM rubber.

of the final product. The maximum value of 21.48 dN·m occurred in compound E4 which contained 20 phr of Modified-CB-Oxi. The value of compound E3 reached 16.41 dN·m when it used 20 phr of Modified-CB instead of Modified-CB-Oxi. The rubber curing process receives significant influence from the covalent

Table 3.10 Curing properties of formulated EPDM rubber mixtures with various fillers.

Samples	t_{s2} (min:s)	t_{90} (min:s)	M_L (dN·m)	M_H (dN·m)	$M_H - M_L$ (dN·m)	CRI (1/min)
E1	4:27	13:40	1.20	18.02	16.82	10.95
E2	4:04	13:11	1.31	17.25	15.94	11.03
E3	4:10	11:40	1.29	16.41	15.12	13.70
E4	4:21	12:33	1.42	21.48	20.06	12.32
E5	4:21	19:56	1.15	16.62	15.47	6.51
E6	4:18	10:18	1.37	16.68	15.31	16.67
E7	4:17	10:32	1.41	16.68	15.27	16.26
E8	4:20	10:49	1.41	15.89	14.48	15.90
E9	4:21	17:38	1.27	18.10	16.83	7.59
E10	4:16	13:43	1.31	16.46	15.15	10.79
E11	4:23	11:13	1.34	15.79	14.45	14.49
E12	4:40	12:02	1.4	17.44	16.04	13.12
E13	4:29	12:55	1.45	18.84	17.39	12.11
E14	4:21	12:45	1.42	20.45	19.03	12.14
E15	4:43	13:31	1.29	17.66	16.37	11.26
E16	4:37	13:40	1.28	16.90	15.62	11.07
E17	4:20	13:56	1.20	15.33	14.13	10.68

bond between TESPT and CB-Oxi in Modified-CB-Oxi. The values of hybrid filler compounds (E6–E17) reached their highest points when Modified-CB-Oxi was combined with nanosilica in compounds E12 and E14.

The delta torque (ΔTorque) measures the difference between maximum torque (M_H) and minimum torque (M_L) to show variations in vulcanized rubber CLD. The ΔTorque values in Table 3.10 follow the same pattern as the M_H values. The maximum and minimum ΔTorque values of 20.06 and 15.12 dN·m, respectively, occurred in compounds E4 and E3 which contained Modified-CB-Oxi and Modified-CB. The covalent bond between CB-Oxi and TESPT in Modified-CB-Oxi enhanced EPDM rubber curing and produced better CLD. The graph in Figure 3.19 shows how ΔTorques change from compound E1 to compound E17. The compounds with Modified-CB-Oxi as filler material showed increased ΔTorque values when the filler amount increased to reach their highest points. The results demonstrate that rubber crosslinking improves during curing when functionalized CB is used together with a coupling agent.

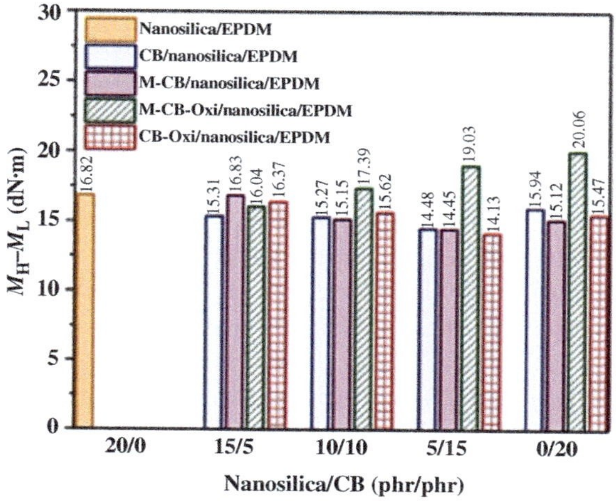

Figure 3.19 The analysis of ΔTorque in relation to nanosilica/CB loading for EPDM rubber compounds filled with fillers.

The CLD values showed a direct relationship with the ΔTorque measurements from the previous compounds. The highest CLD values occurred in compounds E5 and E4 with measurements of 85.78 and 156.19 mol/m^3, respectively. The research demonstrated successful CB surface modification through TESPT which enabled the formation of CB–EPDM macromolecule covalent bonds. The CLD values of compounds E12–E14 showed higher readings than the values from compounds E6–E11 and E15–E17. The CB type proved to be a major factor which increased CLDs to levels that matched the observed crosslinking improvements from using premodified CB (CB-Oxi) and subsequent coupling agent grafting.

The optimum curing time (t_{90}) represents the time needed for rubber formulations to reach 90% of their full cured state during specific testing conditions. The measure helps users determine curing time performance under established conditions to assess how fast the curing process occurs. The scorch time (t_{s2}) represents the time rubber compounds remain workable at specific temperatures before their curing process begins. The cure rate index (CRI) derived from rheometric tests provides a numerical value that shows the speed at which the curing reaction progresses. The curing rate metric helps users understand the chemical reaction speed of the curing process by revealing essential information about the rate of curing. The evaluation of rubber curing methods depends on these factors because they determine the final quality and characteristics of rubber products.

The scorch times for compounds E1–E5 appear in Table 3.10 with E2 reaching four minutes and 4 seconds and E1 reaching four minutes and 27 seconds. The scorch time for E4 with Modified-CB-Oxi filler lasted four minutes and 21 seconds.

The research shows that modifying CB materials leads to better pre-scorch duration results. The scorch times for dual filler compounds E6–E17 reached longer durations than those with individual due to crosslinking reaction blocking which results in extended scorch times. The combination of modified CBs with nanosilica and specific curing agents leads CB resulted in significant changes to optimal cure duration where E5 reached its peak time at 19 minutes and 56 seconds. The t_{90} value of E4 decreased to 12 minutes and 33 seconds after TESPT modification of the CB material. The rubber compound CLD increased after TESPT treatment of CB-Oxi to create Modified-CB-Oxi which resulted in a shorter optimal cure time that benefits rubber curing operations. The nanosilica-only compound E1 required an extended optimal cure duration of 13 minutes and 40 seconds. The CRI measurements for all tested compounds showed identical patterns of results. The CRI value for E5 stood out as the lowest at 6.51 1/min because the CB surface modification created obstacles for curing reaction rates. The CRI values from altered CB samples exceeded those obtained from unmodified CB samples. The parameters for compounds E6–E17 which used silica and CB blends showed different results.

The research findings show how surface modifications of CB and filler additions affect rubber curing parameters including prescorch time and optimal curing time and processing speed. The specific effects depend on the extent of modification and filler properties and their chemical interactions with rubber. The research shows that proper selection and optimization of filler materials and surface treatments require for customized curing processes to achieve desired rubber properties and performance in applications.

The images in Figures 3.20 and 3.21 present a comparative analysis of SEM data which shows the cross-sectional surfaces of EPDM rubber compounds with single fillers (E2, E4, and E5) and hybrid fillers (E8, E14, and E17) at 100 and 20 μm magnifications. The SEM image of compound E4 shows Modified-CB-Oxi particles that distribute evenly throughout the EPDM matrix. The rubber macromolecules bond covalently with CB surface through TESPT coupling agent which produces efficient dispersion. The rubber–filler interaction in compound E4 becomes more effective because of better rubber–filler bonding which leads to improved filler distribution. The poor filler distribution in compounds E2 and E5 becomes more pronounced in E5 because CB surface contains polar functional groups including hydroxyl and carboxyl groups. The SEM image demonstrates that CB particles in E5 exist as large clusters. The nonpolar nature of EPDM rubber causes CB-Oxi particles to form clumps instead of achieving uniform distribution in the matrix. The SEM images from Figure 3.21 demonstrate that compound E14 with Modified-CB-Oxi/nanosilica particles show superior dispersion in the rubber matrix compared to compound E8 with commercial CB/nanosilica. The SEM images demonstrate nanosilica agglomeration which proves

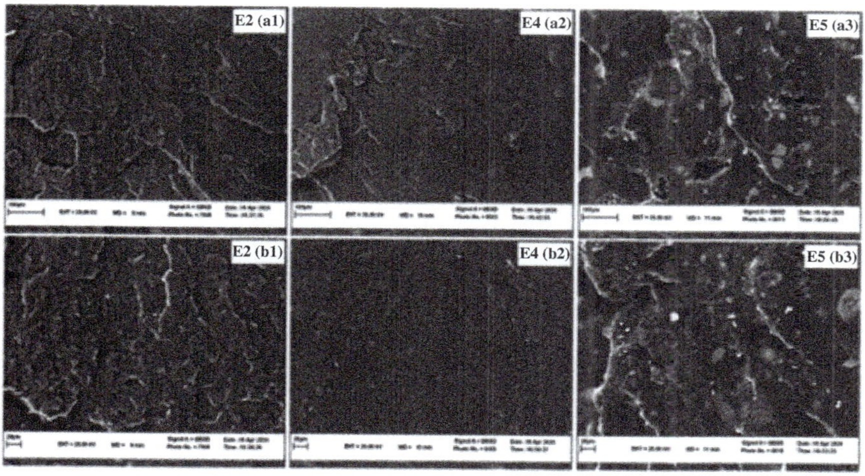

Figure 3.20 The SEM images for chosen individual fillers-filled EPDM rubber compounds, CB/EPDM (E2), Modified-CB-Oxi/EPDM (E4), and CB-Oxi/EPDM (E5) at magnifications of (a) 100 μm and (b) 20 μm.

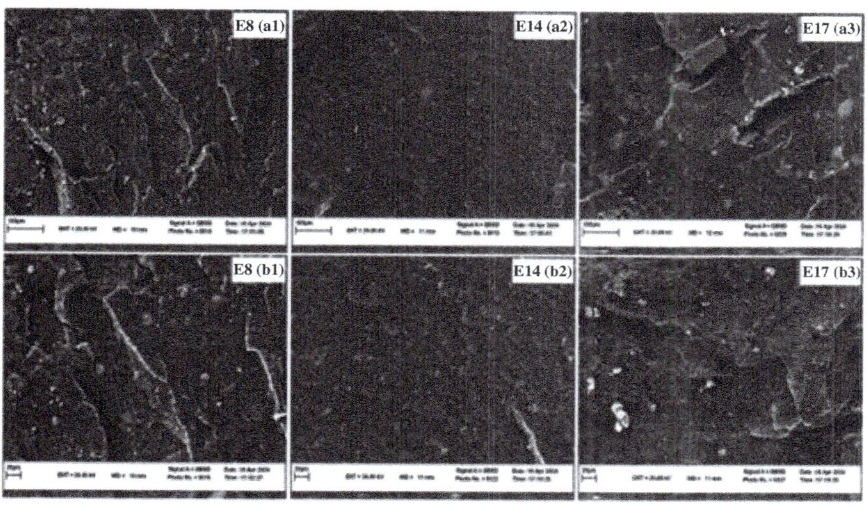

Figure 3.21 The SEM images for specific hybrid filler-filled EPDM rubber compounds, CB/nanosilica/EPDM (E8) and Modified-CB-Oxi/nanosilica/EPDM (E14), along with CB-Oxi/nanosilica/EPDM (E17) at magnifications of (a) 100 μm and (b) 20 μm.

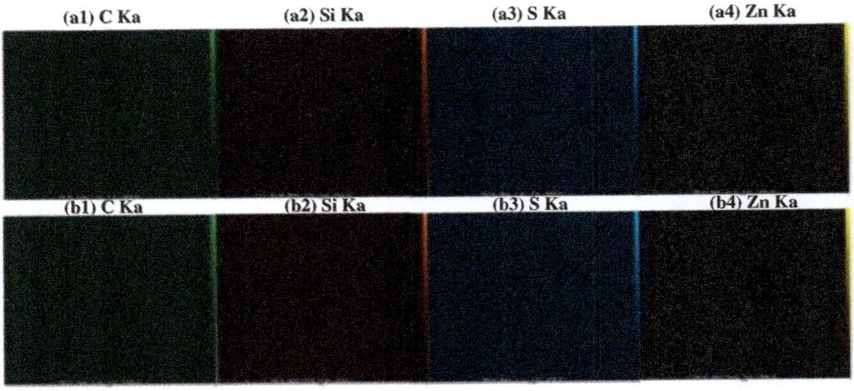

Figure 3.22 EDX elemental mapping outcomes for carbon, silica, sulfur, and zinc particles from compounds (a) E4 and (b) E14.

that filler distribution affects the final properties of rubber compounds after curing. The distribution of fillers in cured rubber affects both mechanical properties and dynamic performance. The uniform distribution of filler particles enables better stress transfer between rubber and filler materials which leads to improved strength and modulus and better tear resistance.

The absence of proper filler dispersion will produce adverse effects. The presence of filler agglomerations creates stress concentration points which leads to increased crack formation and propagation that damages the material's mechanical and dynamic properties. The coupling agent TESPT in compound E4 created strong covalent bonds between Modified-CB-Oxi and EPDM matrix which led to improved filler distribution. The elemental mapping of E4 and E14 compounds shows that sulfur (S) and silicon (Si) elements maintain uniform distribution throughout the materials (Figure 3.22). The surface modification techniques achieved successful filler–rubber compatibility and superior dispersion according to the SEM analysis results. The SEM results match previous research and the mechanical and dynamic property results from next paragraphs of this text demonstrate that filler dispersion directly affects the final performance of rubber compounds after curing.

The EDS analysis in Figure 3.23 shows the weight distribution of elements in compounds E4 and E14. The composition of compound E4 (Modified-CB-Oxi/EPDM) consists of 83.93 wt.% carbon, 5.96 wt.% oxygen, 5.67 wt.% sulfur, 3.94 wt.% zinc, and 0.49 wt.% silica. The composition of compound E14 (Modified-CB-Oxi/nanosilica/EPDM) contains 81.66 wt.% carbon, 6.28 wt.% oxygen, 8.52 wt.% sulfur, 1.97 wt.% zinc, and 1.56 wt.% silica. The Si peak at 1.8 keV in E4 and E14 spectra confirms silica presence in compound E4 which demonstrated

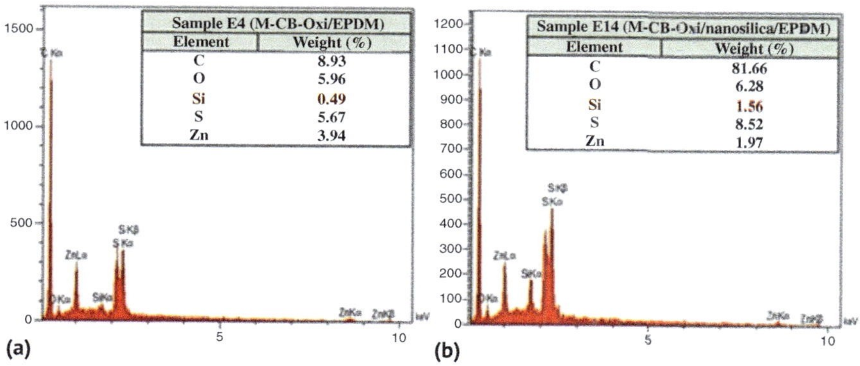

Figure 3.23 The EDS spectra for (a) Modified-CB-Oxi/EPDM (E4) and (b) Modified-CB-Oxi/nanosilica/EPDM (E14).

successful silane coupling agent functionalization on CB surfaces through covalent bonding.

The research into EPDM compound mechanical properties studied how silane treatment and filler incorporation affect the materials when used individually and when nanosilica is added. The research findings appear in Table 3.11. The research focused on tensile strength because this property determines rubber compound performance under tensile stress and maintains structural integrity during operation across different applications. The tensile strength of compound E4 (Modified-CB-Oxi) reached 4.09 MPa which exceeded the values of compound E2 (CB) at 3.34 MPa and compound E3 (Modified-CB) at 3.31 MPa. The lowest tensile strength measurement occurred for CB-Oxi at 1.63 MPa because its polar nature and poor rubber matrix distribution caused this result. The hybrid filler compounds (E6–E17) showed no specific pattern in their tensile strength measurements which ranged from 2.53 MPa in compound E17 to 4.81 MPa E6.

The elongation at break metric is a key indicator of a rubber compound's flexibility, range of motion, compatibility, and resistance to fatigue, all essential for choosing materials in different applications to guarantee durability and dependable performance. Among compounds E1–E5, each comprising distinct fillers, compound E4 reached the highest elongation at break of 162.49%, whereas E5 showed the lowest at 114.51%. Compound E1, featuring nanosilica, demonstrated a significant elongation at break of 153.8%, just after E4. The reduced value for E5 can be associated with inadequate compatibility and weak relations between the rubber and the filler (CB-Oxi). In compounds utilizing hybrid fillers (E6–E17), compounds E15–E17 featuring CB-Oxi/nanosilica displayed better elongation values, whereas compounds E12–E14, E6–E8, and E9–E11 showed progressively

Table 3.11 Selected mechanical properties of the cured EPDM rubber compounds with different fillers.

Samples	Tensile strength (MPa)	Elongation at break (%)	Initial modulus (MPa)	Yield modulus (MPa)	Tear strength (MPa)	Hardness (Shore A)
E1	3.82	153.80	2.48	2.28	17.53	62
E2	3.34	138.14	2.41	2.22	11.82	64
E3	3.31	133	2.48	2.47	16.03	61
E4	4.09	162.49	2.53	2.01	15.94	63
E5	1.63	114.51	1.51	2.83	11.66	59
E6	4.81	169.97	2.83	2.26	18.38	63
E7	3.43	119.83	2.86	2.92	17.97	63
E8	4.52	150.34	3	2.84	17.34	65
E9	3.54	124.40	2.85	2.67	17.03	64
E10	3.75	136.14	2.75	2.56	18.32	64
E11	2.86	118.66	2.41	2.71	17	62
E12	4.53	170.74	2.65	2.21	16.71	62
E13	4.19	152.11	2.76	2.46	18.38	63
E14	3.82	150.63	2.53	2.33	18.09	65
E15	4.22	191.46	2.21	1.80	20.25	62
E16	3.52	198.37	1.77	1.68	17.47	62
E17	2.53	180.30	1.40	2	16.38	60

increased values. These findings indicate that elongation at break in the EPDM rubber compounds is influenced by the interaction between the filler and rubber, the creation of filler networks, and the characteristics of the surface modifications made to the fillers.

The rubber mechanical testing evaluates stiffness through initial and yield moduli which represents distinct material properties. The initial modulus of compound E4 reached 2.53 MPa while E1–E3 compounds showed moduli between 2.41 and 2.48 MPa. The hybrid filler compounds (E6–E17) displayed starting moduli that spanned from 1.4 MPa in E17 to 3 MPa in E8. The single-filler compounds showed their lowest modulus at 2.01 MPa in compound E4 and their highest at 2.83 MPa in compound E5. The hybrid filler compounds reached their minimum modulus at 1.68 MPa in E16 and their maximum at 2.92 MPa in E7. The SEM images demonstrate how CB and CB-Oxi fillers produce lower stiffness than Modified-CB and Modified-CB-Oxi fillers because of their poor compatibility with rubber. The coupling agent TESPT demonstrates high efficiency in enhancing both initial and yield moduli.

Rubber compounds require tearing strength as their main characteristic to fight against punctures and cuts and fractures which determine their operational reliability and safety in different applications. The nanosilica compound E1 achieved the highest tearing strength at 17.53 MPa, while E3 and E4 reached 16.03 and 15.94 MPa, respectively. The surface pre-modification of CB (CB-Oxi) and traditional CB resulted in decreased tearing strength which produced values between 11.66 and 11.82 MPa. The hybrid filler compounds (E6–E17) showed similar results when Modified-CB-Oxi was combined with nanosilica because the tearing strength increased between 16.71 and 18.38 MPa. The combination of Modified-CB-Oxi surface properties with nanosilica network formation and extensive surface area resulted in enhanced tear resistance. The experimental findings match the results that previous research studies have documented.

The specific requirements of particular applications determine the optimal hardness level for their needs. The materials need to achieve exact stiffness levels while maintaining their flexibility and tear resistance properties. The Shore A hardness values of all synthesized compounds appear in Table 3.11. The highest Shore A hardness values of 64 and 63 were achieved by compounds E2 and E4 while E5 showed the lowest value at 59 which proves CB and Modified-CB-Oxi work well as fillers in EPDM rubber production. The hybrid filler combinations in compounds produced consistent results which showed hardness values between 60 and 65.

The compression sets of EPDM rubber compounds appear in Figure 3.24 through three different filler types which include single fillers in E2 and E4 and mixed fillers in E8, E14, and E17. The results indicate that the compression set value of Modified-CB-Oxi/nanosilica hybrid filler compounds decreased, which demonstrates that lower compression set percentages indicate better material recovery from deformation after compressive loading.

Research professionals use DMA as their primary method to evaluate rubber material dynamic properties. The test system provides complete measurement of material dynamic properties through storage, loss modulus tests, tanδ measurements, and glass transition temperature determination. The storage modulus shows how stiff a material is while its elastic behavior and the loss modulus show its viscous properties and energy absorption ability. The damping properties of materials become measurable through tanδ which represents the loss modulus to storage modulus ratio. The glass transition temperature of materials indicates their transition from glassy to rubbery states while showing their thermal behavior. The test method of DMA evaluates material dynamic properties through a broad temperature range which duplicates real-world operational conditions.

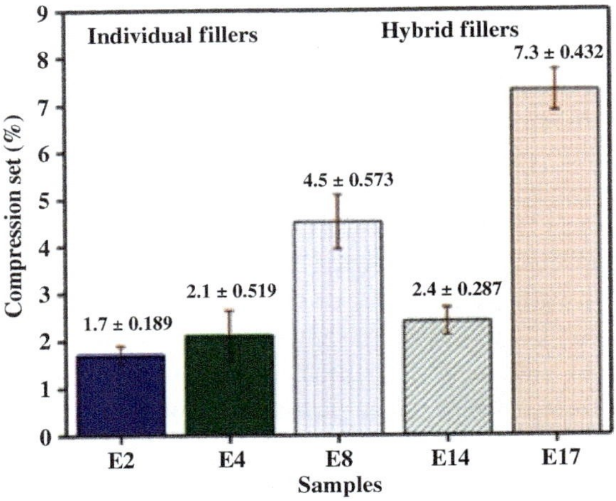

Figure 3.24 Compression characteristics of EPDM rubber compounds with specific individual fillers (E2 and E4) and hybrid fillers (E8, E14, and E17).

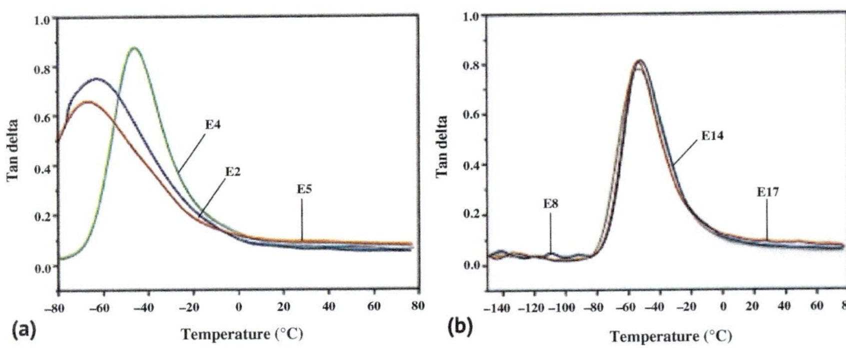

Figure 3.25 Tanδ as a function of temperature for (a) specific individual fillers-filled cured rubber compounds E2, E4, and E5 along with (b) hybrid fillers-filled cured rubber compounds E8, E14, and E17. All measurements were performed at 1 Hz with a temperature range of −150 to 80 °C at a heating rate of 5 °C/min.

The comparison of tanδ values at different temperatures for E2, E4, and E5 compounds and their hybrid versions E8, E14, and E17 appears in Figures 3.25a and b.

The tanδ measurements at different temperatures for E7 and E13 compounds which contain CB/nanosilica and Modified-CB-Oxi/nanosilica fillers, respectively, are shown in Figure 3.26a. The data shows equivalent parameters across different temperatures which include T_g at −20, 0, and 60 °C. The T_g values for compounds E2, E4, and E5 were measured at −62.8, −45.7, and −66.8 °C,

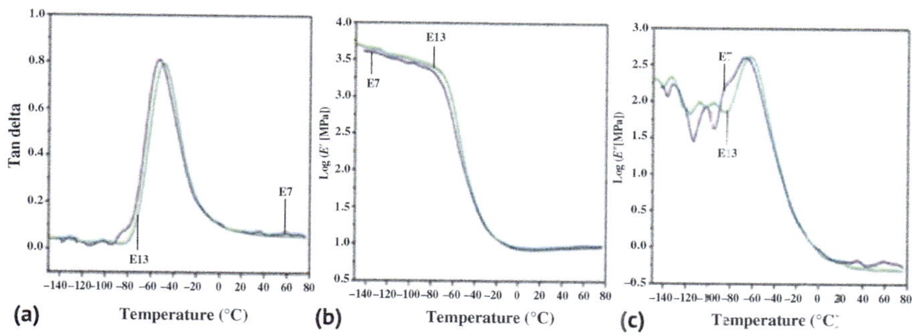

Figure 3.26 (a) Tanδ, (b) storage modulus, and (c) loss modulus as a function of temperature for cured EPDM rubber formulations E7 and E13. All measurements were carried out at 1 Hz with a temperature range from −150 to 80 °C at a heating rate of 5 °C/min.

respectively. The higher CLD of E4 resulted in elevated T_g values which indicate better stiffness and reduced segmental movement. The T_g values for hybrid filler compounds E8, E14, E17, E7, and E13 were measured at −51.8, −53.5, −53.7, −53.3, and −49.2 °C, respectively.

The substitution of commercial CB with Modified-CB-Oxi influenced the glass transition temperature of the cured rubber compound in different ways. In hybrid filler compositions, the utilization of the coupling agent-modified Modified-CB-Oxi decreased the T_g to −53.5 °C, indicating that the improved rubber–filler interaction in E14 might have led to a more pliable rubber matrix, thereby lowering T_g. The coupling agent utilized in E4, E14, and E13 enhanced the distribution of Modified-CB-Oxi throughout the rubber matrix, resulting in a more consistent structure that reduces stress concentrations and alleviates crystallization. Additionally, tanδ usually reaches its highest point near T_g, offering an exact indication of this important temperature. The height of the tanδ peak indicates the degree of energy dissipation in the material at T_g, with greater tanδ values signifying higher energy dissipation. A higher tanδ peak in vulcanizates indicates a less robust filler network for the loss factor (tanδ). The tanδ value at T_g for E2, which contained CB, was 0.75, lower than E4's value of 0.88, suggesting greater energy dissipation in the EPDM rubber mixture filled with Modified-CB-Oxi.

The rubber formulation shows better grip on slippery surfaces when tanδ values reach higher levels at −20 °C. Its increased tanδ value indicates better energy, which produces stronger friction forces between rubber and wet and icy surfaces. The improved friction between rubber surfaces and slippery areas enhances grip but researchers need to evaluate tanδ values together with rubber properties including hardness and tire tread designs to determine winter suitability. The Modified-CB-Oxi-filled rubber compound in compound E4 demonstrated higher tanδ values at −20 and 0 °C with 0.26 and 0.13, respectively, than E2 recorded at 0.22 and 0.10. The Modified-CB-Oxi-filled rubber compound shows the potential to enhance ice and wet surface grip according to its tanδ measurements. The tanδ readings from E14 matched E8 values at 0.21 and 0.10. The rubber compound containing Modified-CB-Oxi/nanosilica (10 phr/10 phr) showed better ice grip performance according to its higher tanδ value at −20 °C compared to E7.

The rubber compound generates less heat during rolling at 60 °C because it loses less energy which results in lower tanδ values. The reduced rolling resistance from this compound requires less energy to fight internal rubber friction in tires and other rubber products. The hybrid filler condition in compounds E14 and E13 shows a tanδ value of 0.06 which is lower than E8 and E7 at 0.07 because their fillers distribute better throughout the rubber matrix. The results from Figure 3.26b and c match the data obtained from the DMA tests.

The analysis in Figure 3.27a and b compares thermal stability between three individual fillers (E2, E4, and E5) and three hybrid fillers (E8, E14, and E17) in EPDM compounds. The thermal properties of cured rubber compounds underwent TGA testing under nitrogen from 25 to 600 °C at a heating rate of 10 °C/min. The thermal degradation of EPDM rubber started between 390 and 500 °C while the maximum weight loss occurred at 460 °C. The formulations contained light hydrocarbons which decomposing between 25 and 390 °C. The hybrid filler compounds E8 and E14 demonstrated identical thermal stability through their 79.1% total mass loss. The individual filler compounds E2, E4, and E5 showed total mass losses of 78.2%, 80.1%, and 85.1%, respectively. The pyrolysis of surface carbonyl and hydroxyl functional groups on CB-Oxi particles caused E5 and E17

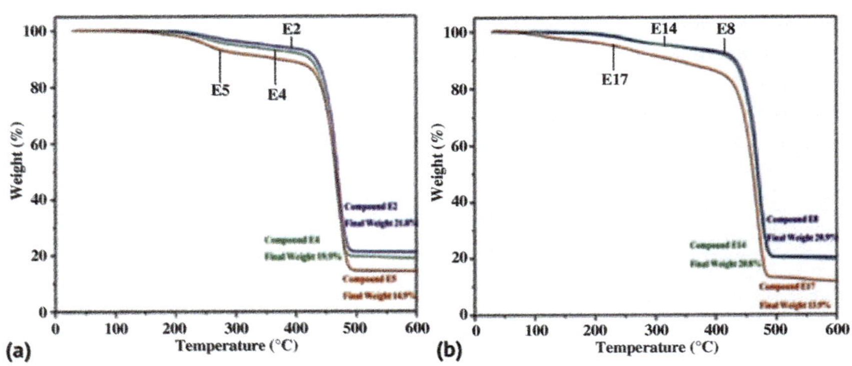

Figure 3.27 The TGA thermograms show (a) individual fillers in cured rubber compounds E2, E4, and E5 and (b) hybrid fillers in EPDM rubber compounds E8, E14, and E17. The TGA measurements ran from 25 to 600 °C at 10 °C/min while using nitrogen as the atmosphere.

to experience increased mass losses of 85.1% and 86.1%, respectively. The remaining materials including CB and zinc oxide showed no weight reduction when temperatures exceeded 500 °C. The TGA results show no substantial variations in thermal stability between the tested compounds.

The evaluation of rubber compound resistance against organic solvents used toluene for both immersion and swelling tests. The swelling ratios of E1–E5 compounds containing separate fillers spanned from 1.20 for E4 to 1.57 for E5. The organic solvent degradation of EPDM rubber compound E5 showed expected results. The treatment of CB (CB-Oxi) with TESPT to make Modified-CB-Oxi resulted in a rubber compound that demonstrated superior resistance to organic solvents according to the E4 test results. The hybrid filler compounds E6–E17 showed E14 and E17 as their most and least resistant to toluene with swelling values of 1.25 and 1.48, respectively. The addition of Modified-CB-Oxi to CB and Modified-CB and CB-Oxi and nanosilica combinations led to improved toluene resistance in E14 and E17 compounds.

The EPDM compounds contain bound rubber as their primary indicator to show filler–rubber chain interactions. The bound rubber percentages for E1–E5 single-filler rubber compounds were measured at 96.10%, 95.23%, 95.28%, 95.58%, and 95.30%, respectively. The silane modification process of CB leads to significant changes in bound rubber content according to the test results. The bound rubber content in hybrid filler compounds showed no pattern between 88.48% and 95.97% while remaining stable across all samples.

The elastic Gibbs free energy (ΔG) and conformational entropy (ΔS) serve as essential thermodynamic properties which affect both the elastic properties and operational performance of rubber materials. The rubber compound requires less work to extend when ΔG reaches more negative values because it becomes more elastic. The rubber compound's polymer chain disorder increases when ΔS values rise which results in better elasticity and flexibility.

The ΔG and $\Delta S \times 10^{-2}$ values for compounds E1–E5 except E5 show small changes between the compounds because ΔG values span from −18.04 to −20.97 J/mol and $\Delta S \times 10^{-2}$ values range from 6.06 to 7.04 J/mol. The stretching properties of compound E4 became more effective because it showed the lowest ΔG value of −20.97 J/mol and the highest $\Delta S \times 10^{-2}$ value of 7.04 J/mol. The stretching properties of compound E5 became more challenging because it showed a less negative ΔG value of −11.12 J/mol and the lowest $\Delta S \times 10^{-2}$ value of 3.73 J/mol. The hybrid filler compositions in compounds E6–E17 produced ΔG and $\Delta S \times 10^{-2}$ values that spanned from −12.79 J/mol for E17 to −19.03 J/mol for E14 and from 4.29 to 6.39 J/mol, respectively. The rubber compounds' elasticity suffered a negative impact because of CB-Oxi addition according to the observed trend. The coupling agent TESPT reduced the negative impact which CB-Oxi had on the rubber compounds.

In summary, the silane-modified CB shows promising potential for use as a filler material in EPDM rubber compounds. The research examined how silane-modified CB interacts with nanosilica as secondary fillers to affect vulcanization and filler dispersion and resulting composite properties. The surface modification process achieved better filler dispersion and improved rubber–filler interaction while simultaneously affecting curing behavior, mechanical properties, dynamic performance, thermal stability, solvent resistance, and thermodynamic properties which determined the final product quality of rubber-based materials. The research produced the following essential findings:

- The highest ΔTorque values of 20.06 and 19.03 dN·m corresponded to Modified-CB-Oxi which produced the highest CLD values of 156.19 and 142.47 mol/m^3.
- The SEM images show that Modified-CB-Oxi particles distribute evenly throughout the EPDM matrix when used alone or with nanosilica and they outperform commercial CB and CB-Oxi in terms of dispersion quality.
- The tear strength of Modified-CB-Oxi/EPDM samples increased substantially when using either single or hybrid filler combinations.
- The TGA results showed that all tested compounds maintained equal thermal resistance during the analysis.
- The Modified-CB-Oxi/nanosilica combination delivers superior wet grip performance while minimizing rolling resistance which leads to better fuel efficiency and reduced environmental pollution compared to commercial CB/nanosilica.
- The thermodynamic study showed that Modified-CB-Oxi/EPDM samples achieved higher ΔS and ΔG values because of their improved dispersion and enhanced material compatibility.

3.10.2 Case Study II

The influence of silane-treated CB and nanosilica, both separately and together, on the properties of NR compounds is discussed.

This case has been written based on author's published paper entitled "Synergistic effects of silane-modified carbon black and nano-silica hybrid fillers on the performance of natural rubber compounds" in *Journal of Applied Polymer Science* (2024).

The research evaluates the operational efficiency of surface-modified CB when used alone and when combined with nanosilica as a hybrid filler to enhance NR compound properties. The research investigates how commercial CB and modified CBs and nanosilica affect the curing process and mechanical and dynamical properties of NR-filled rubber compounds when used separately and together. The study evaluated both the filler distribution and rubber compound morphology. The research results examined how NR rubber compounds could be used to create tire tread formulations.

NR demonstrates superior resilience, strong tensile strength, excellent tear resistance, wear resistance and superior adhesion properties and flexible behavior under cold temperatures. The rubber industry uses NR to create truck tires and aircraft tires and heavy-duty tire formulations. The material shows poor resistance to heat and ozone and sunlight exposure and it fails to protect against oil and gasoline and hydrocarbon solvents. The affordable rubber material known as NR serves as a primary material for producing various rubber products including automotive tires. The addition of different fillers to NR compounds enhances their properties while decreasing the final cost of rubber compounds. The two most widely used fillers in NR compounds consist of CB and silica. The rubber industry relies on CB as its primary filler material. The tire industry consumes more than 90% of all CB products that manufacturers produce worldwide. The rubber industry selects CB as their preferred filler because it offers affordable costs and easy processing and acceptable wear resistance. The addition of CB to NR compounds results in improved mechanical, electrical, and thermal properties.

The rubber matrix of CB shows weak dispersion which leads to multiple defects when the rubber compound performs. The rubber compound receives improved dynamical properties from silica filler which results in better rolling resistance and superior traction on wet and icy roads. The high cost of silica filler combined with its poor wear resistance stems from its inadequate dispersion within NR compounds. The rubber industry now focuses on nanosilica as a filler material because it offers exceptional characteristics.

The current trend in rubber technology involves hybrid filler systems which combine two or more different filler materials. The combination of different fillers in hybrid systems produces better rubber–filler interactions which results in superior reinforcement and dispersion compared to single-filler systems. The use of hybrid fillers enables manufacturers to decrease filler quantities for achieving target properties which results in reduced material expenses and environmental impact. The combination of CB with nanosilica represents a promising approach to achieve better performance through the integration of their individual advantages while minimizing their respective limitations. The combination of CB with silica has become a common subject of research in scientific literature.

The research used NR (Rubber, SMNR 20 GM, Teh Ah Yau Rubber Factory Sdn Bhd, Malaysia) as its raw material which contained 0.16% dirt and 1% ash and 0.60% nitrogen and 0.80% volatile matter at maximum levels. The research used commercial CB N660 (Pars Carbon Ltd., Saveh, Iran) as the primary filler and ULTRASIL® (Evonik Industries AG, Germany) as the secondary filler. The rubber formulation required additional components beyond raw rubber and fillers which included curing chemicals. The rubber formulation required TBBS (accelerator from Flexsys Belgium) and nano-zinc oxide (activator from Robin Arman Nano Technology Co., Iran) and stearic acid (activator from Anchor Chemical Ltd, UK) and elemental sulfur (curing chemical from Solvay Barium Strontium, Hannover, Germany). The physical property measurements of CLD and bound rubber content required toluene (Mojalali, Iran) as the solvent.

The surface modification of CB was performed as described in case study I based on the procedure developed by authors (Mofrad et al., 2024c).

The prepared NR rubber compounds received their formulation from Table 3.12. The two-roll mill operated with 25 cm diameter and 75 cm length rolls to perform the mixing process. The mill operated with a cooling system while maintaining a 3–4 mm gap between the rolls. The rubber compounding sequence followed the steps shown in Figure 3.28. The two-roll mill received preheated NR for mastication before the next step.

The mixture received fillers during a 10–15-minute blending period. The mixing process required all curing chemicals including stearic acid nano-zinc oxide, accelerator, and curing agent (sulfur) until the compounds reached uniform consistency. The complete mixing process required 30 minutes to finish. The rubber compounds underwent compression molding in a hydraulic hot press (SPH-500 from Santam, Iran) at 160 °C and 150 bar pressure after determining their optimal cure times from ODR cure curves. The hot press produced 15 × 15 cm^2 sheet molds from rubber materials to measure their mechanical properties. The hot press produced cylindrical samples with 12.5 ± 0.5 mm thickness and 29 ± 0.5 mm diameter for hardness testing. Figure 3.29 shows the proposed mechanism which explains how Modified-CB-Oxi interacts with NR through chemical reactions.

The formulation details along with vulcanization properties of NR compounds appear in Tables 3.12 and 3.13. The total filler content in each compound reaches 20 phr. The fillers in compounds N1–N5 consist of nanosilica, CB, Modified-CB, Modified-CB-Oxi, and CB-Oxi, respectively. The compounds from N6–N17 contain different nanosilica levels (15, 10, and 5 phr) in combination with CB, Modified-CB, Modified-CB-Oxi, and CB-Oxi to achieve 20 phr total filler content. As stated earlier in this text, the minimum torque value indicates the viscosity level and flowability of rubber compounds before curing.

The M_L values of compounds with various filler types (N1–N5) reached their peak at 1.1 dN·m for compound N1 with 20 phr nanosilica and their lowest point at 0.9 dN·m for compound N5 with 20 phr CB-Oxi. The nanosilica-filled compound showed the highest viscosity among all tested compounds. The viscosity of the compounds decreased when nanosilica was partially replaced by CB and its modified versions in compounds N6–N17. The surface area of nanosilica exceeds that of CBs by a significant margin. The increased surface area of fillers

Table 3.12 Formulation of NR compounds with different fillers (phr).

Compound no.	NR	Fillers					TBBS	Nano-ZnO	Stearic acid	Sulfur
		Nanosilica	CB	Modified-CB	Modified-CB-Oxi	CB-Oxi				
N1	100	20	—	—	—	—	4	6	2	3
N2	100	—	20	—	—	—	4	6	2	3
N3	100	—	—	20	—	—	4	6	2	3
N4	100	—	—	—	20	—	4	6	2	3
N5	100	—	—	—	—	20	4	6	2	3
N6	100	15	5	—	—	—	4	6	2	3
N7	100	10	10	—	—	—	4	6	2	3
N8	100	5	15	—	—	—	4	6	2	3
N9	100	15	—	5	—	—	4	6	2	3
N10	100	10	—	10	—	—	4	6	2	3
N11	100	5	—	15	—	—	4	6	2	3
N12	100	15	—	—	5	—	4	6	2	3
N13	100	10	—	—	10	—	4	6	2	3
N14	100	5	—	—	15	—	4	6	2	3
N15	100	15	—	—	—	5	4	6	2	3
N16	100	10	—	—	—	10	4	6	2	3
N17	100	5	—	—	—	15	4	6	2	3

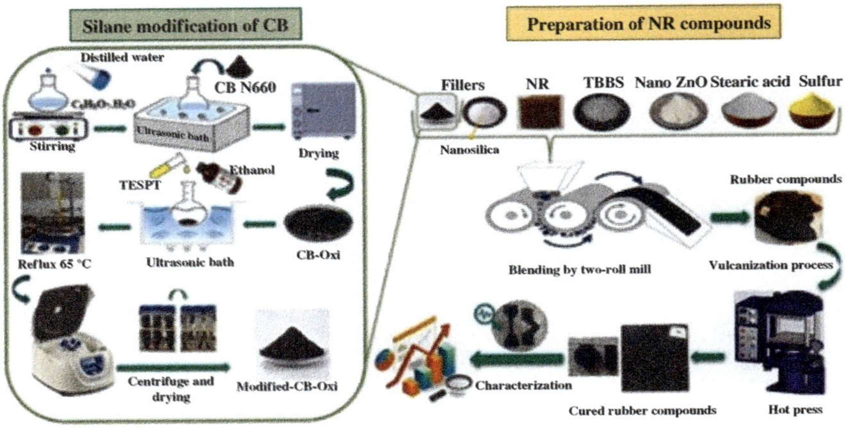

Figure 3.28 The procedure for preparation of Modified-CB-Oxi/NR compounds.

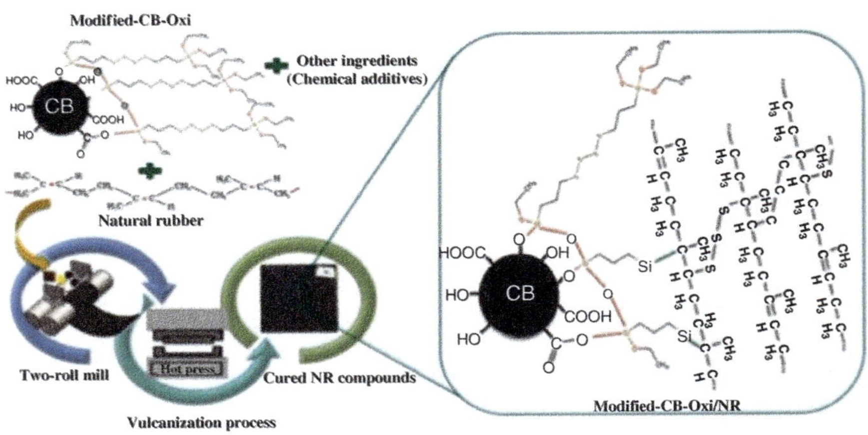

Figure 3.29 The proposed mechanism for reactions between Modified-CB-Oxi and NR.

creates additional points of contact between rubber chains and filler particles which restricts chain movement and causes viscosity to rise.

The small size of nanosilica particles together with their high polarity and surface energy makes them more prone to agglomeration. The formation of agglomerates creates barriers for rubber chain movement which leads to increased viscosity. The maximum torque M_H functions as an indicator of CLD during curing as mentioned earlier to show the stiffness level of cured rubber. The M_H value reached its

Table 3.13 The curing properties of NR compounds with different fillers.

Samples	t_{s2} (min:s)	t_{90} (min:s)	M_L (dN·m)	M_H (dN·m)	$M_H - M_L$ (dN·m)	CRI (1/min)
N1	2:22	4:52	1.1	13.5	12.3	43.5
N2	1:39	3:44	1	15.9	14.9	48.8
N3	1:40	3:57	1	15.8	14.8	46.1
N4	2:01	4:35	1	13.8	12.8	42.7
N5	2:26	3:45	0.9	5.8	4.9	84
N6	2:09	4:25	1	13.3	12.3	46.3
N7	2:04	4:02	0.9	13.6	12.8	50.5
N8	1:51	4:16	1	14.3	13.3	37.7
N9	2:13	4:31	1	14.4	13.4	45.9
N10	2:01	4:09	1.1	15.3	14.2	48.1
N11	1:55	4:01	1	14.3	13.4	40.7
N12	2:18	4:49	1.1	13.9	12.8	43.3
N13	2:18	4:38	1.1	15.3	14.2	45.5
N14	2:12	4:43	1	14.6	13.6	43.3
N15	2:36	5:06	1	11.2	10.2	37
N16	2:31	4:53	1	9.5	8.4	45.1
N17	2:33	4:32	1	8.4	7.5	50.3

highest point at 15.9 dN·m when compound N2 used commercial CB as its single filler. The M_H value of compound 5 reached 5.8 dN·m when it contained 20 phr of CB-Oxi. The rubber curing process becomes less efficient when CB contains grafted functional groups such as carboxyl and hydroxyl groups. The functional groups on CB particles create chemical bonds with TBBS, ZnO, and stearic acid which are present in the compound mixture.

The carboxyl groups in TBBS prevent it from forming ionic bonds with TBBS because TBBS contains available N—H groups which block its acceleration function. The hydrogen bonding between hydroxyl groups and ZnO activator reduces ZnO mobility which prevents complete curing initiation. The compound will experience under-curing because of this effect which produces decreased ΔTorque values. The interaction between stearic acid and hydroxyl groups disrupts the stearic acid–rubber matrix bond which affects its dispersing and lubricating performance. The M_H values of hybrid filler compounds (compounds N6–N17) showed lower values at 11.2, 9.5, and 8.4 dN·m when CB-Oxi content reached 5, 10, and 15 phr in compounds N15–N17.

The results match the previous data obtained for compound 5. The delta torque (ΔTorque) shows the variation between maximum and minimum torque values which indicate changes in vulcanized rubber CLD. The M_H values follow the same pattern as the ΔTorque values which are shown in Table 3.13. The ΔTorque values of compounds N2 and N5 showed the highest and lowest readings at 14.9 and 4.9 dN·m, respectively. The surface addition of carboxyl and hydroxyl groups to CB (CB-Oxi) disrupted NR rubber curing as previously explained. The addition of TESPT as a coupling agent produced a 12.8 dN·m increase in ΔTorque for the Modified-CB-Oxi rubber compound (compound N4). The rubber compound containing only nanosilica (compound N1) showed lower ΔTorque than compound N2 which used CB but lacked a coupling agent. The observation occurs because silica exhibits polarity which enables potential interactions with specific curing agents according to previous discussions. The comparison between compounds N6, N7, and N8 and compounds N12, N13, and N14 shows dual CB and nanosilica in different formulations with different nanosilica to CB ratios. The first set of compounds used CB as their CB material but the second set used Modified-CB-Oxi. The ΔTorques measurements from Modified-CB-Oxi compounds exceeded those from CB compounds by 11% or more based on the nanosilica to CB ratio.

The results demonstrate that functionalized CB with coupling agent grafting enhances rubber crosslinking during curing. The CLD values matched ΔTorque measurements for all tested compounds. The CLD of compound N5 reached 62.33 mol/m^3 which produced the smallest ΔTorque value among all compounds while compound N1 containing only nanosilica showed a CLD of 118.6 mol/m^3. The CLDs of compounds N12–N14 showed higher values than compounds N6–N8 by as much as 35% (compound N14). The observed difference demonstrates that CB type and nanosilica/CB ratios play a major role in determining CLDs which resulted in better crosslinking performance after using functionalized CB and coupling agent grafting.

The rubber compound needs an optimum cure time (t_{90}) to reach 90% of its complete curing state when exposed to particular conditions. The metric helps users determine how fast rubber compounds cure under controlled conditions. The scorch time (t_{s2}) indicates the processing window for rubber compounds at specific temperatures before they start to cure. The CRI derived from rheometric data provides a numerical value that shows the speed at which the curing reaction advances as stated earlier. The curing process speed becomes measurable through this parameter, which delivers essential information about the reaction rate. The rubber curing process efficiency depends on these three parameters which determine the final quality and properties of rubber products. The scorch times for compounds N1–N5 appear in Table 3.13 with compound N2 filled with CB reaching 1 minute and 39 seconds and compound N5 filled with CB-Oxi reaching two minutes and 26 seconds. The scorch time for N4 compound containing

Modified-CB-Oxi reached two minutes and 1 second, which represented a minor decrease from previous results. The results demonstrate that CB modification leads to better prescorch time performance. The scorch time measurement for a rubber compound containing nanosilica reached two minutes and 22 seconds at the same level as N5. The modified CBs and nanosilica surfaces interact more strongly with curing chemicals which prevent early crosslinking reactions from occurring. The surface modification of CB did not affect the optimum cure time because it remained at three minutes and 44 seconds. The application of TESPT coupling agent to CB-Oxi resulted in an extended optimum cure time that reached four minutes and 35 seconds which was considered unacceptable.

The treatment of compound N4 with TESPT coupling agent resulted in longer optimum cure times because the grafted coupling agent created steric hindrance on the CB surface. The steric hindrance from the coupling agent grafts on CB surfaces might block curing chemicals from reaching rubber macromolecules, which leads to delayed crosslinking reactions. The coupling agent creates barriers that block curing chemicals from reaching the rubber matrix, which results in extended curing times. The nanosilica-filled compound N1 reached its highest optimum cure time in four minutes and 52 seconds. The CRI values of all compounds except N5 showed comparable results. The CRI value of N5 exceeded all other compounds by reaching 84, which demonstrated that CB surface functionalization strongly influenced curing reaction speed. The cure time and CRI values for compounds N6–N17 containing hybrid silica–CB mixtures depended on the specific ratios of nanosilica to CB in their formulations.

The research findings conclude that surface treatment of CB and different filler materials affect rubber curing parameters through their impact on prescorch time and optimum cure time and cure rate. The effects on the rubber curing process depend on the extent of modification and the specific filler material and the strength of filler–rubber bonding. The research demonstrates that proper selection and optimization of filler materials and surface modifications lead to optimized curing processes for rubber-based applications.

The analysis in Figures 3.30 and 3.31 shows SEM images of cured NR compounds with single (N2, N4, and N5) and hybrid (N8, N14, and N17) fillers through cross-sectional surface observations at 100 and 20 μm magnifications. Figure 3.30 shows compound N4 with Modified-CB-Oxi particles that maintain a uniform distribution pattern throughout the NR matrix. The rubber macromolecules form covalent bonds with CB surface particles through TESPT coupling agent which results in this uniform distribution.

The improved filler–rubber interaction in compound N4 results in better filler distribution. The compounds N2 and N5 show poor filler dispersion because they contain CB and CB-Oxi as their fillers. The surface of CB-Oxi contains more polar functional groups which leads to increased agglomeration of CB-Oxi particles

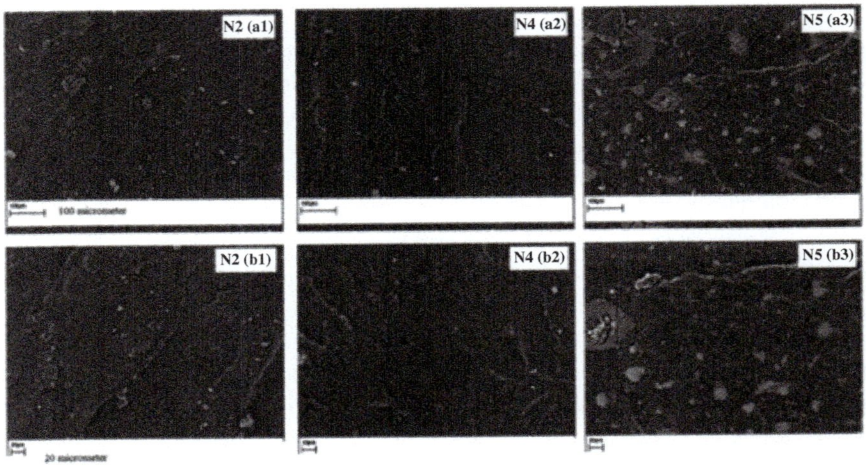

Figure 3.30 The SEM images show individual filler-filled rubber compounds at 100 and 20 μm magnifications for CB/NR (N2), Modified-CB-Oxi/NR (N4), and CB-Oxi/NR (N5).

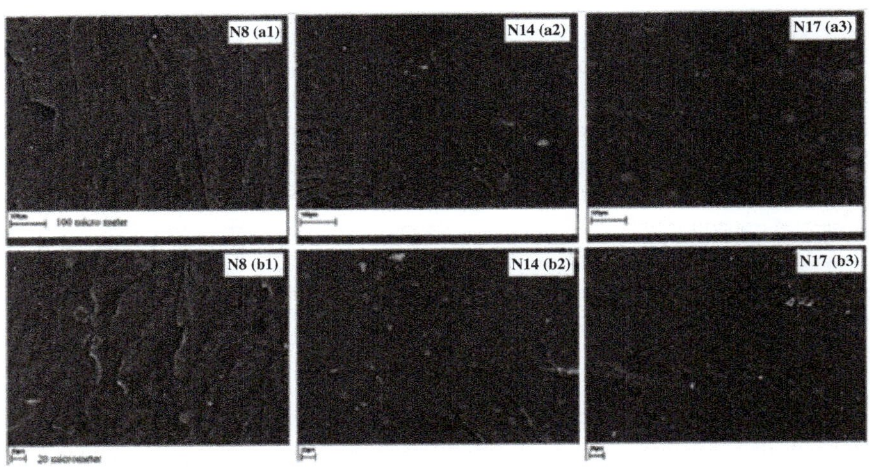

Figure 3.31 The SEM images for selected hybrid filler-filled rubber compounds, Modified-CB-Oxi/nano-Si (N14) and CB-Oxi/nanosilica/NR (N17) and CB/nano-Si/NR (N8), Modified-CB-Oxi/nanosilica/NR (magnitudes of (a) 100 μm and (b) 20 μm).

in NR because NR has nonpolar properties. The rubber matrix fails to distribute CB-Oxi particles because of their nonpolar nature, which causes them to form large agglomerates instead of dispersing evenly. The rubber matrix shows better distribution of Modified-CB-Oxi/nanosilica particles in compound N14 compared to the commercial blends of CB/nanosilica in compound N8 and CB-Oxi/nanosilica in compound N17. The SEM images show nanosilica agglomerates in the latter compounds which demonstrate that filler dispersion affects the final properties of cured rubber compounds. The mechanical and dynamic properties of cured rubber compounds improve when their fillers achieve proper dispersion throughout the material. The rubber-to-filler stress transfer becomes more effective when filler particles distribute evenly throughout the material, which leads to better strength and modulus dispersion. The presence of agglomerated filler particles creates stress concentrations that lead to increased crack formation and propagation, which results in decreased mechanical and dynamic performance. Surface modification methods show promise to enhance filler–rubber interface quality, which results in better filler distribution. The SEM image results matched the mechanical addition of TESPT coupling agent to compound N4 resulted in better filler distribution because TESPT and dynamic property data presented in oncoming phrases. The consistent results between SEM images and mechanical and dynamic property data confirm that filler dispersion directly affects the final performance of rubber compounds after curing.

The research team studied NR compounds through tests which evaluated silane modification and filler addition as standalone elements and when nanosilica was added to the mixture. The research findings appeared in Table 3.14. The tensile strength of rubber compounds represents a vital material property because it determines their ability to handle stretching forces and retain their form during different operational scenarios. The addition of a coupling agent in Modified-CB-Oxi, N4, resulted in better tensile strength than Modified-CB, N3, because it improved both filler–rubber bonding and filler distribution.

The tensile strength measurements for N1 and N5 compounds showed expected results because they reached 2.5 MPa which indicates weak rubber–filler bonding because of filler polarity differences. The CB-filled compound (N2) achieved the highest tensile strength at 3.1 MPa because it demonstrated the best rubber–filler bonding among all tested compounds. The tensile strength measurements of hybrid filler compounds (N6–N17) did not follow any specific pattern. The tensile strength measurements for compounds N13 and N17 showed the lowest value at 2 MPa and the highest value at 3.3 MPa. The elongation at break parameter provides essential information about rubber compound flexibility and motion range and compatibility and fatigue resistance which determines material selection for various applications to achieve reliable long-term performance. The elongation at break values reached their highest point at 114.1% in compound N5 and their lowest point at 70% in compound N3. The nanosilica filler in compound N4 produced the highest elongation at break measurement, which reached 104.7%. The elongation at break values of compounds N15–N17 reached their highest points

Table 3.14 Selected mechanical properties of prepared NR compounds with different fillers.

Samples	Tensile strength (MPa)	Elongation at break (%)	Initial modulus (MPa)	Yield modulus (MPa)	Tear strength (MPa)	Hardness (Shore A)
N1	2.5	104.7	2.6	2.8	18.2	61
N2	3.1	84.5	3.7	3.7	17.9	64
N3	2.4	69.9	3.4	3.9	15.5	63
N4	2.5	79.7	3.1	3.2	18.9	63
N5	2.5	114.1	2.2	2.5	13.3	54
N6	2.1	69.2	3	3.7	14	68
N7	2.7	87.1	3.1	3.4	14	68
N8	2	63.8	3.2	3.6	11.5	64
N9	2.5	74.6	3.3	3.8	11.3	68
N10	2.2	72.1	3.1	3.6	13.3	67
N11	2.3	79.9	2.9	3.4	11.8	63
N12	2.2	70.5	3.2	4	16.4	68
N13	2	65.1	3.1	3.9	18.5	68
N14	2.3	75.4	3	3.9	15.2	64
N15	2.5	115.7	2.1	2.6	22.3	63
N16	3	157.1	1.9	2	18.9	60
N17	3.3	190	1.8	1.8	19.6	56

when they used CB-Oxi/nanosilica hybrid fillers which produced results between 115.7 and 190%. The elongation at break values of the other compounds spanned from 64.8% in N8 to 80% in N11 based on their CB/nanosilica filler ratios. The test results show that filler–rubber bonding strength, filler network development, and surface treatment methods of fillers determine the test results.

Rubber mechanical testing uses initial modulus and yield modulus to measure stiffness, but these values represent different stages of material behavior. The initial modulus represents the material stiffness when it experiences small deformations because it measures the slope of the stress–strain curve during its first linear section. The yield modulus represents the stress value which causes permanent plastic deformation in materials, thus marking their transition from elastic to plastic behavior. The material loses its ability to return to its original form after reaching the yield point. The initial modulus values for compounds

N5 and N1 measured at 2.2 and 2.6 MPa, respectively, while N2–N4 compounds with single-filler materials showed values between 3.1 and 3.7 MPa. The parameter showed values between 1.8 MPa for N17 and 3.3 MPa for N9 when hybrid fillers were used in compounds N6–N17. The addition of nanosilica and CB-Oxi to materials resulted in decreased values of this parameter, but TESPT treatment of CB-Oxi helped reduce the negative impact on this parameter. The yield modulus data showed an identical pattern to the tensile strength results. The rubber compounds showed their minimum and maximum yield modulus values at compounds N5 and N3 which measured 2.5 and 3.9 MPa, respectively. The lowest and highest yield modulus values for hybrid filler compounds occurred in compounds N17 and N9 at 1.8 and 3.8 MPa, respectively. The rubber compounds containing nanosilica and CB-Oxi showed decreased initial and yield modulus values compared to individual filler compounds because these fillers create matrix incompatibility as shown in SEM images. The results demonstrate that TESPT functions as an effective coupling agent to enhance both initial and yield modulus values. Rubber compounds require tearing strength as their main property to prevent punctures and cuts and cracks which affect their operational performance and safety and reliability in different applications. The highest tearing strength value of 18.9 MPa occurred in compound N4 which used Modified-CB-Oxi as its formulation ingredient. The surface modification of CB through functionalization and subsequent treatment with Modified-CB-Oxi creates an effective solution to enhance tearing strength in NR compounds beyond commercial CB standards. The tearing strength of compound N1 reached 18.2 MPa because it contained nanosilica as a key ingredient. The hybrid filler compounds N15–N17, which used Modified-CB-Oxi with nanosilica at different ratios, achieved the highest tearing resistance values between 23.2 and 31.5 MPa. The combination of Modified-CB-Oxi with nanosilica produces an enhanced effect in tear resistance. The rubber–filler bond improves through Modified-CB-Oxi while nanosilica creates a strong network structure that enhances matrix strength to achieve better tear resistance.

The required hardness level for applications depends on specific needs and desired operational characteristics. A material requires a combination of hardness with additional properties which include flexibility and tear resistance. The Shore A hardness values of all prepared compounds appear in Table 3.14. The hardness values of compounds N1–N5 showed N5 as the lowest at 54 and N2 as the highest at 64. The hybrid filler parameter showed no consistent pattern between 56 and 68.

Rubber compounds achieve their performance through dynamic properties which handle different and recurring loads while static properties measure material behavior under steady stress conditions. The material response to oscillating forces and deformations becomes measurable through dynamic properties because these applications need specific behavior during vibration, impact, repeated stretching, and compression cycles. The effective vibration absorption and dissipation of rubber compounds depend on their dynamic properties including hysteresis and loss modulus which makes them suitable for vehicle engine mounts that reduce noise and improve ride quality. The friction coefficient of rubber determines its grip and traction performance which tires need to maintain safe traction on different road surfaces. The scientific rubber community recognizes DMA as an essential method to evaluate rubber compound dynamic properties. The testing method enables researchers to measure multiple dynamic material properties through storage and loss modulus measurements and tanδ and glass transition temperature (T_g) assessments. The storage modulus shows how stiff a material is while measuring its elastic behavior, but the loss modulus shows its viscous characteristics and energy absorption ability as stated earlier in this text. The loss to storage modulus ratio known as tanδ delivers vital information about material damping properties which determine their performance evaluation. The T_g measurement through DMA reveals the point where materials shift from glass to rubbery states while showing their thermal response. The testing method enables researchers to measure dynamic properties at various temperatures while creating conditions that match actual operational environments.

The comparison between tanδ values at different temperatures for compounds N8, N14, and N17 with CB, Modified-CB-Oxi, and CB-Oxi fillers appears in Figure 3.32. The data show the parameters at different temperatures which include T_g (maximum tanδ) and −20, 0, and 60 °C. The T_g values for compounds N8, N14,

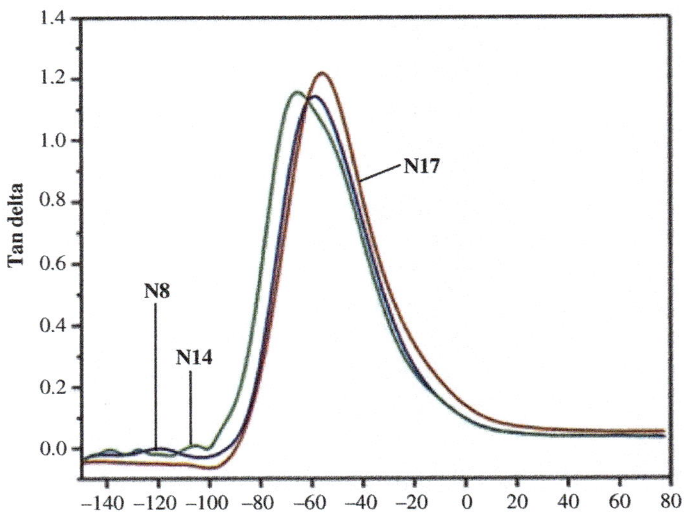

Figure 3.32 Tanδ versus temperature for cured rubber compounds N8, N14, and N17. All measurements were conducted at 1 Hz with a temperature sweep between −150 and 80 °C at a heating rate of 5 °C/min.

and N17 reached −56, −65, and −52 °C, respectively. The T_g of cured rubber compounds changed differently when commercial CB was replaced by modified CBs CB-Oxi and Modified-CB-Oxi. The addition of CB-Oxi to the compound resulted in a T_g increase to −52 °C from its original value of −56 °C. The T_g of N17 decreased to −65 °C when Modified-CB-Oxi replaced CB-Oxi. The observed increase in T_g results from stronger filler–filler interactions in N17 which form rigid filler networks and agglomerates.

The rubber matrix of N14 became more homogeneous because the coupling agent improved Modified-CB-Oxi dispersion throughout the material. The improved dispersion of the material network became more uniform, which decreased stress concentrations and prevented crystallization from occurring, thus resulting in a lower T_g. The T_g determination becomes possible through precise measurements of tanδ peaks which occur at this temperature. The peak value of tanδ indicates the amount of energy that materials absorb during T_g testing with higher values showing increased energy absorption. The T_g values of N8 showed the lowest tanδ peak at 1.14 compared to N14 and N17, which had peak values of 1.16 and 1.22, respectively. The different results between N17 and N18 indicate that N17, with its poor filler distribution pattern, experienced the highest amount of energy dissipation. The rubber compound's ability to perform well on icy and wet surfaces becomes evident through its high tanδ values at −20 and 0 °C temperatures. The rubber compound demonstrates better energy dissipation at these temperatures, which produces stronger frictional forces between the rubber surface and ice or wet surfaces. The improved friction between the rubber surface and the ice or wet surface results in enhanced traction and grip. A complete assessment of rubber compound suitability for winter conditions requires evaluating tanδ data together with rubber hardness measurements and article geometry characteristics including tire tread patterns. The tanδ values of compound N17 reached 0.34 at −20 °C and 0.14 at 0 °C. The rubber compound with CB-Oxi shows potential to deliver enhanced ice and wet traction performance. A rubber compound which shows lower tanδ values at +60 °C will produce less heat during rolling because it dissipates less energy. The rubber articles such as tires experience decreased rolling resistance because they generate less internal friction which requires less energy to overcome. The tanδ values of N8 and N14 reached 0.04, while N17 showed a value of 0.05. The two compounds demonstrated lower rolling resistance because their fillers distributed more evenly throughout the rubber material than the third compound. The results from Figure 3.33 showed complete agreement with the observations made about loss and storage modulus.

The thermal stability comparison of NR compounds N8, N14, and N17 appears in Figure 3.34. The TGA test evaluated thermal properties of cured rubber compounds through nitrogen exposure from 25 to 600 °C at 10 °C/min temperature increments. The TGA results show NR rubber starts degrading between 320 and

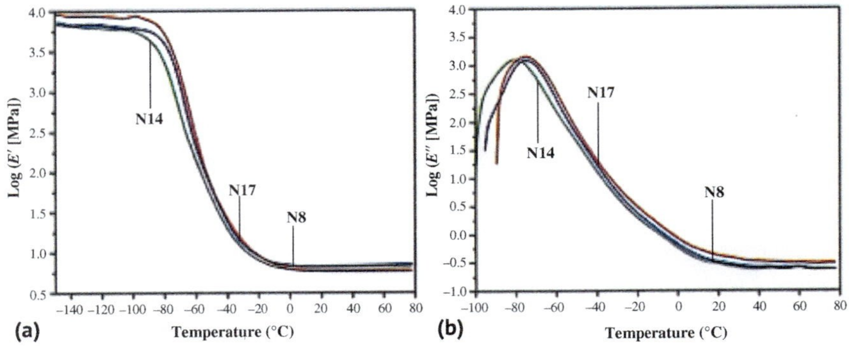

Figure 3.33 Storage and loss modulus versus temperature for compounds N8, N14, and N17 at 1 Hz. Measurements were conducted at 1 Hz with a temperature sweep between −100 and 80 °C at a heating rate of 5 °C/min.

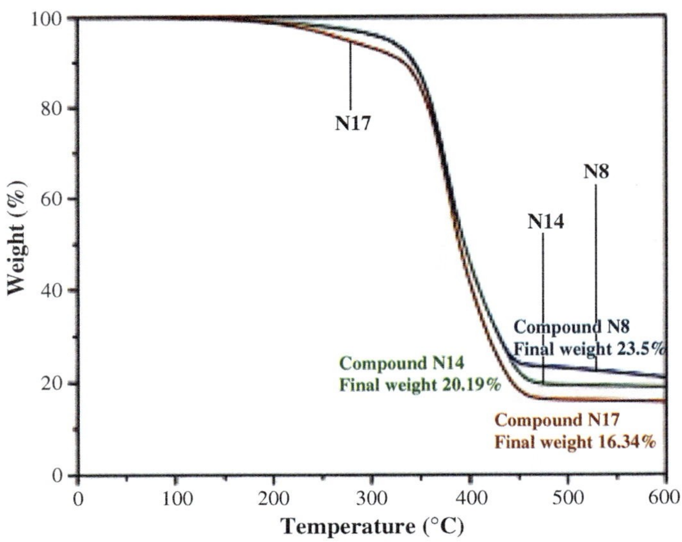

Figure 3.34 TGA thermograms for cured rubber compounds N8, N14, and N17. TGA was performed in the range of 25–600 °C at a heating rate of 10 °C/min under a nitrogen atmosphere.

480 °C. The compound formulations underwent mass loss between 25 and 320 °C because of light hydrocarbon decomposition. The total weight losses for N8, N14, and N17 compounds reached 76.5%, 79.8%, and 83.7%, respectively. The pyrolysis of surface carbonyl and hydroxyl groups on CB-Oxi particles caused N17 to experience the highest total mass loss. The temperature range above 480 °C showed no

mass loss because the materials included unburned CB and zinc oxide. The TGA thermographs of all studied compounds showed identical thermal stability patterns according to the results.

The evaluation of organic solvent resistance in prepared rubber compounds used toluene for immersion and swelling tests. The swelling ratios of compounds N1–N5 with individual fillers stayed at 1.4 except for compound N5 which showed a value of 2 because it demonstrated higher susceptibility to organic solvent degradation. The swelling ratio results for hybrid fillers compounds N6–N17 showed no pattern based on filler type or nanosilica/CB ratio. The swelling ratio reached 1.6 in N15 and N17 compounds which contained 5 and 15 phr CB-Oxi, respectively. The bound rubber measurement serves as a standard method to evaluate how well fillers interact with rubber chains in rubber compounds. The bound rubber contents of N1–N5 samples reached 95%, 95.2%, 95.1%, 95.1%, and 94.2%, respectively. The silane treatment of CB did not affect bound rubber levels because all NR compounds showed identical results. The bound rubber content in rubber compounds with hybrid fillers stayed between 94.7% and 95.5% with small changes between samples. The compound N16 showed the lowest bound rubber value at 90% because its CB-Oxi and nanosilica particles demonstrated exceptional bonding.

Rubber compounds experience their elasticity and performance through the elastic Gibbs free energy (ΔG) and conformational entropy (ΔS) which serve as essential thermodynamic properties as described previously. The work needed to extend a rubber compound reaches its minimum value at ΔG negative values which indicate better stretchability and higher elasticity. The rubber compound's polymer chain disorder level determines its ΔS value which shows higher values for better elasticity and flexibility. The ΔG and $\Delta S \times 10^{-2}$ values for compounds N1–N5 show no change except for N5 which has different values of −7.8 and 2.6 J/mol. The stretching process becomes more challenging for N5 because its ΔG value is less negative (−7.8 J/mol) and its $\Delta S \times 10^{-2}$ value is at its minimum (2.6 J/mol). The hybrid filler containing compounds N6–N17 showed ΔG values between −12.3 and −27.5 J/mol and $\Delta S \times 10^{-2}$ values between 4.1 and 9.2 J/mol. The addition of CB-Oxi results in decreased rubber compound elasticity according to the observed trend. The coupling agent TESPT successfully reduced this negative effect.

From this study, the authors concluded that the surface treatment of CB as a single component or with nanosilica as a hybrid filler system produces better results for NR compounds in multiple industrial applications. The surface modification of CB through coupling agent treatment led to better CLD and curing properties in rubber compounds. The addition of nanosilica as a single component caused negative effects on curing behavior because of its polar nature and tendency to form aggregates. The surface modifications led to significant improvements in the mechanical properties of rubber compounds after vulcanization because they enhanced tear resistance. The combination of nanosilica with Modified-CB-Oxi modified CB produced the best tear resistance because the two separate effects worked together to create better filler distribution in the rubber material. The glass transition temperature response of CB surface modification depended on both the specific modification method and its degree of application. The surface modification of CB led to better tanδ values at low temperatures, which indicates better ice and wet grip performance and reduced rolling resistance because of decreased energy loss after coupling agent treatment. The studied compounds showed no difference in thermal stability, but their swelling resistance depended on the specific filler combination and proportions used. The measurements of bound rubber content showed positive interactions between fillers and rubber in most of the rubber compounds. The rubber compounds showed better elasticity and chain mobility because of CB surface modification through coupling agent treatment according to thermodynamic analysis. The selection of fillers and modification techniques based on performance requirements leads to maximum success in rubber compound development and application.

3.10.3 Case Study III

The influence of silane-treated CB and nanosilica, both separately and together, on the properties of butyl rubber (IIR) compounds is discussed.

This case has been written based on author's published paper in *Polymers and Polymer Composites* (2024) entitled as the "Effect of modified carbon black and nano-silica, individually and in combination, on the curing, mechanical, and dynamical properties of butyl rubber (IIR)." The research assesses the performance of surface-modified CB as a standalone filler and when used with nanosilica as a hybrid filler for butyl rubber (IIR) compounds. The discovery of butyl rubber, also known as isobutylene–isoprene rubber (IIR), occurred during the early 20th century. The discovery of this exceptional elastomer material occurred in 1931 when Standard Oil (now ExxonMobil) Scientists William J. Sparks and Frank J. Soday accidentally created it during their search for synthetic gasoline production techniques. The scientists discovered a highly gas-resistant rubber through their unintentional polymerization of isobutylene which became a pivotal achievement in scientific and industrial development. The distinct chemical composition of butyl rubber sets it apart from typical elastomers including NR and SBR. The main components of butyl rubber consist of isobutylene at 98% and isoprene at 2%.

The unique properties of butyl rubber stem from this particular makeup. The chain structure of isobutylene contains two methyl groups which attach to every other carbon atom to create a hindered structure. The restricted chain movement because of steric hindrance results in lower crystallinity and better gas barrier properties than other materials. The small number of isoprene units with their double bonds enables rubber chain crosslinking during the vulcanization process. The inner tubes of tires require butyl rubber with superior air retention properties for their construction. The material prevents gas leakage because it blocks air from escaping, which helps maintain tire pressure. The combination of chemical resistance and fluid tolerance with high flexibility makes butyl rubber suitable for seal and gasket applications across different industries. The biocompatible nature of butyl rubber makes it suitable for use in automotive parts, appliances, plumbing systems, and pharmaceutical packaging. Scientific literature shows that butyl rubber made from isobutylene–isoprene requires different fillers to improve its properties and performance while reducing production expenses. The rubber industry uses two main fillers in IIR compounds, which are CB and silica. The rubber industry uses CB as its primary filler because it represents 90% of worldwide CB manufacturing for tire production. The rubber industry selects CB as its preferred filler because it offers affordable costs and easy processing and acceptable abrasion protection. The addition of CB to butyl IIR compounds improves multiple mechanical, electrical, and thermal properties. The rubber compound develops multiple defects when CB fails to disperse properly throughout the rubber matrix. The rubber compound benefits from silica addition because it improves dynamic properties which results in better traction on wet and icy surfaces and reduced rolling resistance as described in previous sections. The high cost of silica stands against its poor wear performance because it fails to disperse well in rubber compounds.

The high cost of silica and its poor rubber compound wear performance result from insufficient filler dispersion. The use of nanosilica as a rubber filler material has become popular because it provides better performance than traditional fillers. The development of rubber technology now includes hybrid filler systems which combine two or more different fillers. Hybrid filler systems consisting of multiple fillers demonstrate the ability to produce enhanced rubber–filler interactions which result in better reinforcement and dispersion than single-filler systems. The use of hybrid fillers enables manufacturers to decrease the amounts of required fillers which results in lower production expenses and reduced environmental impact from rubber compounds. The development of a hybrid filler system that combines CB with nanosilica presents a promising approach to leverage their individual benefits while overcoming their individual limitations. The scientific community has shown strong interest in CB and silica hybrid systems based on previous research findings.

This research investigates how commercial CB and modified CBs and nanosilica affect rubber compounds made from IIR when used separately and together. The study assessed both the curing behavior and mechanical strength and dynamic performance of rubber compounds. The research investigated how the fillers distributed throughout the rubber compounds. The research findings focus on determining IIR rubber compound suitability for tire inner liner formulation.

The main raw material used in this study was IIR which included IIR-1675N with 1.4–1.8 mol.% unsaturation and ML(1+4) at 125 °C Mooney viscosity between 46 and 56. The butyl rubber variant IIR-1675N came from SIBUR based in Russia. The research used commercial CB N660 from Pars Carbon Ltd. in Saveh, Iran as the reinforcing filler. The general-purpose furnace CB serves as the main filler in this study. The secondary filler used in this study was nanosilica which Evonik Industries AG from Germany provided as ULTRASIL®. The rubber formulation received its curing chemicals from various components. The rubber formulation received TBBS from Flexsys in Belgium as an accelerator and nano-zinc oxide from Robin Arman Nano Technology Co. in Iran as an activator and stearic acid from Anchor Chemical Ltd in the UK as an activator and elemental sulfur from Solvay Barium Strontium in Hannover, Germany as a curing chemical. The physical property measurements of CLD and bound rubber content required toluene as a solvent which Mojalali in Iran provided. The coupling agent TESPT came from Evonik Industries AG based in Germany.

The research methodology from previous studies (Mofrad et al., 2024c) guided the first step which involved CB oxidation through citric acid monohydrate solution immersion with distilled water under ultrasonic vibration (Sono Swiss, Switzerland, 280 W, 50/60 Hz). The oxidation process produced CB-Oxi through chemical surface modification of commercial CB (N660) by adding hydroxyl (—OH) and carboxyl (—COOH) functional groups. The researchers added functional groups to the CB surface because these groups would help the silane coupling agent bond during subsequent reactions. The ultrasonic bath treated CB-Oxi with ethanol for 30 minutes before adding TESPT to the solution. The reflux process operated at 65 °C for three hours. The solution reached room temperature after the grafting reaction was finished. The treated CB-Oxi required centrifugation to separate it followed by three ethanol washing cycles. The TESPT-functionalized CB-Oxi underwent drying at 50 °C for 12 hours to become "Modified-CB-Oxi." The researchers performed the same treatment on unmodified commercial CB to create "Modified-CB" (Figure 3.17).

The prepared IIR rubber compounds received their formulation from Table 3.15. The two-roll mill operated with 25 cm diameter and 75 cm length rolls to perform the mixing process. The mill operated with a cooling system while maintaining a 3–4 mm gap between the rolls. The rubber compounding sequence followed the steps shown in Figure 3.35. The two-roll mill received preheated IIR for mastication before the process began. The mixture of IIR with fillers underwent blending for 10–15 minutes. The rubber compounds received all curing chemicals including stearic acid and nano-zinc oxide and accelerator and curing agent (sulfur) until they reached complete homogeneity. The entire mixing process

Table 3.15 Formulation of IIR compounds with different fillers (phr).

		Fillers								
Compound no.	**NR**	**Nanosilica**	**CB**	**Modified-CB**	**Modified-CB-Oxi**	**CB-Oxi**	**TBBS**	**Nano-ZnO**	**Stearic acid**	**Sulfur**
A1	100	10	—	—	—	—	4	6	2	3
A2	100	—	10	—	—	—	4	6	2	3
A3	100	—	—	10	—	—	4	6	2	3
A4	100	—	—	—	10	—	4	6	2	3
A5	100	—	—	—	—	10	4	6	2	3
A6	100	5	5	—	—	—	4	6	2	3
A7	100	5	—	5	—	—	4	6	2	3
A8	100	5	—	—	5	—	4	6	2	3
A9	100	5	—	—	—	5	4	6	2	3

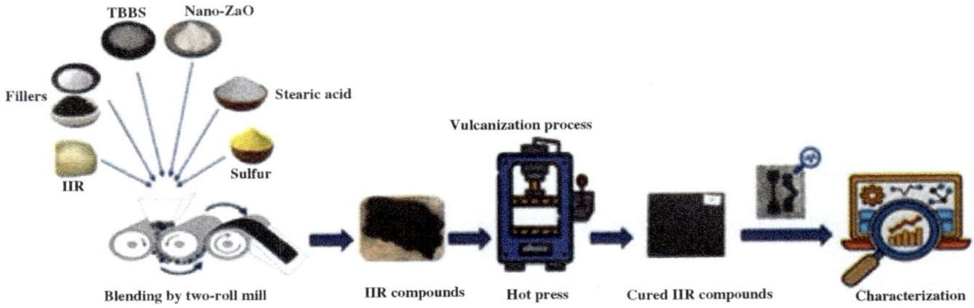

Figure 3.35 The procedure for preparing Modified-CB-Oxi/IIR compounds.

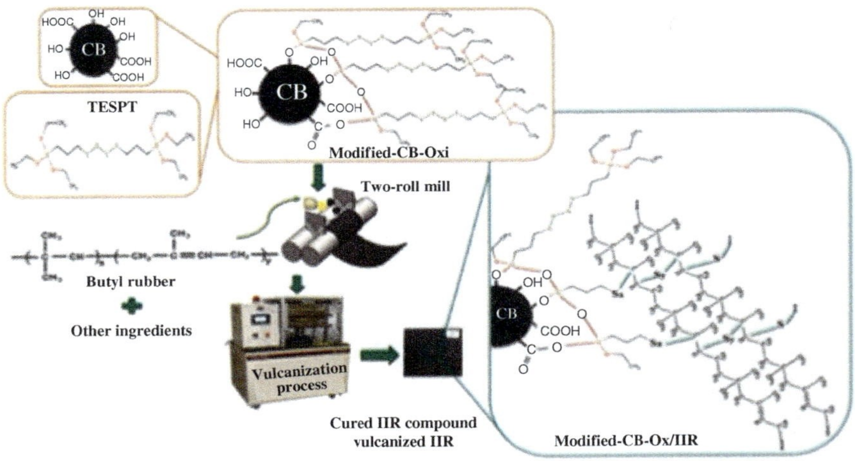

Figure 3.36 The proposed mechanism for reactions between Modified-CB-Oxi and IIR rubber.

required 30 minutes to finish. The rubber compounds underwent compression molding in a hydraulic hot press (SPH-500 from Santam, Iran) at 160 °C and 150 bar pressure after determining their optimal cure times from ODR cure curves. The hot press produced sheet molds for rubber testing which measured 15×15 cm^2 with 2 mm thickness. The hot press produced cylindrical samples with 12.5 ± 0.5 mm thickness and 29 ± 0.5 mm diameter for hardness testing. The experiments ran three times with median results reported for each test.

Figure 3.36 shows the proposed mechanism which demonstrates how Modified-CB-Oxi interacts with butyl rubber through chemical reactions.

The following information appears in Tables 3.15 and 3.16 about the preparation methods and vulcanization characteristics of IIR compounds. The total filler content in each compound reaches 10 phr. The compounds A1–A5 contain

Table 3.16 Curing characteristics of prepared IIR compounds with different fillers.

Samples	t_{s2} (min:s)	t_{90} (min:s)	M_L (dN · m)	M_H (dN · m)	$M_H - M_L$ (dN · m)	CRI (1/min)
A1	11:08	18:53	0.9	4.3	3.4	13.4
A2	11:24	16:10	0.9	4	3.1	20.5
A3	11:56	16:44	0.9	4.5	3.6	20.5
A4	12:04	19:55	1	4.4	3.5	13.3
A5	11:46	15:40	0.9	3.6	2.7	25.4
A6	12:54	16:45	0.9	3.9	3	25.6
A7	12:57	16:35	1	4.2	3.2	26.5
A8	12:00	18:33	0.9	4.3	3.4	15.3
A9	13:05	19:20	0.9	3.9	3	16.3

separate fillers which include nanosilica, CB, Modified-CB, Modified-CB-Oxi, and CB-Oxi, respectively. The compounds A6–A9 contain 5 phr nanosilica with 5 phr of CB or Modified-CB or Modified-CB-Oxi or CB-Oxi to reach a total filler amount of 10 phr. The curing chemical amounts of TBBS and nano-Zn and 2 and 3 phr for all compounds.

The compound A4 with 10 phr Modified-CB-Oxi achieved the highest M_L value (viscosity) of 1 dN·m. The other compounds containing individual fillers (A1–A3 and A5) showed no change in their M_L value which remained at 0.9 dN·m. The physical barrier between IIR macromolecules and TESPT grafts on Modified-CB-Oxi particles limited rubber macromolecule movement which produced higher viscosity. The viscosity of compounds A6–A9 stayed constant at 0.9 dN·m when nanosilica received partial replacement by CB and modified CBs except for compound A7 which used equal nanosilica/Modified-CB amounts to achieve 1 dN·m viscosity through its unique filler interactions.

The M_H value reached its highest point at 4.5 dN·m when compound A3 used grafted TESPT commercial CB (Modified-CB) as its single filler. The M_H value of compound A5 reached 3.6 dN·m when it contained 10 phr of CB-Oxi. The rubber curing process becomes less efficient because CB particles with grafted functional groups such as carboxyl and hydroxyl groups interfere with the process. The functional groups on CB particles create chemical bonds with TBBS and ZnO and stearic acid in the compound mixture which hinders their ability to accelerate the curing process. The carboxyl groups in TBBS form ionic bonds with available N—H groups in TBBS which blocks TBBS from performing its accelerator function. The hydrogen bonds between hydroxyl groups and ZnO activator particles decrease ZnO mobility and prevent it from starting the curing reaction properly. The process results in insufficient curing of the compound which produces lower M_H and delta torque (ΔTorque) values. The presence of hydroxyl groups in stearic acid allows it to interfere with rubber matrix–stearic acid interactions which affects its dispersing and lubricating performance. The M_H values of hybrid filler compounds (compounds A6–A9) reached their lowest points at 3.9 dN·m when CB/nanosilica and CB-Oxi/nanosilica were used together in compounds A6 and A9. The curing process of rubber becomes negatively affected when nanosilica particles are present in the compound. The polar surface of nanosilica particles creates interactions with polar curing chemicals which prevents these materials from bonding with rubber macromolecules. The M_H values of compounds A7 and A8 reached 4.2 and 4.3 dN·m, respectively, through the addition of Modified-CB and Modified-CB-Oxi despite the initial deterioration. The M_H values correspond to the ΔTorque values which follow a systematic pattern in Table 3.16. The highest and lowest ΔTorque values among compounds A1–A5 belonged to A3 and A5 which measured 3.6 and 2.7 dN·m, respectively. The curing process of IIR rubber became worse when carboxyl and hydroxyl groups were added to the CB surface through CB-Oxi modification. The addition of TESPT as a coupling agent reduced the negative effects on curing which resulted in compound A4 achieving a higher ΔTorque value of 3.5 dN·m. The rubber compound curing process received negative effects from nanosilica filling in compound A1 because it produced a 3.4 dN·m ΔTorque value. The curing process of the rubber compound suffered from silica's polar nature which might interfere with specific curing chemicals according to previous discussions. The research compares ΔTorques between compounds A6–A9 which use dual CB and nanosilica additives with different CB types including CB and Modified-CB and Modified-CB-Oxi and CB-Oxi. The ΔTorque value of compound A8 with Modified-CB-Oxi reached 13% higher than all other compounds. The research findings demonstrate that rubber crosslinking during curing becomes more effective when using functionalized CB followed by coupling agent grafting. The CLD measurements showed similar patterns to the ΔTorque results except for A3. The lowest CLD value of 87.8 mol/m^3 occurred in compound A5 which produced the smallest ΔTorque reading. The CLD values for compounds A2 and A4 showed the highest and lowest readings at 65.7 and 165.1 mol/m^3, respectively. The surface modification of CB through TESPT created effective covalent bonds between IIR macromolecules and CB which resulted in better performance. The nanosilica-based compound A1 showed a high CLD value of 156.3 mol/m^3. The CLDs of compounds A6–A9 showed higher values than compounds A1–A5 because nanosilica addition resulted in better CLD performance for dual filler IIR compounds. The CLD of A8 reached its highest value at 163.1 mol/m^3 when using Modified-CB-Oxi/nanosilica as the filler material. The CLD showed substantial dependence on CB type which led to better crosslinking results when using functionalized CB and subsequent coupling agent grafting. The rubber compound needs a cure time (t_{90}) to reach 90% of its complete curing state when tested under established conditions. The measurement helps determine curing speed by establishing a standard for evaluating processing time. The scorch time (t_{s2}) represents the processing window for rubber compounds at specific temperatures before they start to cure. The CRI derived from rheometric data provides a numerical value that shows how fast the curing reaction advances. The curing process speed becomes measurable through this parameter, which delivers essential information about the reaction rate. The rubber curing process efficiency depends on these three parameters which determine the final quality and properties of rubber products as described earlier. The scorch times for compounds A1–A5 extended from 11 minutes and 8 seconds for compound A1 nanosilica filled to 12 minutes and 4 seconds for A4 Modified-CB-Oxi filled. The scorch time of A5 CB-Oxi filled compound decreased to 11 minutes and 46 seconds. The research shows that CB modification leads to shorter prescorch times. The scorch times of dual filler compounds A6–A9 became longer than those containing single fillers. The modified CBs and nanosilica surfaces with specific curing chemicals create synergistic interactions which slow down crosslinking reactions according to previous research.

The surface modification of CB through functionalization resulted in significant changes to the optimum curing time which remained at 15 minutes and 40 seconds for A5 but extended to 19 minutes and 55 seconds for A4. The nanosilica-filled compound A1 required an optimum curing time of 18 minutes and 53 seconds. The rubber curing industry faces an undesirable situation because the CLD of the rubber compound increases when CB-Oxi receives TESPT treatment to produce Modified-CB-Oxi. The observed longer optimum cure time for compound A4 after TESPT treatment can be explained by the blocking effect of the coupling agent molecules that attach to CB surfaces. The blocking effect from steric hindrance might block curing chemicals from reaching rubber macromolecules which would slow down crosslinking reactions. The coupling agent creates obstacles which prevent the curing chemicals from reaching the rubber matrix properly, thus extending the curing duration. The curing process of compound A1 required the longest time because it contained only nanosilica. The CRI values of the mentioned compounds showed significant differences between each other. The CRI value of A5 exceeded all other compounds because it reached 25.4 1/min which demonstrated that CB surface functionalization enhanced curing reaction speed. The parameters of compounds A6–A9 containing hybrid silica-CB materials depended on the specific CB type used in their formulations. The research findings show that surface treatment of CB and selection of different fillers affects rubber curing parameters through prescorch time, optimum cure time, and cure rate measurements. The observed effects stem from the combination of modification methods and their extent and the choice of filler materials and their bonding

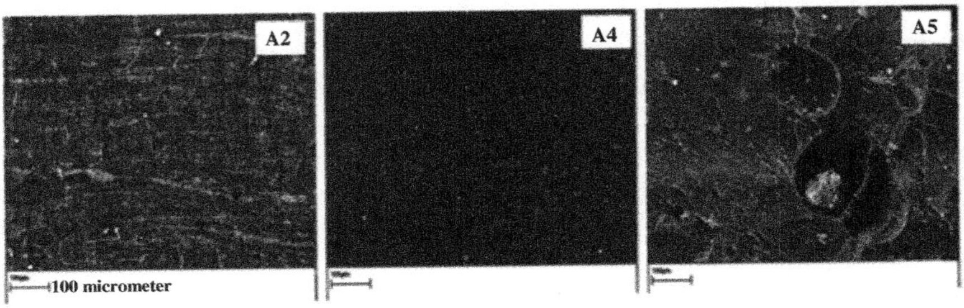

Figure 3.37 The SEM images for selected individual filler-filled rubber compounds, CB/IIR (A2), Modified-CB-Oxi/NR (A4), and CB-Oxi/NR (A5) at the magnitude of 100 μm.

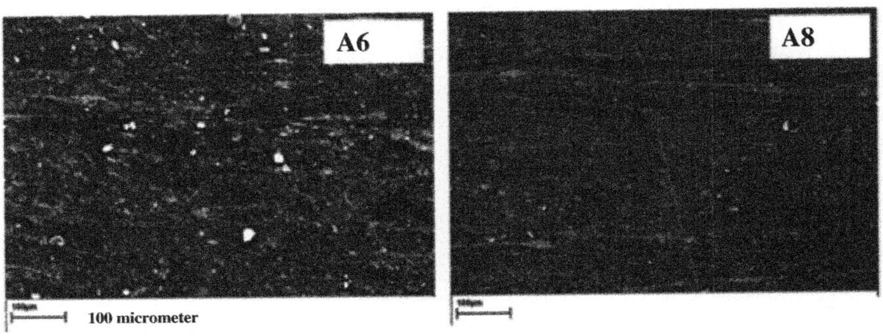

Figure 3.38 The SEM images for selected hybrid filler-filled rubber compounds, CB/nanosilica/IIR (A6) and Modified-CB-Oxi/nanosilica/IIR (A8).

with rubber. The research findings demonstrate that proper selection and optimization of filler materials and surface modifications lead to optimized curing processes for rubber-based applications. The SEM images in Figures 3.37 and 3.38 show the cross-sectional surfaces of IIR compounds after curing with single (compounds A2, A4, and A5) and hybrid fillers (A6 and A8) at 100 μm magnification. The SEM image of compound A4 shows Modified-CB-Oxi particles that distribute evenly throughout the IIR matrix. The rubber macromolecules form covalent bonds with the CB surface through TESPT coupling agent which leads to uniform filler distribution. The better interaction between fillers and rubber in compound A4 results in superior filler distribution. The filler distribution in compounds A2 and A5 shows poor results because they contain CB and CB-Oxi, respectively. The CB surface contains more polar functional groups which leads to enhanced poor filler dispersion in compound A5. The SEM image of A5 shows large CB clusters. The nonpolar nature of IIR causes CB-Oxi particles to form big clusters instead of spreading throughout the rubber matrix. The rubber matrix shows better distribution of Modified-CB-Oxi/nanosilica particles in compound A8 compared to commercial CB/nanosilica in A6 according to Figure 3.38.

The SEM images of these compounds demonstrate nanosilica agglomerates which demonstrate why filler dispersion affects the final properties of rubber compounds. A rubber compound with properly dispersed fillers will achieve better mechanical and dynamic performance according to research studies as described earlier. The rubber material benefits from stress transfer efficiency when filler particles are distributed evenly throughout the compound, which leads to better strength and modulus and tear resistance.

The presence of poorly distributed filler particles creates stress concentration points which weakens the material structure and makes it more prone to crack development and propagation. The mechanical and dynamic performance of materials deteriorates when filler particles fail to distribute evenly throughout the material. The addition of TESPT coupling agent to compound A4 resulted in better filler distribution because it formed chemical bonds between Modified-CB-Oxi and IIR matrix components. Surface modification methods show promise to enhance filler–rubber interface quality which results in better filler distribution. The SEM image results matched the data obtained for most mechanical and dynamic properties which are presented in following phrases. The consistent results between SEM images and mechanical and dynamic property data confirm that filler dispersion plays a key role in determining the final performance of rubber compounds after curing.

Also, the research investigated IIR compounds through mechanical property tests which evaluated silane modification, filler addition, and nanosilica integration separately and together. The tensile strength of CB-Oxi (compound A5) showed a small improvement compared to CB (compound A2) with measured values of 3.2 and 3.1 MPa, respectively. The tensile strength measurements for A3 and A4 compounds showed unexpected lower values at 2.2 and 2.3 MPa, respectively. The highest tensile strength measurement occurred in the CB-Oxi sample. The rubber compound's microstructure likely experiences changes because of grafted functional groups which lead to this observed tensile strength result. The tensile strength of nanosilica-filled IIR rubber (compound A1) showed the lowest value at 2.5 MPa because nanosilica particles have polar properties and do not disperse well within the rubber matrix. The hybrid filler compounds (A6–A9) showed no pattern in their tensile strength results, which were lower than the values obtained from individual rubber compounds (A1–A5). The tensile strength of compounds A9 and A8 showed the lowest and highest values at 1.9 and 2.2 MPa, respectively. The rubber matrix becomes less homogeneous because of filler incompatibility in dual filler compounds, which results in decreased tensile strength values. The elongation at break parameter delivers essential information about rubber compound flexibility and motion range and filler compatibility and fatigue resistance which determine material selection for various applications to achieve reliable long-term performance. The elongation at break test revealed A2 as the compound with the highest value at 576.3% while A1 showed the lowest value at 325.7%. The elongation at break measurement for compound A5 with CB-Oxi reached 560.2% which surpassed A2's value. The rubber matrix in A1 shows poor compatibility because the rubber–filler (nanosilica) interaction remains weak, which results in its lowest elongation at break value. The elongation at break measurements for hybrid filler compounds A6–A9 showed A9 with CB-Oxi/nanosilica at 422.1% and A8 with Modified-CB-Oxi/nanosilica at 334%. The elongation at break measurements of A7 through A6 compounds ranged from 346.7% to 393.7% based on their CB selection. The results show that elongation at break depends on rubber–filler bonding strength and filler network development and surface treatment methods used on the fillers in IIR rubber compounds. The initial modulus represents the material stiffness through its slope in the stress–strain curve during small deformation. The yield modulus represents the stress value which causes permanent plastic deformation in materials, thus marking their transition from elastic to plastic behavior. The material loses its ability to return to its original form after experiencing deformation past the yield point. The initial modulus of compounds A1–A5 except compound A1 maintained a constant value of 0.8 MPa while the other compounds showed values between 0.5 and 0.6 MPa. The parameter showed a small variation between 0.5 MPa for A6 and A9 and 0.7 MPa for A8 in hybrid filler compounds A6–A9. The yield modulus followed an identical pattern to the initial modulus. The lowest and highest values among individual rubber compounds belonged to compounds A1 and A2 and A5 with 0.6 and 0.3 MPa, respectively. The lowest and highest values among hybrid filler compounds occurred in A6 and A9 and A8 with 0.4 and 0.6 MPa, respectively. The rubber compounds containing nanosilica and CB and CB-Oxi showed lower initial and yield modulus values than Modified-CB and Modified-CB-Oxi compounds because these fillers do not match well with the rubber matrix as shown in SEM images. The coupling agent TESPT demonstrates its ability to enhance both initial and yield modulus values. The nanosilica formulation of compound A1 produced the highest tearing strength which reached 17.2 MPa while A2 reached 16 MPa. The surface modification of CB resulted in a decrease of tearing strength to 12.4–13.5 MPa. The hybrid filler compounds in compounds A6–A9 showed identical patterns of behavior. The combination of surface-modified CB with nanosilica resulted in decreased tearing strength which fell between 12.4 and 12.9 MPa. The combination of CB with nanosilica in A6 produced a synergistic effect which decreased this mechanical property because the destructive interaction between CB and nanosilica fillers increased. The optimal hardness level for applications depends on specific requirements which determine the desired performance characteristics. A material requires hardness to exist alongside other essential properties which include flexibility and tear resistance. The highest hardness value of 45 appeared in compound A4 while compound A5 demonstrated the lowest hardness value of 32 which makes Modified-CB-Oxi suitable for IIR rubber compound applications. The hardness values of hybrid fillers followed a pattern where nanosilica and CB-Oxi and Modified-CB-Oxi compounds showed hardness values between 39 and 45.

Rubber compounds achieve their performance through dynamic properties which handle different and recurring loads since static properties measure material behavior under steady and constant stress conditions. The material response to oscillating forces and deformations becomes measurable through dynamic properties because these applications need specific performance during vibration, impact, repeated stretching, and compression cycles. The effective vibration absorption and dissipation of rubber compounds depend on their dynamic properties including hysteresis and loss modulus which makes them suitable for vehicle engine mounts that reduce noise and improve ride quality. The friction coefficient of rubber materials determines their grip performance, which directly affects tire traction on various road surfaces to ensure safe driving operations. The comparison between tanδ values at different temperatures for compounds A6 and A8 with CB and Modified-CB-Oxi fillers appears in Figure 3.39. Also, it shows the same parameter at different temperatures which include

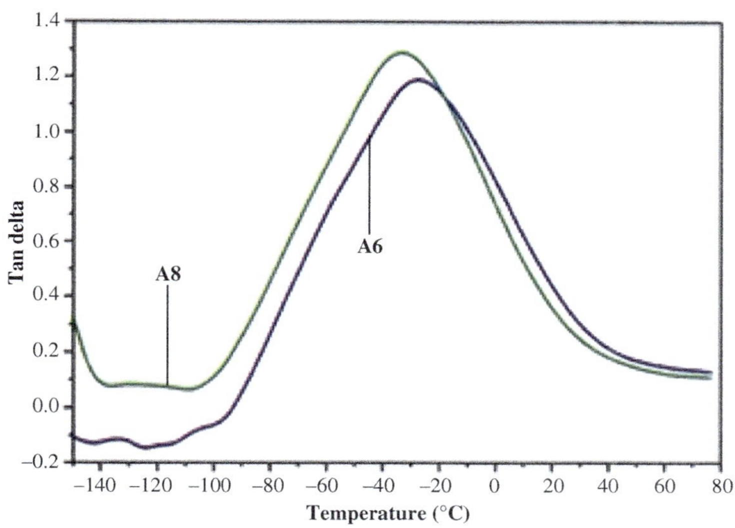

Figure 3.39 Tanδ versus temperature for cured rubber compounds A6 and A8. All measurements were conducted at 1 Hz with a temperature sweep between −150 and 80 °C at a heating rate of 5 °C/min.

T_g (maximum tanδ), −20, 0, and 60 °C. The T_g values for compounds A6 and A8 were found to be −24 and −30 °C, respectively. The glass transition temperature of cured rubber compounds changed differently when commercial CB was replaced by modified CB (Modified-CB-Oxi). The rubber compound containing 10 phr CB showed a T_g value of −24 °C, but the use of Modified-CB-Oxi with coupling agent modification resulted in a T_g value of −30 °C. The rubber–filler interaction strength in the A8 rubber matrix became stronger, which led to a more flexible rubber structure that decreased the T_g value. The coupling agent in A8 improved the distribution of Modified-CB-Oxi throughout the rubber matrix. The better dispersion of Modified-CB-Oxi throughout the rubber matrix created a more consistent network structure which decreased stress points and blocked crystallization processes to produce a lower T_g value.

The T_g temperature appears as a peak in tanδ measurements which provides both sensitive and precise methods to determine this essential temperature point. The peak height of tanδ measurements shows the amount of energy dissipation which occurs in materials at T_g temperatures. The energy dissipation of A8 material exceeds A6 material because A8 shows a higher tanδ value of 1.29 compared to A6 which has a value of 1.19 after Modified-CB-Oxi filling of the IIR rubber compound. The rubber compound's ability to perform well on icy and wet surfaces becomes evident through its higher tanδ values at −20 and 0 °C temperatures as stated earlier. The rubber compound shows better traction and performance on icy and wet surfaces because its tanδ values reach higher levels at −20 and 0 °C. The rubber compound demonstrates better energy dissipation at these temperatures, which produces stronger friction forces between the rubber surface and ice or wet surfaces. The increased friction between the rubber surface and the ice or wet surface results in better traction and grip. A complete evaluation requires rubber properties including hardness and article geometry such as tire tread patterns for tires together with tan δ data to determine rubber compound suitability for winter conditions. The tanδ value of Compound A8 reached 1.16 at −20 °C, while A6 reached 1.14. The rubber compound containing Modified-CB-Oxi shows better ice grip performance according to these test results. The tanδ value at 0 °C for A8 (0.74) was lower than A6 (0.82) which suggests A8 might have reduced wet grip performance compared to A6.

The rubber compound shows lower tanδ values at +60 °C which means it dissipates less energy and produces less heat during rolling operations. The rubber-made articles experience decreased rolling resistance because they need less energy to fight internal friction as mentioned earlier in this text. The tanδ value of compound A8 reached 0.12 while A6 showed a value of 0.15. The compound A8 demonstrated better rolling resistance because its rubber matrix contained better filler dispersion than compound A6. The results from Figure 3.40 supported the observations made about loss and storage modulus.

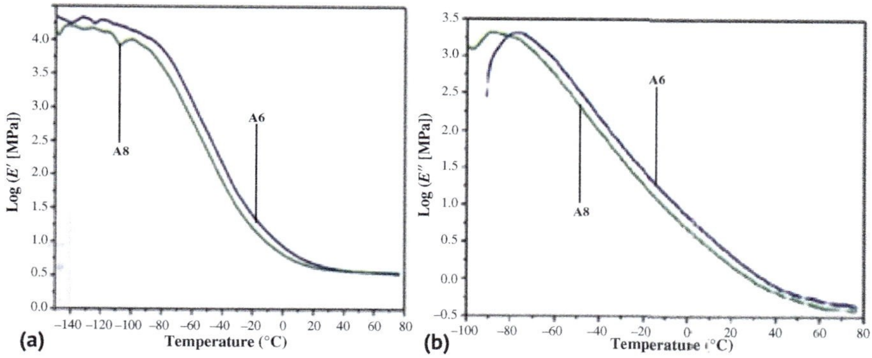

Figure 3.40 Storage and loss modulus versus temperature for compounds A6 and A8 at 1 Hz. Measurements were conducted at 1 Hz with a temperature sweep between −100 and 80 °C at a heating rate of 5 °C/min.

The thermal stability comparison between IIR compounds A6 and A8 appears in Figure 3.41. The TGA method analyzed rubber compound thermal responses through nitrogen gas exposure from 25 to 600 °C while the heating rate reached 10 °C/min. The research data shows IIR rubber degrades between 300 and 430 °C while reaching its highest degradation rate at 406 °C. The compound formulations underwent mass loss between 25 and 300 °C because of light hydrocarbon decomposition. The total mass loss of A6 and A8 compounds reached 86% without showing any distinction between them. The temperature range above 430 °C showed minimal mass loss because the compounds contained unburnable materials including CB and zinc oxide. The TGA thermographs of the studied compounds demonstrated identical thermal stability performance.

The prepared rubber compounds underwent organic solvent resistance testing through toluene immersion and swelling assessment. The swelling ratios of compounds A1–A5 with individual fillers showed values ranging from 1.6 for A1 and A4 to 2.1 for A2. The swelling test results for compound A5 showed a value of 1.9 which indicated higher solvent degradation than expected. The treatment of CB (CB-Oxi) with TESPT to create Modified-CB-Oxi resulted in better solvent resistance for the IIR rubber compound as shown by A4 test results. The swelling results for A6–A9 hybrid filler compounds did not show any specific pattern because the values stayed similar when different CB types were combined with nanosilica in IIR rubber formulations. The swelling test results showed that A8 had the lowest swelling value of 1.6 which indicated better toluene resistance than the other hybrid compounds with 1.8 swelling values. The IIR compounds contain bound rubber data which serves as a metric

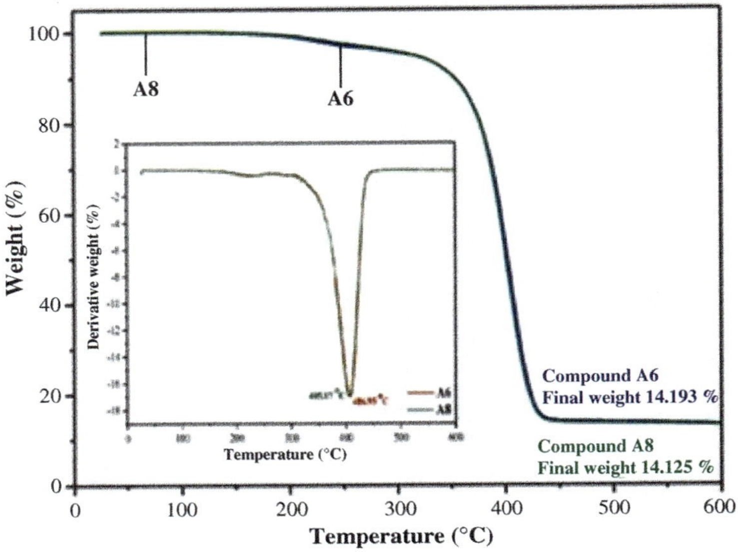

Figure 3.41 TGA thermograms for cured rubber compounds A6 and A8. TGA was performed in the range of 25–600 °C at a heating rate of 10 °C/min under a nitrogen atmosphere.

to evaluate filler–rubber chain interactions. The bound rubber content of A1–A5 samples reached 94.9% and 94.4% and 91.9% and 94% and 93.1%, respectively. The silane treatment of CB resulted in no substantial change to bound rubber levels which stayed consistent throughout the CB-filled IIR compound A2. The bound rubber content in rubber compounds with hybrid fillers showed minimal change between 92.6% and 94.8%. The compound A6 showed the lowest bound rubber value at 92.6% because its CB and nanosilica particles had the strongest destructive interaction. Rubber compounds experience elasticity through the elastic Gibbs free energy (ΔG) and conformational entropy (ΔS) which function as essential thermodynamic properties. The minimum work needed to extend a rubber compound reaches its maximum point when ΔG shows negative values which indicate better elasticity as stated earlier in this text. The rubber compound's polymer chain disorder level becomes measurable through ΔS which produces higher values when chains move freely for better elasticity and flexibility. The ΔG and $\Delta S \times 10^{-2}$ data show minimal variation between compounds A1–A5 except for A1 and A4 which display values between −11.9 and −8.8 J/mol for ΔG and between 3 and 4 J/mol for $\Delta S \times 10^{-2}$. The parameters of A4 showed the most negative values at −22.8 J/mol and the highest value at

7.7 J/mol which demonstrated enhanced stretching ability and flexibility and superior elasticity and chain mobility. The hybrid filler-containing compounds A6–A9 displayed ΔG and $\Delta S \times 10^{-2}$ values between −22.5 and −14.2 J/mol and between 7.6 and 4.8 J/mol, respectively. The addition of CB-Oxi leads to rubber compound elasticity degradation according to the observed trend. The negative effects of CB-Oxi on rubber compound elasticity became less pronounced after TESPT treatment.

As concluding remarks, the research demonstrates that surface treatment of CB and filler material choices determine multiple curing properties which affect rubber compound processing and final product performance. The curing properties of rubber compounds showed substantial changes because of different filler materials and surface treatments which affected their viscosity and stiffness and their curing time. The visual inspection of cured rubber compounds through SEM showed how different filler distributions affect their structural appearance. The surface treatment of CB led to better filler distribution and superior rubber–filler bonding, which resulted in better mechanical properties. The mechanical properties of rubber compounds depended on three factors which included filler material selection and surface treatment and rubber–filler bonding strength. The addition of nanosilica and surface-treated CB particles resulted in compounds with different stiffness levels and tear resistance and flexibility properties. The DMA results showed how different filler materials and surface treatments affect storage modulus and loss modulus and tanδ values. The glass transition temperature and tanδ measurements at different temperatures revealed performance characteristics of the compounds under different operational conditions including ice and wet grip and rolling resistance. The thermal stability results from TGA tests showed that all compounds maintained equal thermal stability, but their solvent resistance depended on the filler materials and their surface treatments. The surface treatment process led to better solvent resistance because the swelling ratios showed significant improvement. The bound rubber content measurements showed minimal changes between different compounds because they indicated weak filler–rubber interactions. The thermodynamic evaluation demonstrated that surface-coupling agent-modified fillers help reduce negative effects which untreated fillers create in rubber compounds. The selection of filler materials and their surface modifications require precise optimization to achieve desired rubber compound properties for specific industrial applications. The surface treatment process improves filler distribution and rubber–filler bonding while simultaneously affecting curing behavior and mechanical properties, dynamic characteristics, thermal stability, solvent resistance, and thermodynamic properties which determine rubber product quality and performance.

References

Ansarifar, A., Saeed, F., Wang, L., Ostad Movahed, S., and Yasin, K. A. Curing rubber compounds efficiently and cost effectively. *Tire Technol. Int.* 2009, *1*, 110–114.

Jourabchi, S. and Ostad Movahed, S. Effect of microwave irradiation on the molecular diffusion of ethyl acetate into SBR/EPDM rubber blends. *Prog. Rubber Plast. Recycl. Technol.* 2024, *40*, 201–217.

Mofrad, F. J., Ahmadpour, A., and Ostad Movahed, S. Effect of modified carbon black and nano-silica, individually and in combination, on the curing, mechanical, and dynamical properties of butyl rubber (IIR). *Polym. Polym. Compos.* 2024a, *32*, 09673911241256553.

Mofrad, F. J., Ahmadpour, A., and Ostad Movahed, S. Synergistic effects of silane-modified carbon black and nano-silica hybrid fillers on the performance of natural rubber compounds. *J. Appl. Polym. Sci.* 2024b, *141*(28), e55640.

Mofrad, F. J., Ahmadpour, A., and Ostad Movahed, S. Effect of silane-modified carbon black and nano-silica, individually and in combination, on the performance of ethylene–propylene–diene monomer rubber. *Polym. Compos.* 2025, *46*(1), 547–566.

Mofrad, F. J., Ostad Movahed, S., and Ahmadpour, A. Surface modification of commercial carbon black by silane coupling agents for improving dispersibility in rubber compounds. *J. Appl. Polym. Sci.* 2024c, *141*, e55155.

Ostad Movahed, S. Nanofiller in aircraft and heavy vehicles tyre rubber compounds. In *Proc. Third Natl. Conf. Mod. Technol. Chem. Chem. Eng.*; Iran, May 2014.

Ostad Movahed, S., Ansarifar, A., and Song, M. Comparing the dynamic behaviour of several rubbers filled with silanized silica nanofiller. *Polym. Int.* 2009a, *58*(2), 209–217.

Ostad Movahed, S., Ansarifar, A., and Song, M. Using modulated-temperature differential scanning calorimetry to study interphases in blends of styrene–butadiene and polybutadiene rubbers filled with silanized silica nanofiller. *J. Appl. Polym. Sci.* 2009b, *111*(2), 1644–1652.

Ostad-Movahed, S., K. Ansar Yasin, A. Ansarifar, M. Song, and S. Hameed. "Comparing effects of silanized silica nanofiller on the crosslinking and mechanical properties of natural rubber and synthetic polyisoprene." *Journal of Applied Polymer Science* 109, no. 2 (2008): 869-881.

Sabeti Meybod, J. and Ostad Movahed, S. Effects of several parameters on the storage stability of melamine–formaldehyde polymers using a design of experiments approach. *J. Adhes. Sci. Technol.* 2020, *34*(16), 1705–1719.

4

Introduction to Rubber Recycling and Classifications of Waste Rubbers

4.1 Introduction

The recycling of rubber materials creates both financial prospects and essential environmental requirements. The increasing global rubber consumption in automotive and industrial applications creates an urgent need to create environment-friendly methods for handling postconsumer and postindustrial rubber waste. Scientists have dedicated substantial research efforts to improve rubber recycling methods throughout the last few years. Ostad Movahed et al. (2016) conducted a detailed analysis of waste rubber reclamation, regeneration, and devulcanization methods. The authors identified three main recycling approaches which include thermal methods, mechanical approaches, and chemical processes. The thermal recycling process for scrap rubber uses a jacketed autoclave system that operates similarly to traditional heater processing. The early reclamation methods that used alkaline aqueous solutions for rubber treatment have become obsolete because of their negative effects on the environment. The mechanical recycling process through grinding and size reduction includes three sub methods which are ambient grinding, cryogenic grinding, and wet grinding. The two-roll cracker mill processes large rubber pieces into 10–40 mesh particles which eventually become 80 mesh or smaller fine crumbs through multiple processing stages. The process requires sequential operations that start with coarse reduction followed by ultrafine sizing and then metallic and fiber contaminant removal and precise classification. The process of cryogenic grinding uses liquid nitrogen to freeze rubber scraps until they become brittle before impact milling produces smaller particles. The processed crumb requires drying followed by mesh size and bulk property-based sorting. The grinding process with water-cooled wheels operates as the basis for wet grinding. The rubber crumb undergoes drying and separation procedures before scientists assess its particle size distribution and physical characteristics.

4.1.1 The Significance of Rubber Recycling Within a Circular Economic Framework

The circular economy depends on rubber recycling to achieve its essential goals of waste reduction and material resource optimization and continuous material reuse. The new economic system operates against the traditional linear "take–make–dispose" approach. The industrial recycling of used rubber materials helps extend product lifespans while protecting essential resources from depletion. The worldwide rubber market, especially in tire manufacturing produces enormous amounts of waste materials from postconsumer products. The recycling process helps decrease environmental damage from waste while it decreases the need for virgin materials which stem from petroleum resources. The combination of mechanical, thermal, and chemical because it aligns with their circular economy initiatives. The implementation of extended producer responsibility (EPR) insulation materials and fresh rubber parts. Multiple nations across the world have established national policies which support rubber recycling methods enables the conversion of discarded rubber into various products including flooring materials and road surfacing products and sound regulations and recycling technology subsidies and landfill diversion requirements form part of these policies. The combination of deforestation and climate change effects on natural rubber production creates an urgent need for recycled rubber as a strategic resource. The development of devulcanization and pyrolysis technologies has made it possible to extract valuable compounds including carbon black and liquid fuels from waste rubber according to Emami Tajodin and Ostad Movahed (2024). The market demand for premium uses of recycled rubber materials continues to expand because of industry developments in devulcanization and pyrolysis technologies.

4.1.2 Plastic Recycling Methods Differ From Those Used for Rubber Recycling

The recycling approaches for plastics and rubbers differ substantially because their molecular structures and processing characteristics are distinct. Thermoplastics maintain their ability to undergo multiple heating and reshaping cycles because they do not contain permanent crosslinks. The recyclability of thermoplastics makes them the focus of worldwide large-scale recycling operations. Rubbers exist as thermoset materials which require different recycling approaches than plastics. The vulcanization process creates permanent covalent bonds between molecules that stop them from melting which prevents heat-based reshaping. The crosslinked structure of rubber makes it impossible to process through standard thermal or mechanical plastic recycling methods. The recycling of rubber requires devulcanization and cryogenic grinding and thermochemical depolymerization and pyrolysis methods to break down its crosslinked matrix structure. The process of sorting rubber waste materials becomes more complicated than plastic waste sorting operations. The recycling of rubber products including tires requires separation of their complex composition which includes natural and synthetic rubber blends (NR, SBR, and BR) together with steel cords, textile reinforcements, processing oils, and fillers. The diverse composition of rubber waste makes it difficult to achieve proper separation and treatment of specific materials. The recycling industries of rubber and plastics face similar contamination issues although they deal with different types of contaminants. The processing of rubber becomes more difficult because rubber contains fibers and wire reinforcements and processing additives which decrease the quality of recycled materials.

The plastic recycling industry's established methods including spectroscopic sorting, density-based separation and design-for-recyclability principles are being implemented in rubber recycling operations. The achievement of rubber recycling circularity depends on technological advancements in recycling methods together with regulatory support and economic benefits. Table 4.1 compares plastics and rubber recycling.

4.2 Classification of Waste Rubber Sources

This section provides detailed classification of waste rubber origins together with national and regional quantity assessments which focus on ELTs and rubberish articles, industrial process scrap, municipal/composite rubber waste, and medical/household items. The recycling paths for rubber depend heavily on the waste material's origin and its level of contamination and chemical makeup because rubber exists as a chemically bonded substance. The framework aims to establish a decision system which connects waste streams to their best recovery methods while offering preliminary life-cycle assessment (LCA) and techno-economic analysis (TEA) results. The values used in this study come from industry reports, NGO publications, and governmental data released between 2020 and 2024 (Castillo et al., 2025).

The production of waste rubber occurs throughout the world because people buy vehicles and construct buildings and industrial facilities. The majority of waste rubber mass consists of discarded tires which amount to 70–75% of total waste rubber. The EU, Japan, and the United States have achieved significant improvements in their organized ELT collection systems during the past 20 years, while China and India have experienced rapid industrial growth that has raised their total tire and industrial rubber waste amounts. The environment faces multiple threats because of unregulated stockpiles, open fires, and toxic substances

Table 4.1 Plastics versus rubber recycling.

Feature	Plastics recycling	Rubber recycling
Dominant type	Thermoplastics (e.g., PE, PP, PET) that soften on heating and can be remelted	Crosslinked elastomers (vulcanized rubber such as NR, SBR, EPDM) that do not melt under heat
Recyclability	Generally high, especially for clean, single-resin streams with minimal contamination	Low due to vulcanization; crosslinked structure prevents remelting
Mechanical recycling	Widely applicable to thermoplastics, effective for simple streams	Limited; used primarily for size reduction and crumb rubber production
Chemical recycling	Depolymerization, pyrolysis, and solvolysis, particularly useful for mixed plastic waste	Devulcanization, pyrolysis; still under research for cost-effectiveness
Energy recovery	Moderate calorific value; often used in waste-to-energy systems	High calorific value; tire-derived fuel (TDF) is widely used in cement kilns
Infrastructure availability	Well-developed systems in many developed countries	Less widespread and less standardized globally
Market size	Larger and more diversified, involving packaging, textiles, electronics, etc.	Primarily focused on automotive and industrial sectors (especially tire recycling)
Policy maturity	More advanced regulations in place (e.g., EU Waste Framework, EPR)	Still evolving; more recent regulatory focus on end-of-life tires (ELTs) and microplastics
Contamination sensitivity	Highly sensitive to food residues, dyes, adhesives, etc.	Complications include textile, steel reinforcements, and compound diversity

that leak from landfills and tiny rubber particles that come from road deterioration. The process of separating waste sources between industrial scrap, ELTs, and municipal composites stands as the best method to enhance material recovery rates and decrease environmental impacts throughout the life cycle. Figure 4.1 shows ELTs recycling flowchart.

4.2.1 Industrial Resources

Industrial rubber waste consists of two categories which include preconsumer process scraps and postindustrial components. The preconsumer scraps which

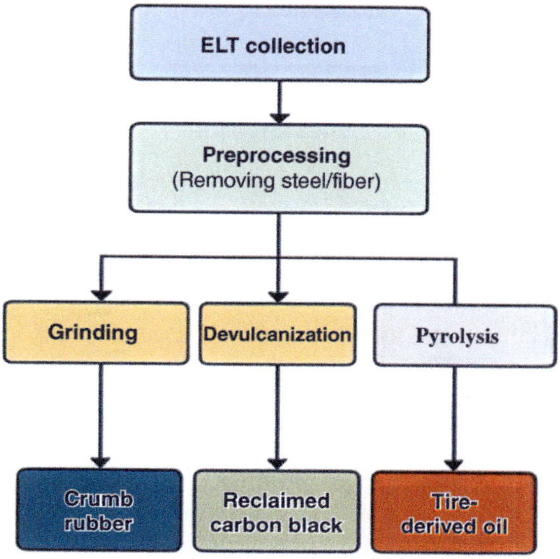

Figure 4.1 The ELTs recycling flowchart.

include trim, flash, and off-spec sheets maintain their cleanliness and known composition which allows for direct reuse or controlled devulcanization. The postindustrial parts which include hoses, seals, belts, and membranes need to undergo metal and fabric extraction followed by washing and solvent treatment before they can be reduced in size and devulcanized. The collection of specific waste streams results in high-value feedstocks with defined compositions, e.g., EPDM roofing membranes, pharmaceutical butyl stoppers, and halogenated elastomers.

The data in Table 4.2 presents representative industrial sectors together with their corresponding recycling considerations.

The production of process scraps happens when compounds are made and when materials undergo extrusion and molding operations, e.g., tread/sidewall trim and off-spec calendaring sheets. Masterbatch production accepts green stock scraps which need mastication before reuse. The process of size reduction becomes necessary for cured scraps before they can undergo partial decrosslinking through mechano-chemical or microwave-assisted devulcanization to achieve property recovery. Tire plants achieve internal recycling rates between 2% and 5% of their total production which decreases their need for new virgin polymers. The sorting process for end-of-life industrial parts (postindustrial) organizes materials based on their polymer family between diene rubbers, saturated rubbers, and

Table 4.2 Representative industrial sectors together with their corresponding recycling considerations.

Industrial sector	Typical components	Dominant rubber(s)	Recycling considerations
Automotive and transport	Tires, hoses, mounts, seals	NR/SBR/BR; NBR; EPDM	Large volume; remove steel/textile; grinding/devulcanization
Oil and gas/petrochemistry	O-rings, gaskets, linings	NBR; IIR; EPDM; FKM	Hydrocarbon contamination; controlled reclamation; pyrolysis for mixed streams
Construction/civil	Expansion joints, membranes	NR; EPDM; CR	Often laminated; mechanical delamination; size reduction
Mining and conveyor	Conveyor belts, liners	NR; SBR; BR; NBR	Steel/textile reinforcement; ideal for ambient/cryogenic grinding
Electrical/cables	Sheaths, insulations	EPDM; CR; CPE	Halogens/fillers affect emissions in thermolysis; prefer mechanical routes
Pharma/medical	Butyl stoppers, peristaltic tubes	IIR; SI; NBR	High purity; sterilize → devulcanize or reprocess
Roofing systems	EPDM sheets	EPDM	Clean removal yields high-value granulate for secondary sheets

specialty elastomers. The removal of metal inserts and textiles occurs through mechanical or cryogenic processing methods.

The process of prewashing followed by solvent extraction becomes necessary for NBR/IIR materials which have exposed to oil during refinery operations. The processing methods include three main paths which start with ambient granulation to produce powders and then proceed to controlled devulcanization through mechanical, chemical, microwave, and or a combination method and finally, pyrolysis for streams with high contamination levels to extract steel and carbonaceous materials. Table 4.3 presents estimated 2023–2024 waste generation data for each country based on their industrial and tire-rubber waste production.

4.2.2 Home-made and Municipality Resources

The municipal waste stream contains postconsumer rubber products which include ELTs, footwear, floor mats, gloves, toys/sports goods, and small automotive parts. The recyclability index of these items decreases because of their diverse

Table 4.3 Estimated 2023–2024 waste generation data for each country based on their industrial and tire-rubber waste production.

Country/region	ELTs generated (Mt/yr)	Industrial rubber waste (kt/yr)	ELT collection rate (%)	Material recovery (%)	Energy recovery (%)
European Union (27)	3.2–3.5	900–1,000	≈ 95–99	50–60	35–45
United States	3.5–4	1,000–1,200	≈ 85–90	45–55	35–45
China	8–9.5	2,300–2,600	≈ 75–85	35–45	30–40
India	1.3–1.7	700–900	≈ 60–70	25–35	20–30
Japan	0.9–1.1	250–300	≈ 95–99	40–50	45–55
Turkey	0.35–0.45	120–150	≈ 70–80	30–40	25–35
Iran	0.25–0.35	90–120	≈ 55–65	20–30	20–30

Table 4.4 The ELT recycling mass balance for ambient grinding route.

Product stream	Mass yield (wt.%)	Primary use	Notes
Crumb rubber (0.5–3 mm)	40–55	Playgrounds, sports, asphalt	Particle size distribution dictates use
Fine powder (< 500 μm)	5–15	Polymer modifiers, coatings	Surface functionalization improves compatibility
Steel	10–15	Scrap markets	Removed by magnets
Textiles	5–10	Composite filler	Energy recovery or composite cores
Losses and fines	5–10	Dust control/filters	Controlled collection required

composition and contamination levels which require pretreatment including washing, sorting, and delamination. ELTs represent the largest amount of waste in municipal streams so they should receive first priority for collection.

The ELT recycling mass balance for ambient grinding route appears in Table 4.4.

The small household rubber articles consist of gloves made from NR/NBR materials and shoe soles from SBR/EVA materials and hoses and gaskets from EPDM materials and toys made from NR/TPE/PU blends. The sorting process faces significant difficulties. The separation of rubber from plastic materials can be achieved through methods which included but not limited to triboelectric separation of rubber–plastic blends, selective dissolution to extract elastomer fractions and compatibilizer-assisted blending of rubber powders into thermoplastics, and float–sink technique.

Public infrastructure implements rubber composites for playground surfaces through crumb rubber and PU binder combinations, sound barrier construction, and expansion joint applications which classifies as municipal and construction rubber composites. The optimal treatment method for these composites involves size reduction followed by secondary forming into mats and underlays or thermochemical processing when binder materials prevent recycling. The practice of urban mining for rubberized asphalt during resurfacing operations allows civil engineering projects to achieve partial closed-loop recycling.

Waste medical rubbers (NR/NBR gloves, catheters, silicone tubes, and IIR stoppers) as healthcare and medical rubbers experienced increased demand because of the COVID-19 pandemic. The process of sterilization through steam autoclave, microwave, or plasma followed by shredding makes medical rubbers suitable for industrial recycling. The successful implementation of sterilized glove powder into asphalt and elastomer blends becomes possible when organizations establish proper collection systems. The recyclability index based on source categories appears in Table 4.5.

Table 4.5 The recyclability index based on source categories.

Waste source	Contamination	Polymer uniformity	Recyclability index (1–5)	Preferred route
Industrial process scrap	Low	High	5	Direct reuse; devulcanization if cured
ELTs (passenger/truck)	Moderate	Medium	4	Grinding; devulcanization; pyrolysis
Industrial end-of-life parts	Moderate–high	Medium	3	Pretreat; grind; targeted devulcanization
Household goods	High	Low	2	Sorting; compatibilized blends; energy recovery
Municipal composites	High	Low	2	Secondary forming; thermochemical routes
Medical rubber	Variable	Medium–high	3–4	Sterilize; shred; blend/modify

4.3 Economic and Environmental Assessment (TEA/LCA)

The economic results of processes depend on both the quality of their raw materials and their operational intensity. The process of grinding rubber into crumb form requires minimal capital investment while producing stable operating expenses. The process of devulcanization enhances product value because it enables the recovery of mechanical properties in materials. The pyrolysis process requires significant capital investment to produce marketable oil and steel and recovered carbon black (rCB). The LCA assessment shows that material recovery systems prevent the need for new virgin polymer and carbon black materials during production while energy recovery systems replace fossil fuels but maintain atmospheric carbon levels.

The scoping studies require the TEA/LCA metrics, which Table 4.6 presents as indicative values.

4.4 Global Regulations and Future Outlook

The combination of EPR schemes and ELT regulations creates conditions which boost waste collection numbers and establish market stability. The EU operates a near-perfect ELT collection system which maintains equal material and energy

Table 4.6 The LCA/TEA metrics.

Route	Typical initial investment cost scale	Main products	Unit cost (USD/Mt feed)	LCA note (greenhouse gas or GHG trend)[a]
Ambient/cryogenic grinding	Small–medium	Crumb/powder; steel; textiles	80–150	Avoids virgin filler, low process emissions
Devulcanization	Medium	Reclaimed rubber with restored elasticity	150–350	Higher value retention; energy moderate
Pyrolysis (batch/continuous)	Medium–large	Tire-derived oil; rCB; steel	180–350	Substitutes fossil fuel and virgin CB; process emissions to control
Coprocessing in cement kilns	Large (host site)	Thermal energy	50–120	Offsets coke/coal; no material recovery

[a] The LCA note (GHG trend) shows the quantitative effect of recycling methods on total CO_2-equivalent emissions relative to landfilling or incineration.

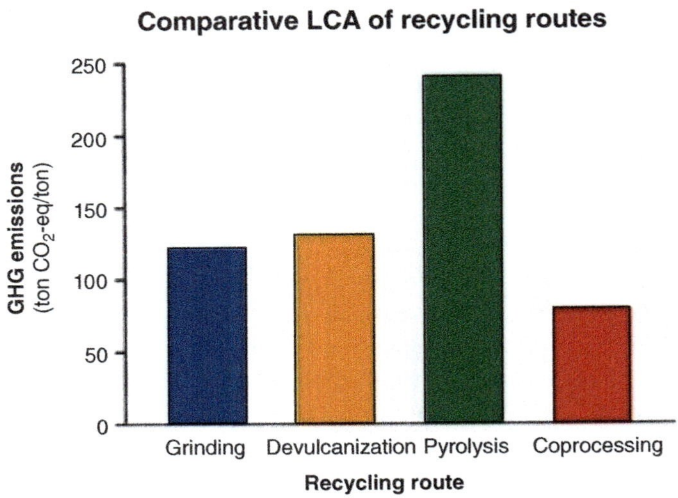

Figure 4.2 Comparative LCA of recycling routes.

recovery rates while Japan reaches similar collection levels through its strong energy conversion capabilities. The United States demonstrates inconsistent ELT collection rates between states while rubberized asphalt markets continue to grow. The implementation of digital tracking systems with QR/RFID technology and AI-based sorting systems will enhance both waste segregation at source and final product quality. The market adoption of reclaimed rubber products will increase through standardized properties and certified rCB products.

Figure 4.2 shows comparative LCA of recycling routes.

References

Castillo, R., Alva, A., Viviana, O., and Bosch, M. A systematic bibliometric review analysis of research on the use of waste rubber tyres in building and construction materials and their applications. *Polymers* 2025, *17*(18), 2480.

Emami Tajodin, S. and Ostad Movahed, S. Effect of 1,3-(bis-citraconimidomethyl) benzene on the reversion of cured ethylene–propylene–diene monomer rubber using a semi-efficient vulcanisation system and nano zinc oxide as activator. *Prog. Rubber Plast. Recycl. Technol.* 2024, *40*, 144–163.

Ostad Movahed, S., Ansarifar, A., and Estaji, S. Review of the reclaiming of rubber waste and recent work on the recycling of ethylene–propylene–diene rubber waste. *Rubber Chem. Technol.* 2016, *89*(3), 54–78.

5

Strategies and Methods in Rubber Recycling

5.1 Strategies for Rubber Recycling

The recycling of rubber waste stands as a complex technical problem because elastomeric materials maintain their thermoset properties through crosslinking bonds. The vulcanization process creates permanent covalent bonds in rubbers (elastomers) which makes them impossible to melt and reshape like thermoplastics. The sulfur and peroxide crosslinks in vulcanized rubber materials including natural rubber (NR), styrene–butadiene rubber (SBR), butadiene rubber (BR), and ethylene–propylene–diene monomer (EPDM) create permanent bonds that stop chain movement and block melting and reprocessing in standard recycling methods. The recycling of waste rubber demands unique mechanical, chemical, and thermo-mechanical approaches which target crosslinked networks for destruction or transformation to maintain original material characteristics.

The circular economy framework enables rubber recycling strategies to be organized into four distinct levels based on their hierarchical structure. The material or product receives direct reemployment through minimal reprocessing at this level of primary recycling and reuse. The material reduction process through size reduction enables rubber particle recovery for compounding and blending operations in secondary mechanical recycling. The chemical recycling process of tertiary recycling uses devulcanization, depolymerization, and thermal cracking including pyrolytic conversion to extract monomers, oils, and fillers from waste rubbers. The last stage of recycling known as quaternary recycling or energy recovery uses rubber waste as a substitute for traditional fuels because of its high energy value. The selection of recycling methods depends on waste origin, state, and required product quality and operational expenses of the process. The current industrial practice focuses on mechanical grinding and devulcanization but new methods like microwave devulcanization (Ostad Movahed et al., 2016), supercritical fluid treatment, and catalytic pyrolysis show potential to achieve better material recovery with lower environmental impact.

The recycling of rubber in comparison with plastics recycling requires more integrated system which unites chemical engineering with surface science and process optimization methods. The combination of partial devulcanization with mechanical reuse through hybrid strategies produces the best sustainability results according to recent techno-economic analyses (TEAs) and life-cycle assessments (LCAs) because it reduces energy usage and increases material recovery value. The annual production of end-of-life tires (ELTs) and rubber waste exceeds 30 MMT so recycling methods need to become efficient because they represent both waste management requirements and strategic material engineering sustainability needs.

5.2 Reusing or Primary Strategy

The primary or reuse strategy in rubber recycling involves using rubber materials or components directly for their original purpose without modifying their chemical composition. The method preserves the original polymer structure of materials while skipping demanding reprocessing steps that require significant energy consumption. The method works only when waste rubber maintains enough structural strength and flexibility to serve in a new product or service. The direct reuse method stands as the most sustainable recycling approach because it requires minimal energy for processing and produces the fewest possible emissions while preserving the highest material worth. The fundamental hierarchy guides primary reuse which elastomeric materials call "direct reuse" or "remanufacturing." The complete reuse of rubber-made products proves difficult because their performance deteriorates through time due to mechanical stress, chemical reactions, and heat exposure. The recovery of undamaged components from waste rubber products including inner tubes and conveyor belts and seals and gaskets continues to offer practical solutions for specific markets.

Multiple industrial operations demonstrate that rubber products can be reused through basic treatment methods including the following.

5.2.1 Retreading of Tires

The practice of tire retreading stands as the leading economic example of primary rubber material recycling. The tire retreading process involves tire restoration through tread layer replacement while maintaining the tire casing structure. The tire refurbishment process involves removing the existing tread from tires with intact casings through buffing before adding a new vulcanized tread layer.

The retreading process enables tire manufacturers to conserve 70% of tire weight while decreasing their energy requirements by 50% compared to producing new tires. The environmental benefits of tire retreading become evident through LCAs which demonstrate that this process reduces CO_2 equivalent (CO_2-eq) emissions by 20–25 kg per passenger tire.

The tire retreading process involves tire restoration through tread layer replacement while maintaining the tire casing structure. The retreaded tire maintains the majority of original weight (about 70%) from the original material and achieves performance levels similar to new tires through contemporary retreading methods. The worldwide retreading industry serves all tire types including passenger vehicles, trucks, aircraft, and off-the-road (OTR) tires while achieving 40–60% energy savings compared to producing new tires. The worldwide tire industry uses three main retreading methods which include precure (cold) retreading, mold-cure (hot) retreading, and bead-to-bead (full-casing) retreading. The preparation process for retreading tires includes inspection followed by buffing, skiving, cementing, tread application, and curing, but the method differs based on tread stock and curing system selection.

The precure or cold process requires a prevulcanized tread strip known as precured tread rubber to be attached to a prepared casing through a thin application of un-crosslinked cushion gum. Process sequence may include the inspection process which uses visual methods and X-ray technology to check for separations, blisters and ply damage in the casing. The buffing process creates a specific tread radius and surface roughness (Ra between 0.7 and 1 mm) to achieve proper adhesion of tread to carcass. The process of skiving and repair involves using repair compound to fix defective areas which need additional curing steps. For accomplishment, the bonding cement receives application. The precured tread strip receives precise tension control during its circumferential wrapping process while joint stitching removes air pockets. The tire assembly receives an envelope covering before being placed in an autoclave at 100–110 °C and 5–6 bar for two to three hours to vulcanize the cushion gum. This process has several advantages, including that the process requires minimal investment for equipment because it needs only small autoclaves. The prevulcanized process enables perfect control of tread design and compound quality. The process requires less energy than mold-cure methods because it consumes 25–30% less power. The process has several major drawbacks. The strength of bond lines depends on how well the surfaces are prepared and how well the adhesion promoter works. The appearance of sidewalls from this process will not match the appearance of branded new tires. The process fails to work properly when using thin or heavily deteriorated tire casings. The worldwide use of this method in truck and bus tire production originates from its ability to work with flexible materials and its moderate vulcanization temperature requirements.

The hot process of mold-cure requires an uncured tread rubber strip or tread-on-barrel layer to be placed on the casing before vulcanization occurs in a pressurized mold with tread pattern engravings. The precure process requires identical mentioned buffing and repair steps. The buffed surface receives a raw tread compound application which consists of a NR/SBR blend. The tire undergoes high pressure (15–20 bar) while being heated to 150–155 °C for 30–45 minutes inside the mold cavity. The rubber cures during this process to create both new tread and sidewall designs. The finished product looks identical to a branded new tire. The bonding strength becomes superior because tread and cushion gum vulcanize as a single unit. The process operates best in automated facilities that produce large quantities of tires. The production process requires elevated energy usage together with expensive equipment. The high process temperature leads to accelerated casing deterioration when the curing time exceeds recommended levels. The process demands strict maintenance of mold cleanliness and proper venting systems. Mold-cure retreading serves as the primary method for aircraft and heavy-duty truck tire maintenance because it delivers exceptional mechanical performance and visual appearance.

The bead-to-bead process extends mold-cure retreading by replacing all exterior rubber components including tread and sidewalls while keeping the internal structure of plies and beads. The process uses a big mold that covers the entire tire casing for premium retreads and aviation tires which can be retreaded five to seven times. The process brings back the complete tire surface. The process creates tires that look identical to new ones. The process enables casings to operate for 100,000 km. The process has two main disadvantages. The process needs perfect condition of the casing structure. The process needs specialized equipment which makes it difficult for small workshops to implement.

From environmental and economic significance, the European Tire & Rubber Manufacturers' Association (ETRMA) reports that retreading one truck tire results in the following environmental benefits: the production of new tires requires 70% more raw materials than retreading while generating 44 kg of CO_2-eq emissions and using 57 L of crude oil equivalents. The worldwide practice of retreading prevents the disposal of millions of tires each year which helps decrease landfill waste and decreases tire incineration particulate emissions. The cost of retreading tires amounts to 30–50% of what new tires cost while extending their service life to 80–90% of their original duration.

As a future trend, the development of microwave-assisted bonding and plasma activation techniques aims to improve the bonding strength between existing rubber layers and newly applied rubber. Also, AI-powered inspection systems that use machine vision technology have achieved better results in detecting defects. The EU Circular Economy Action Plan 2.0 has created new opportunities for circular business models which enable manufacturers to recover and reprocess their own branded tire casings. The latest technological developments prove that retreading serves as a fundamental sustainable practice for tire production because it achieves waste minimization, financial benefits, and improved resource utilization. The new technologies have brought down retread failure numbers to match new tire performance in controlled fleet tests.

5.2.2 Reuse of Industrial Rubber Articles

Industrial rubber articles undergo reuse after their local deterioration becomes apparent while their main structure remains operational. The process of trimming, resizing, and repair enables these products to serve as shock-absorbing mats, dock bumpers, and floor materials. The rubber industry produces off-cuts which developing countries use to create low-pressure products including shoe soles and floor tiles, thus prolonging the useful life of their original materials. The postindustrial and postservice elastomer waste contains significant amounts of industrial rubber articles which include conveyor belts, hoses, vibration mounts, seals, gaskets, expansion joints, and protective linings. The products made from specialized rubbers (NR, SBR, NBR, EPDM, CR, and IIR) with fabric reinforcements, metallic inserts, and chemical-resistant fillers create composite materials with distinct structures. The discarded components maintain their mechanical strength, chemical resistance, and elastic properties which makes them suitable for direct reuse or alternative applications instead of full recycling.

Categories of industrial rubber waste suitable for reuse may include the conveyor belts and transmission belts. The material composition of these belts consists of multiple layers which include NR/SBR rubber and nylon or steel cords. Belts that show surface deterioration but maintain their core structure can be transformed into new applications for light material transportation, agricultural equipment, and protective lining purposes. The operations of mining and port facilities use worn belt sections to create shorter conveyors which handle light materials including fertilizer, grain, and sand. Another industrial rubberish articles are hoses and tubing. The failure points of industrial hoses that transport fuel and hydraulic fluids and air occur at specific areas near couplings and bends. The process of segmental cutting followed by recoupling helps extend the lifespan of these products which can then be used as drainage lines and irrigation hoses and or suction ducts. The flexible nature of reinforced hoses with fabric braids enables users to flatten them into gasket materials or wrapping strips.

The chemical resistance properties of thick tank liners made from NR, EPDM, and butyl rubber materials survive beyond the point when their surfaces become hard. The process of peeling off these materials followed by trimming and rebonding with new adhesive layers enables their use in acid-proof flooring and corrosion-resistant barrier applications.

The reuse of large molded seals with dimensional tolerance failures occurs in prototype testing and noncritical sealing operations because they still demonstrate elastic properties. The aerospace and power industries use discarded gaskets to create training and calibration parts which helps minimize waste generation.

Vibration and shock absorbers used in heavy machinery and railway components show deterioration only at their connection points. The process of remachining mounting holes or adding metallic sleeves enables these products to continue their operational service.

Research shows that industrial rubber component reuse leads to a 35–45% decrease in material usage and a maximum 2-ton reduction in CO_2 emissions for every ton of new rubber material. The process of refurbishing a 1 m × 50 m conveyor belt costs between 40% and 50% of what it would cost to produce a completely new belt while preserving more than 85% of its operational strength. The economic benefits of waste avoidance become more significant when EU regions implement disposal taxes that exceed €120 per ton.

The reused components need to fulfill three essential requirements which include mechanical standards, dimensional specifications, and safety protocols. The Tire & Rubber Manufacturers' Association (TRMA) provides guidelines for conveyor belt resplicing operations. The testing standards for tensile strength and hardness evaluation follow ASTM D412 and D2240, respectively. The ISO 7623 standard defines the testing method for measuring the adhesive strength between different plies of the belt. The belt maintains its reliability through regular nondestructive testing (NDT) which includes ultrasonic flaw detection methods. The failure to achieve these performance standards limits reuse to situations where safety is not a concern.

5.2.3 Architectural and Civil Applications

The practice of using vulcanized rubber products in architectural and civil engineering projects has proven to be both technically sound and financially beneficial for extending the useful life of industrial rubber waste. The applications of this type utilize the specific viscoelastic, damping, and weather-resistant characteristics of cured rubbers which stay operational after their automotive or mechanical service period ends. The chemical structure of rubber remains unchanged when plastic manufacturers use it to create architectural elements and civil structures through cutting and laminating and bonding processes. The practice of recycling rubber waste serves two purposes because it replaces new elastomeric and polyurethane materials while removing substantial amounts of rubber waste from landfills. The direct application of large rubber sheets and slabs from defective products leads to the creation of sound-insulation panels, bridge bearings, and railway vibration dampers. The applications depend on vulcanized elastomer properties for damping and resilience while accepting cosmetic flaws to achieve maximum material reuse.

The main materials for civil engineering reuse come from the end-of-life conveyor belts, which contain NR/SBR rubber material with steel or fabric reinforcement. Also, the rubber materials from mining operations, port facilities, and heavy industrial sites are processed into thick sheets and mats. The tire sidewalls and treads of trucks get cut into panels and pads for reuse. The vibration isolators and bridge bearings from abandoned structures get recovered for future use. The materials show suitable mechanical properties for structural damping applications because they maintain elongation at break above 200% and Shore A hardness between 60 and 80 and compressive modulus between 3 and 8 MPa. They may reuse in structural and functional applications:

a. The flooring industry uses cut sections of conveyor belts and thick rubber sheets to create tiles for gyms and stable covers and walkway pads. The material provides excellent resistance to wear and tear and maintains safe traction while reducing noise levels by 20–30 dB. The process of water-jet cutting and surface cleaning, and adhesive lamination serves these materials without needing thermal processing, which reduces their environmental impact.
b. Machinery vibration isolators made from NR or CR rubber pads with steel shims get refurbished for use as bridge and building base isolation systems. The elastomeric bearings demonstrate flexibility when structures move laterally while supporting heavy vertical loads of multiple hundred kN. The reused bearings undergo remachining followed by shear modulus testing (0.8–1.2 MPa) before receiving weather-resistant sealant applications. The reuse of bearings in India, China, and Eastern Europe leads to 20–40% cost savings when proper strain history and fatigue cycle verification is performed.
c. The entire tire tread and sidewall sections get transformed into guardrail buffers, railway-track under-pads, and level-crossing mats. The material demonstrates its ability to absorb vibrations and withstand repeated stress through its damping properties. The use of recycled rubber pads in railway applications leads to vibration amplitude reduction of 30–50% which helps decrease urban noise levels. Municipal projects use stacked tire segments as lightweight embankment foundations which replace gravel and reduce construction weight by 40%.
d. Acoustic façade systems that use rubber sheets with aluminum or composite panels get repurposed for installation along highways and industrial facilities. The internal damping properties of vulcanized rubber effectively reduce acoustic energy transmission between 200 and 800 Hz. The outdoor performance of reused rubber panels exceeds polyurethane foams because they maintain better aging stability and moisture resistance.

e. The marine and transport sectors use large tire sidewalls and molded rubber blocks to create dock fenders, bumpers, and collision absorbers. These installations need only basic preparation which includes cleaning and bolting before they can operate for 10–15 years under 0.5–2 MPa compressive loading.

The embodied energy of reused rubber flooring and bearing components amounts to 10–15 MJ/kg while new elastomeric products require 70–80 MJ/kg for production. The TEA shows that bridge-bearing refurbishment facilities and recycled flooring production lines will generate capital returns within less than two years and one year, respectively. The practice of reusing materials decreases construction waste output which results in a 30% reduction of landfill requirements for infrastructure projects that incorporate rubber-derived panels.

Multiple international standards exist to verify elastomeric components used in civil structures which enable organizations to obtain reuse certification: EN 1337-3 establishes requirements for structural bearings that include elastomeric elements. AASHTO M251 provides specifications for plain and laminated elastomeric bridge bearings. ASTM C1028/D2240 standards measure friction coefficients and material hardness. ISO 22391-5 defines testing procedures for thermoplastic and rubber pipe systems. The standards do not specifically address reused materials but organizations can accept materials that pass physical tests which demonstrate modulus, compression set, and shear fatigue performance within established thresholds.

5.2.4 Design-for-reuse and Emerging Innovations

Industrial designers now use design-for-reuse (DfR) concepts as their primary design approach: the design uses modular assemblies with detachable rubber segments instead of molded parts that cannot be replaced. The bolt-on design of linings allows users to detach and refurbish them easily. Smart identification tags with RFID or QR code technology track service hours to help predict maintenance needs and plan reuse operations. The surface treatment methods of CO_2-laser ablation and ozone-based oxidation enable the restoration of adhesive bonding capabilities for seals and liners which extends their operational life without needing complete revulcanization.

As a conclusion, the practice of reusing industrial rubber products stands as a vital yet frequently disregarded stage in the rubber recycling hierarchy. The method connects direct product reemployment to full mechanical reprocessing through a low-energy, high-value process that works best for big structurally sound components. The approach becomes more practical because of advances in inspection methods, surface activation techniques and modular design systems despite the challenges from different compound formulations and reinforcement materials. The combination of industrial-article reuse with retreading and selective repair operations helps achieve zero-waste elastomeric manufacturing through the circular economy framework. In conclusion, the process of using rubber products in civil and architectural projects represents material upcycling instead of traditional recycling methods. The applications utilize built-in properties of rubber to transform waste materials into permanent infrastructure elements. The environmental and economic data show promising results because the process achieves significant carbon reduction while requiring minimal initial expenses and delivering extended operational periods. The path to widespread adoption in sustainable construction requires developers to create standardized designs and concrete and composite bonding systems and implement life cycle tracking protocols.

5.3 Mechanical Recycling or Secondary Strategy

The industrial practice of mechanical recycling of rubbers known as secondary recycling uses this method to transform waste rubber materials especially ELTs and production scrap. The rubber network structure remains intact during this process because the physical structure transforms through size reduction, surface activation, and partial remixing with virgin or devulcanized material. The mechanical recycling process depends on mechanical energy to convert large cured rubber blocks into powder and particle forms which can be used for recompounding through cutting, grinding, shearing, and milling operations.

The main advantages of mechanical recycling exist because of three essential benefits. The process uses less energy than devulcanization or pyrolysis methods. The process needs minimal chemical substances which prevents additional environmental contamination. The recycling process produces crumbs and powders that can be used directly to make rubber products, plastic materials, asphalt materials, and composite materials. The main obstacle in recycling vulcanized rubbers arises from their thermoset properties. The crosslinked network structure prevents polymer chain movement so the ground particles function as inert fillers instead of meltable polymers. The process of secondary recycling produces trade-offs between waste minimization and partial material recovery because it reduces mechanical performance to achieve environmental advantages and lower production costs.

The mechanical recycling process of rubber requires specific sequential operations which depend on the initial materials and desired product outcomes. The following diagram shows the basic sequence of operations which starts with collection followed by presorting and then shredding, grinding, sieving, and surface treatment before ending with reutilization.

Collection → Presorting → Shredding → Grinding → Sieving → Surface treatment → Reutilization.

5.3.1 Pretreatment and Shredding

The rubber waste collection process includes tire and industrial off-cut and rubber sheet cleaning followed by drying and mechanical size reduction into 10–50 mm chips. The processing plants use rotary shear shredders and twin-shaft crushers which run at 20–60 rpm to shred materials. The process of steel wire separation from textile fibers occurs through magnetic separator operations and air classifier systems. The ground material functions as the starting material for producing recycled rubber powder (RRP).

5.3.2 Grinding and Pulverization Techniques

The grinding process shapes both the final shape of RRP particles and their ultimate product characteristics. Three principal techniques essentially steer the course of industrial practice:

- The process of ambient grinding remains affordable but faces restrictions because it generates excessive heat which leads to local devulcanization and oxidation.
- The process of cryogenic grinding requires more expenses because it needs liquid nitrogen, but it generates uniform powder sizes while preserving material properties so it works best for expensive elastomer products and polymer mixtures.
- The production of sub-100-μm particles through wet grinding has become more popular because it enables the creation of masterbatch formulations

Table 5.1 compares and summarizes the grinding techniques.

Table 5.1 The grinding techniques characteristics.

Method	Temperature regime	Typical particle size	Remarks
Ambient grinding	20–60 °C	0.3–1 mm	Conventional; friction mills and granulators; surface roughness high → good adhesion
Cryogenic grinding	−80–−120 °C (using liquid N_2)	75–300 μm	Brittle fracture yields clean, angular particles; less fiber damage
Wet grinding/ hydro-milling	<50 °C (water slurry)	<100 μm	Reduces dust, improves heat dissipation; preferred for fine powders

Table 5.2 Mesh sizes.

Mesh size (US standard sieve)	Nominal opening (μm)	Opening (mm)
20 mesh	850	0.85
30 mesh	600	0.60
40 mesh	425	0.425
60 mesh	250	0.25
80 mesh	180	0.18
100 mesh	150	0.15
120 mesh	125	0.125
200 mesh	75	0.075

Source: Adapted from ASTM D5603.

The classification process of powders occurs through vibratory sieves (ASTM D5603) which produce crumb sizes at 20 mesh (850 μm), 40 mesh (425 μm), and 80 mesh (180 μm). The above nominal value obtained from the ASTM E11/ISO 3310-1 sieve series which serves as the official reference standard for rubber-powder and crumb-rubber characterization according to ASTM D5603. As shown in Table 5.2, the rheological properties of recompounded products depend on particle size because it determines their surface area and interfacial bonding strength. The chemically inert nature of ground rubber particles requires multiple surface modification techniques to achieve better bonding with new material matrices. The surface treatment of particles occurs through three methods which include HNO_3, $KMnO_4$, and H_2O_2 chemical oxidation to create polar functional groups such as —OH and —COOH. The surface treatment of rubber particles through plasma or corona discharge creates reactive radicals which enhance their bonding with polymeric materials. The application of γ-aminopropyl-triethoxysilane as a silane coupling agent enables the formation of connections between rubber particles and polymeric materials. The application of devulcanization coatings through mild chemical treatments breaks some sulfur crosslinks at the surface which leads to better revulcanization bonding. The surface energy of these treated particles rises above 40 mN/m from their initial 20 mN/m value which leads to better bond formation with NR and SBR and thermoplastic elastomer materials.

Mechanical recycling produces diverse materials which find application in rubber-based and nonrubber industrial applications including:

- Recompounding in rubber formulations
  The replacement of virgin elastomer with crumb rubber occurs between 5 and 30 wt.% for automotive mats, footwear, and molded parts. The success of this process depends on achieving proper particle-size distribution

and curing system compatibility. The dispersion of fine powders smaller than 200 μm results in better distribution and minimal strength reduction. The vulcanization process requires specific peroxide or sulfur system adjustments to achieve proper vulcanization. As an illustration, adding 10% of RRP powder ground by cryogenic methods to NR/SBR blends preserves more than 85% of their tensile and tear strength.

- Rubber-modified asphalt (RMA)
  The largest market for mechanically recycled rubber exists in rubberized asphalt applications. The process of blending crumb rubber (20–40 mesh) with bitumen at 160–190 °C results in improved elasticity and enhanced rutting resistance and noise absorption. The rubber particles experience partial swelling during high-temperature mixing which leads to plasticizer release that improves binder flexibility. The addition of 60 kg of waste rubber to 1 ton of binder material (RMA) extends pavement lifetime between 25% and 40%. Multiple national road authorities including US FHWA and EUPAVE have established rubberized asphalt as their official standard application.
- Polymer composites and thermoplastic elastomers
  The application of RRPs functions as impact modifiers for polypropylene (PP), polyethylene (PE), and polyurethane composite materials. The combination of maleic-anhydride-grafted polymers (PP-g-MA) with reactive compatibilization methods produces materials that exhibit enhanced damping properties and resilience. The hybridization approach enables the recycling of rubber waste by uniting elastomer and plastic waste streams through thermoplastic matrix integration.
- Civil engineering and construction fillers
  The addition of coarse crumb (1–5 mm) to concrete, mortar, and railway ballast serves as a lightweight material for filling purposes. The addition of coarse crumb reduces compressive strength by 10–15%, but it provides superior impact resistance and freeze–thaw protection. Rubber granules serve as drainage layer materials and artificial turf components because they create porous structures that maintain flexibility.

The total energy requirements for mechanical recycling amount to 300–600 kWh/MT which represents 30% of pyrolysis energy needs while achieving carbon footprint reductions of 0.8–1.2 MT CO_2-eq per ton of rubber processed. It means the mechanical recycling of 1 metric ton (1,000 kg) of waste rubber results in a reduction of 0.8–1.2 metric tons of CO_2-eq emissions when compared to traditional rubber production and disposal methods including incineration and virgin rubber manufacturing. However, the following challenges and limitations exist in mechanical recycling of waste rubbers:

- The processing of waste materials becomes difficult because different types of waste materials with different fillers make it challenging to handle them.

- The absence of chain mobility results in decreased elongation and fatigue resistance because of property deterioration.
- The current low oil prices make virgin rubber more competitive than recycled rubber products in the market.
- The production of dust during processing creates two major problems which affect both the environment and workplace safety of workers.
- The market expansion for crumb rubber faces obstacles because there are no established worldwide standards for product quality or surface characteristics.

The recent LCA and TEA studies from ETRMA (2024) show the following results:

- The recycling process achieves 50–65% better energy performance than producing new rubber from scratch.
- The recycling process reduces CO_2 emissions by approximately 1 ton for every ton of rubber that gets recycled.
- The production cost for 20-mesh crumb stands at €200–300 per ton while virgin NR/SBR compounds cost between €1,000 and €1,200 per ton.
- The market value of recycled rubber products reaches €400–600 per ton based on their particle size and purity levels.
- The research shows mechanical recycling provides the best economic solution for big waste-rubber management operations because it works best when there are no energy recovery or chemical recycling facilities available.

As concluding remarks, the rubber circular economy hierarchy depends on mechanical recycling to connect basic reuse operations with advanced chemical recovery processes. The process of grinding, classification, and surface activation enables vulcanized rubber to become valuable materials which can be used for pavement construction, composite manufacturing, and product molding. The process of mechanical recycling maintains its industrial value because it operates at low energy costs and provides simple operations that scale up well. The development of hybrid mechano-chemical methods will unite grinding advantages with devulcanization control to boost powder reactivity. The combination of AI-based sorting systems with energy-efficient cryogenic equipment and standardized quality control measures will transform secondary recycling into a fundamental element of sustainable elastomer technology.

5.4 Chemical Recycling or Tertiary Strategy

The process of rubber recycling through chemical or tertiary methods transforms the crosslinked polymer (rubber) network into smaller molecular pieces which become available for producing new elastomeric and petrochemical materials.

The process of chemical recycling focuses on breaking down the covalent bonds between crosslinks and main-chain bonds through thermal or chemical reactions to produce devulcanized rubber (DVR, reclaim), oils (chemical and petrochemical plants feedstocks) and monomers. The method achieves maximum product value before reaching complete energy conversion by uniting material recycling with feedstock production. The method requires precise bond cleavage of sulfur and carbon–sulfur bonds to produce reusable rubber materials while preventing backbone damage. The sustainability benefits of chemical recycling include elastomeric feedstock regeneration and reduced emissions but require precise process management to optimize energy usage and product formation.

The main objective in chemical recycling involves achieving controlled depolymerization through thermal or chemical triggers to break down crosslinks and polymer chains into smaller liquid or soft-solid molecules. The extent of degradation ranges from partial devulcanization which breaks crosslinks while preserving long rubber chains to complete pyrolytic cracking which produces gases and oils.

The main reaction pathways are:

- S—S and C—S bond scission in which vulcanization links undergo selective cleavage.
- C—C main-chain scission in which high temperature causes extensive depolymerization of the rubber chain.
- Dehydrogenation and aromatization which produces volatile oils and carbon black.

The rubber type (NR, SBR, EPDM, and IIR) along with crosslinking system and filler or catalyst presence determines these transformation processes.

The process of waste rubber chemical recycling exists in three main categories as shown in Table 5.3. The three pathways operate with distinct temperature

Table 5.3 The classification of rubber chemical recycling.

Route	Mechanism	Typical products
Devulcanization	Selective cleavage of crosslinks (S—S, C—S) without full chain scission	Reclaim or partially restored rubber (can be revulcanized)
Depolymerization/solvolysis	Chain scission under chemical solvents or supercritical fluids	Soluble oligomers, process oils, low-molecular-weight rubbers
Pyrolysis/gasification	Thermal decomposition (400–800 °C) in inert or reducing atmosphere	Gas, pyro-oil, rCB, steel

profiles and catalyst and produce different products while working to recycle carbon-based elastomer materials.

The process of devulcanization stands as the fundamental step in tertiary rubber recycling because it produces flexible material which can be transformed into new vulcanized products. For this purpose, several approaches are used:

a. Mechanical–thermal devulcanization
The process of waste rubber crumb treatment involves using twin-screw extruders, open mills, and internal mixers to apply high shear forces and heat between 180 and 250 °C (Ostad Movahed et al., 2015a). The process of mechanical and thermal energy application breaks S—S and C—S bonds while maintaining the integrity of the polymer structure. The addition of chemicals, namely, zinc oxide, disulfides, and phosphines serve as bond-cleavage promoters which select specific bonds for breaking. The process reduces crosslinking by 60–80% while causing 10–20% main-chain scission (Ostad Movahed et al., 2015b).

b. Chemical devulcanization
The process of devulcanization uses specific chemical agents which include disulfides, amines, thiols, and ionic liquids to target sulfur crosslinks. The process uses diphenyl disulfide (DPDS) and benzoyl peroxide systems to react with hydrocarbon-based materials at temperatures between 180 and 220 °C. The process produces rubber that becomes soft and workable during the reaction at 180–220 °C. The combination of microwave or ultrasonic activation with chemical devulcanization methods leads to better selectivity in the process (Ostad Movahed et al., 2016).

c. Microwave-assisted devulcanization
The rubber matrix experiences uniform crosslink breakdown when carbon black in the material absorbs microwave energy to produce localized temperatures between 250 and 300 °C (Mollanorouzi and Ostad Movahed, 2016). The process achieves energy efficiency of 0.6 kWh/kg while completing operations within less than five minutes. The process generates high-quality DVR that maintains stable mechanical properties and shows excellent revulcanization capabilities. The EU and East Asian regions have adopted this method for processing tire-derived rubber.

d. Ultrasonic and supercritical fluid devulcanization
The process of bond rupture at crosslink points occurs through ultrasonic wave application at 20 kHz with 10–20 μm amplitude which creates cavitation and shear forces in molten rubber. The process of devulcanization occurs at 160–180 °C when CO_2 acts as a supercritical fluid to improve chemical reagent diffusion. The methods achieve high bond selectivity but need complex equipment and precise pressure management systems.

The depolymerization process achieves complete molecular fragmentation which exceeds the level of devulcanization. The process of vulcanized rubber conversion into soluble low-molecular-weight fractions occurs through solvent treatment with aromatic hydrocarbons and alcohols, and catalytic systems that use acidic or basic catalysts. The thermal depolymerization process takes place between 300 and 450 °C under an inert gas environment to produce process oil and gel fractions. The process of catalytic solvolysis requires supercritical toluene or ethanol or acetone as reagents which sometimes need transition-metal oxide catalysts for activation. The hydrogenolysis process involves heating rubber chains at 350–400 °C with H_2 gas in the presence of Ni or Mo catalysts to produce saturated hydrocarbons and light oils. The processes show potential for recovering process oils, plasticizers, and carbon black but the extensive chain breaking makes it impossible to achieve direct revulcanization.

Pyrolysis and gasification represent the complete chemical recycling method. The process of pyrolysis breaks down rubber through heat treatment in nitrogen gas at temperatures between 400 and 800 °C to produce three main products:

- Gas (10–20%) containing H_2, CH_4, CO, CO_2, and C_2–C_4 hydrocarbons.
- Liquid (pyro-oil, 40–60%) including aromatic and aliphatic hydrocarbons (C_6–C_{20}).
- Solid (rCB + ash + steel, 3–40%) including recovered carbon black, fillers, inorganic residues.

The heating process occurs at rates between 5 and 20 °C/min. The process duration ranges between 15 and 60 minutes. The process operates through fixed bed, rotary kiln, fluidized bed, and continuous auger reactors at temperatures between 400 and 800 °C to produce three mentioned main products.

The use of zeolites (HZSM-5, Y, β) and metal oxides (Fe_2O_3, Al_2O_3) in catalytic pyrolysis processes enables temperature reduction and produces more aromatic compounds. The pyro-oil product from this process can replace heavy fuel or get processed into chemicals through chemical refining while the recovered carbon black (rCB) gets transformed into new rubber materials, ink, and pigment products after de-ash and activation.

The gasification process at 900–1,200 °C with restricted oxygen or steam supply produces syngas (H_2 + CO) which can be used for power production or Fischer–Tropsch synthesis.

Table 5.4 compares and summarizes the three main rubber chemical recycling techniques.

TEA evaluates that the cost of building 3,000 MT/annum devulcanization plants ranges between €1 and €3 million while 10,000 MT/annum pyrolysis units require €8–12 million for construction. The operating expenses for processing feedstock amount to €250–400 per ton. The market value of DVR reaches €800 per MT while pyro-oil sells for €500 per MT and rCB reaches €600 per MT. The operational

Table 5.4 The three main rubber chemical recycling techniques comparison.

Process	Temperature (°C)	Main product	Reusability	Energy demand (kWh/MT waste rubber)	CO_2-eq balance (MT/MT rubber)[a]
Devulcanization	180–300	Soft rubber reclaim	Revulcanizable	600–900	−0.7–−1
Depolymerization	300–450	Oligomers/oils	Limited	900–1,200	−0.5–−0.8
Pyrolysis	400–800	Oil + rCB + gas	Feedstock	1,200–1,600	−0.4–−0.6

[a]Negative CO_2-eq values indicate emission savings compared with incineration.

period for mid-scale facilities with proper energy integration and product quality standards will take two to four years to achieve profitability. The implementation of heat recovery systems together with catalyst recycling methods enables organizations to decrease their energy requirements by 20–30%. New microwave and plasma pyrolysis systems demonstrate potential to enhance operational efficiency while producing better gas quality.

From environmental and regulatory aspects viewpoint, the chemical-recycling facilities need to follow emission rules that control volatile organic compounds (VOCs), polycyclic aromatic hydrocarbons (PAHs), and heavy-metal residues in their recovered carbon black production. The EU regulates these operations through Directive 2008/98/EC (Waste Framework Directive) and EU Regulation 2022/1626 which sets end-of-waste standards for pyrolysis waste products. Research on LCAs demonstrates that recycling methods produce 40–70% lower greenhouse gas emissions than incineration and reduce hazardous waste by 70–80%.

As concluding remarks, it should be mentioned that the rubber life cycle requires chemical recycling or tertiary recovery as its technologically advanced yet valuable solution. The material-recycling loop receives its connection through devulcanization which generates compounds for reprocessing while depolymerization and pyrolysis convert waste rubber into chemical feedstocks for circular petrochemical operations. Research and commercialization of these processes should continue because their higher resource recovery rates and product flexibility outweigh their increased energy needs and investment costs. The following factors will determine future development:

- The development of specific catalysts which perform devulcanization at low energy costs.
- The development of hybrid reactors which unite mechanical shearing with microwave energy applications.
- The development of complete LCA and TEA models which will help establish actual sustainability performance.
- The development of policy-based incentives which will establish devulcanized and pyrolytic products as official secondary raw materials.

The recycling of rubber through tertiary methods has evolved into a vital sustainable materials engineering component which supports the development of a net-zero circular elastomer economy.

5.5 Energy Recovery or Quaternary Strategy

The quaternary strategy known as energy recovery becomes the last waste rubber method when materials reuse, mechanical recycling, and chemical regeneration become impossible. The intrinsic calorific value of crosslinked rubber polymers gets converted into usable thermal or electrical energy through combustion and coincineration and gasification processes. The process of energy recovery transforms rubber waste into a solid hydrocarbon fuel which does not maintain the original polymer structure. The heating value of vulcanized rubbers matches bituminous coal because they contain 85–90 wt.% carbon and hydrogen with the heating values range between 28 and 33 MJ/kg. The high-temperature requirements of cement kilns, power plants, and steel industries (>1,400 °C) make ELTs and industrial rubber scraps valuable alternative fuels (AFs). The implementation of strict emission controls for SO_x, NO_x, zinc, and PAH emissions remains crucial for achieving environmental sustainability.

The thermochemical process of rubber combustion during controlled conditions leads to multiple stages of thermal transformation. The process of drying and devolatilization occurs between 200 and 400 °C to release light volatiles and plasticizers. The main-chain hydrocarbons undergo exothermic decomposition into CO and CO_2 and H_2O during the oxidative decomposition stage between 400 and 600 °C. The char oxidation process occurs when temperatures exceed 700 °C to fully oxidize the remaining carbon-based material. The complete reaction produces about 30 MJ of energy per kg of rubber which translates to 0.8 kg of fuel oil for every kg of rubber. Modern incinerators operate at 80–85% efficiency to generate substantial energy recovery while minimizing waste volume by more than 95%.

The cement industry operates as the world's leading market for tire-derived fuel (TDF) consumption. The standard rotary kiln operates at 1,450 °C which provides enough heat to achieve complete organic combustion and complete inorganic material integration into clinker. The thermal process uses waste heat to generate clinker production. The entire mineral content from ash gets incorporated into cement material. The process produces no solid waste products which eliminates the requirement for landfills. The use of TDF as coal replacement in cement production results in a 10–20% reduction of CO_2 emissions per ton of cement. The fuel substitution rate for calorific value should range between 10% and 25% of total fuel consumption. The system accepts three different rubber waste forms which include whole tires and two size ranges of shreds and chips. The rubber waste generates 2–3 kg CO_2-eq/kg emissions, but the avoided fossil fuel usage helps reduce these emissions. The European Commission (BREF, 2019) accepts cement-kiln coprocessing as the Best Available Technique (BAT) for nonrecyclable rubber waste rubber management.

Also, rubber waste undergoes processing at waste to energy plant (WtE) facilities through blending with municipal solid waste (MSW) before being burned in grate or fluidized-bed furnaces. The combustion process produces superheated steam which operates turbine generators to generate electricity. The EU Directive 2010/75/EU requires WtE systems to use flue-gas recirculation and selective noncatalytic and catalytic reduction (SNCR/SCR) systems for NO_x control and activated-carbon injection to capture dioxins and heavy metals. In the other words, the European Union requires waste-incineration plants including rubber and polymer energy-recovery facilities to implement particular emission-control systems which reduce dangerous pollutants before releasing exhaust gases into the air.

Alternative thermochemical energy recovery methods may be included as:

a. Gasification
The process of rubber partial oxidation between 850 and 1,200 °C with steam or oxygen produces syngas ($CO + H_2$) that contains 8–12 MJ/m of energy. The produced syngas can operate internal-combustion engines while Fischer–Tropsch synthesis converts it into methanol or synthetic diesel fuel. The use of Ni/Al_2O_3 or CaO catalysts in gasification processes decreases tar production while producing more hydrogen gas.

b. Copyrolysis with Biomass or Plastics
The combination of rubber with biomass or polyolefins produces better hydrogen-to-carbon ratios and higher oil yields while allowing partial energy recovery without generating solid clinker waste. The hybrid system enables carbon-neutral operations because it uses biomass as its cofeed material.

c. Plasma Arc Conversion
The high-temperature plasma process at temperatures above 2,000 °C breaks down rubber polymers into syngas (H_2, CO) while transforming inorganic materials into glassy slag. The process requires significant energy consumption results in complete emission elimination while generating a fuel cell-compatible gas stream.

The high CO_2 emission intensity of the process results in a net GHG reduction of 0.3–0.5 t CO_2-eq per ton of rubber when compared to uncontrolled landfilling or open burning because fossil-fuel substitution helps reduce carbon emissions. Table 5.5 summarizes energy and emission balances of energy recovery methods.

The construction cost for 10,000 MT/annum coincineration units ranges between €4 and €8 million and WtE plants with 50,000 MT/annum capacity require €15–25 million. The operating expenses for rubber waste treatment amount to €60–90 per ton because they mainly consist of energy costs for air system operations and flue-gas scrubbing systems. The facility generates €100–150 per ton through waste disposal fees and power distribution sales. The investment period for cement or steel plant integration amounts to three to five years. The system achieves its best energy performance when waste heat recovery generates district heating or process steam through combined heat and power (CHP) operations. The presence of sulfur (1–2 wt.%) and zinc oxide in rubber during vulcanization process leads to SO_2 and Zn particulate emissions when flue gases remain

Table 5.5 Energy and emission balances of energy recovery methods.

Process	Net energy output (MJ/kg)	Thermal efficiency (%)	Typical CO_2 emission (kg CO_2-eq/kg rubber)
Cement kiln coprocessing	24–26	85–90	2.5–3
Dedicated incineration + power generation	20–22	70–80	3–3.5
Gasification → syngas	15–18 (equivalent)	60–70	2–2.5
Plasma arc	12–15	50–60	1.5–2 (but high electric input)

unprocessed. The current regulatory standards require facilities to meet specific emission standards. EU Directive 2010/75/EU established the following emission standards:

- The EU Directive 2010/75/EU sets emission standards at 50 mg/Nm^3 for SO_2 and 150 mg/Nm^3 for NO_x and 10 mg/Nm^3 for dust.
- The US EPA 40 CFR Part 60 regulates dioxin emissions at 0.1 ng TEQ/Nm^3.

The following best practices need to be implemented for emission control: the process of SO_x control requires injecting Ca- or Mg-based sorbents into the system. Heavy metals and organics require treatment through activated carbon filter systems. The system requires continuous emission monitoring (CEMS) to track CO and NO_x and SO_x emissions. The facility needs to perform ash leaching tests at regular intervals to confirm its nonhazardous waste status.

It was concluded that the hierarchy of rubber-recycling strategies reaches its end with energy recovery because it transforms waste materials into useful energy resources after all material recovery options have been exhausted. The complete breakdown of polymeric structure in rubber leads to an efficient fossil fuel substitution because rubber contains stable combustion properties mineral compatibility, and high calorific value. The operation of energy-recovery facilities with advanced emission-control systems enables them to meet strict environmental standards while they reduce large amounts of CO_2 emissions from coal and oil combustion. The complete circular economy framework requires mechanical and chemical recycling to occur before combustion but energy recovery enables every nonrecyclable rubber kilogram to support resource efficiency and energy security. The environmental benefits of this essential recycling method will increase through the implementation of carbon-capture technologies and high-efficiency CHP systems and biomass coprocessing. Table 5.6 compares mentioned four strategies for rubber recycling in brief.

Table 5.6 The comparison between four main strategies for rubber recycling.

Recycling strategy	Material recovery (%)	Energy recovery (%)	Typical GHG reduction (t CO_2-eq/t)	Relative economic value
Reuse (primary)	100	0	−1−−1.5	High
Mechanical (secondary)	70–80	10–20	−0.8−−1.2	Medium
Chemical (tertiary)	40–60	40–60	−0.5−−0.9	Medium–high
Energy (quaternary)	0	70–90	−0.3−−0.5	Low–medium

5.6 Future Directions and Emerging Technologies, i.e., Bio-based Reclamation

The recycling industry for polymers and rubbers is experiencing a fundamental transformation because of environmental needs and faster regulatory changes and improved materials and biological and chemical engineering methods. The current traditional recycling methods which include mechanical size reduction, thermo-mechanical shearing, and chemical devulcanization have reached advanced technical levels, yet they continue to encounter major obstacles when processing strongly crosslinked rubbers, composite materials, and waste materials with contaminants. The development of new recycling technologies now focuses on bio-based, chemo-enzymatic, and hybrid catalytic systems which perform selective bond cleavage to maintain polymer structure. The development of new methods will create improved material recycling systems which will decrease the environmental effects of reclamation activities.

5.6.1 Bio-catalyzed Devulcanization of Rubber

The most promising development employs bio-catalysts which include sulfur-metabolizing microorganisms together with enzymes (sulfur oxygenases and sulfatases and laccases) to identify and break particular sulfur bonds in vulcanized elastomers. The bio-based reclamation process follows natural sulfur cycling patterns which enables scientists to reduce crosslink density without triggering the extensive chain degradation that hot temperature treatments would cause. The process requires *Thiobacillus ferrooxidans* and *Rhodococcus* spp. together with other sulfur-reducing bacteria according to research findings. The combination of enzymes with lignin-degrading fungi enables rubber matrix devulcanization at temperatures below 60 °C and pH levels close to neutral which produces better mechanical properties when the materials are recompounded with virgin polymers (Wiśniewska et al., 2022). The biological methods provide scientists with their main scientific benefit because they naturally select specific targets. The enzyme and microbial systems operate with substrate specificity which allows them to break down poly-sulfidic bonds precisely without causing damage to the polymer backbone structure. The system provides enhanced control which enables producers to create high-quality reclaimed rubber materials for tire manufacturing, industrial seal production, and automotive component manufacturing. The process faces multiple obstacles because microbial breakdown occurs at a slow rate and bioreactors need improvement and additives such as antioxidants, accelerators, and carbon black might slow down the process. The development of enzyme immobilization techniques and directed protein evolution methods and metabolic engineering approaches has resulted in the development of new solutions which boost industrial enzyme performance. Scientists achieve better results with biological devulcanization through the use of enzymes which become more effective at breaking sulfur bonds after scientists apply either gentle mechanical activation or microwave-based pretreatment methods.

5.6.2 Bio-based Solvents and Green Chemistry Pathways

The development of bio-derived solvents which include terpenes, γ-valerolactone, ethyl lactate, and deep eutectic solvents shows promise for recycling rubber and plastic materials through dissolution techniques. The use of green solvents allows scientists to move away from petroleum-based aromatic compounds such as toluene, xylene, and THF while they create adjustable solubility parameters which help researchers achieve specific swelling effects and extract particular additives and manage the degradation of crosslinked structures. The combination of choline chloride with biodegradable hydrogen-bond donors in deep eutectic solvents demonstrates ability to break sulfur crosslinks at temperatures ranging from moderate to high while maintaining the integrity of the polymer backbone. The research investigates thiol and phosphine functionalized ionic liquids which demonstrate potential for sulfur bond targeting applications. The solvent systems follow circular economy principles because they can be reused multiple times while minimizing toxic substances and decreasing environmental impacts throughout their entire life cycle.

5.6.3 Enzyme-assisted Recycling of Plastics and Composites

The process of biological recycling has reached higher levels for polyesters (PET [polyethylene terephthalate] and PLA) but scientists now show that engineered enzymes including cutinases and PET hydrolases can break down rubber–plastic combinations and waste materials containing different materials. The bio-enzymatic treatment process in hybrid tires enables researchers to eliminate textile fibers and polyester components which makes it possible to extract rubber material for devulcanization. The development of enzyme engineering will need selective disassembly methods to establish new recycling facilities for multimaterial systems.

5.6.4 Microbial Conversion of Polymer Waste into Value-added Products

The process of microbial conversion enables the transformation of depolymerized polymer fragments into three different valuable metabolites which include PHAs, bio-surfactants, and organic acids. Engineered microbes can use breakdown products from partially oxidized or solubilized rubber and plastic waste as their carbon source after pretreatment methods including chemical, photolytic, and microwave-assisted processes. The method transforms waste rubber and plastics into bio-manufacturing materials which enable the production of valuable products for packaging, medical devices, and biodegradable plastics.

5.6.5 Advanced Physical Technologies: Plasma, Supercritical Fluids, and Ultrasound

Physical technology development now exists as an independent system which operates independently from biological systems. The application of atmospheric plasma treatment to rubber surfaces enables faster devulcanization processes and improved material bonding during recompounding operations. The swollen polymer matrix from supercritical CO_2 treatment eliminates mass-transfer barriers which allows devulcanizing agents to penetrate the material more effectively. High-power ultrasound produces cavitation micro-jets which create two effects: they break down crosslink regions in the material and they help release the filler from its position.

The physical enhancements can be combined with green chemical or biological routes to achieve reduced energy usage and enhanced selection performance.

5.6.6 Digitalization and Intelligent Recycling Systems

The future reclamation plants will use artificial intelligence and machine learning and digital twins to predict devulcanization behavior and optimize reagent dosages and enhance process control. The relationship between polymer composition and filler morphology and crosslink structure allows data-driven models to identify recycling methods which achieve maximum performance for various waste streams. The three main applications of AI technology in biological reclamation consist of enzyme design, metabolic pathway simulation, and reactor optimization. The advancement of intelligent systems will speed up laboratory findings because they perform experiments faster with lower error rates which enables industrial use of laboratory results.

The future technologies will operate as a unified system to establish a complete circular economy which will handle rubber and polymeric materials. The bio-based reclamation process enables both particular substance extraction and sustainable operational practices. The methods of green chemistry enable organizations to perform their processing activities in sustainable ways. Organizations can reduce their energy consumption and reagent requirements through the implementation of sophisticated physical processing techniques. AI-based systems help organizations achieve better operational results when they perform extensive operations. The researchers aim to establish a long-term plan which involves moving past basic recovery methods to achieve controlled molecular construction of materials from waste rubber and plastics into high-performance materials and bio-products while preserving their mechanical strength.

The world requires these innovative technologies to reach sustainability targets through the reduction of petroleum-based materials and prevention of toxic plastic waste from entering the environment.

5.6.7 Case Study

A useful case study by Hosseini and Ostad Movahed et al. (2024) will be discussed here in detail. In this case study, the potential of superworm to remove certain plastics from waste stream has been noted. To do this, aerobic bacterial strains were isolated from the gut of *Zophobas morio* larvae which were fed with PE, PET, PP, and polystyrene (PS) polymers.

In this study, *Z. morio* larvae (approximately 5 cm in length) were purchased from Cricket Superworm (Tehran, Iran). Larvae were subjected to 48-hour starvation prior to the experiment, divided into nine groups (15 larvae per group), and weighed. Each group was fed on one of the 2×2 cm^2 films (PE, PP, and PET) and cube (PS) (3 g) from the commercial polymers of low-density PE (plastic bag), PP (disposable glass), PET (disposable bottle), and expanded PS (foam) alone or with wheat bran (3 g). Larvae only fed on wheat bran were considered as controls. The experimental containers with the superworm larvae were kept at constant humidity (60%) and temperature (25 °C). Scientists studied 10 larvae during the fourth week of their experiment which were randomly selected from each treatment, weighed, and exposed for dissection.

The larvae were suddenly killed by immersion in 70% ethanol for one minute and then rinsed twice with sterile distilled water. The larvae's guts were extracted and suspended in a tube con. The researcher maintained 5 mL sterile phosphate-buffered saline (PBS) followed by vortex and pipetting to obtain an intestinal homogenate. The suspension was used as a microbial inoculum. Bacterial strains were grown overnight in TSB broth media at 30 °C. The results showed that used *Z. morio* superworm larvae as its subject for this investigation, reared on synthetic plastic polymers. The gut bacterial resident of these worms was isolated and surveyed for plastic degradation. Strains designated as *Leminorella* sp. P2, *Citrobacter* sp. P5, *Bacillus* sp. P6, and *Bacillus* sp. P10 revealed the highest degradation rate of PS (2.7%), PP (0.42%), PE (1.3%), and PET (1.1%), respectively. Pretreatment of polymers by thermal, acid, and weathering enhanced the degradation rate. Further analysis on the most degraded polymer, i.e., PS, revealed that pretreatment promoted double bond, as well as carbonyl and hydroxyl groups formation but did not affect the surface topography and wettability.

The biodegradation of PS film was confirmed by SEM and AFM (formation of heavy cracks and surface irregularities), FTIR (increases in oxidized function group), contact angle (increased hydrophilicity), and GPC (diminished polymer molecular weight) analyses. The results highlight the potential of bacterial residence of the plastivor larvae for the remediation of plastics in the integrated procedure with abiotic pretreatment. The research needs further investigations to fulfill its original research objectives which were established at the beginning. The research needs to determine which enzymes remain active and describe their enzyme catalytic processes to enhance the plastic biodegradation process.

References

Ardolino, F., Boccia, C., and Arena, U. Environmental performances of a modern waste-to-energy unit in the light of the 2019 BREF document. *Waste Manag.* 2020, *104*(2020), 94–103.

Hosseini, F. S., Asoodeh, A., Ostad Movahed, S., and Makhdoumi, A. An integrated approach for plastic polymer degradation by the gut bacterial resident of superworm (*Zophobas morio*). *Environ. Sci. Pollut. Res.* 2024, *30*(50), 60359–60370.

Mollanorouzi, M. and Ostad Movahed, S. Reclaiming waste tire rubber by an irradiation technique. *Polym. Degrad. Stab.* 2016, *128*, 115–125.

Ostad Movahed, S., Ansarifar, A., Ghanei, R., and Kermani, Y. Devulcanization of ethylene–propylene–diene waste rubber by microwaves and chemical agents. *J. Elastomers Plast.* 2016, *48*(1), 122–144.

Ostad Movahed, S., Ansarifar, A., Karbalaee, S., and Atharifar, S. Devulcanization and recycling of waste automotive EPDM rubber powder using shearing action and chemical additive. *Prog. Rubber Plast. Recycl. Technol.* 2015a, *31*, 87–116.

Ostad Movahed, S., Ansarifar, A., Karbalaei Nezhad, S., and Atharifar, S. A novel industrial technique for recycling ethylene–propylene–diene waste rubber. *Polym. Degrad. Stab.* 2015b, *111*, 114–123.

Wiśniewska, P., Wang, S., and Formela, K. Waste tire rubber devulcanization technologies: State-of-the-art, limitations and future perspectives. *Waste Manag.* 2022, *150*, 174–184.

6

Waste Rubber Devulcanization, Fundamentals, and Mechanisms

6.1 Introduction

The devulcanization process uses chemical engineering to break specific crosslinks in vulcanized rubber through targeted sulfur–sulfur (S—S) and carbon–sulfur (C—S) bond cleavage without damaging the polymer chain structure (Ostad Movahed et al., 2016a). The process of devulcanization maintains the elastomeric network structure of rubber while breaking crosslinks to create a material that can be revulcanized into new rubber products. The process creates devulcanized rubber (DVR) or reclaim rubber which shows thermoplastic properties and can be combined with fresh elastomers and revulcanized. The technique connects mechanical and chemical recycling methods to create a material with soft thermoplastic characteristics that can be used as a blend with virgin elastomers. The technique has gained significance because the world produces more than 30 million tons of waste tires annually, while manufacturers seek sustainable production methods that preserve material worth.

The sulfur-cured rubber vulcanization network consists of three main bond types. The poly-sulfidic (—S_x—) bonds represent the weakest bonds in the network with bond energies between 250 and 280 kJ/mol. The di-sulfidic (—S—S—) bonds show intermediate strength with bond energies ranging from 270 to 300 kJ/mol. The mono-sulfidic (C—S—C) bonds represent the strongest bonds in the network with bond energies between 285 and 350 kJ/mol The process of devulcanization needs to break down weak sulfur bonds without damaging the primary carbon–carbon backbone which has a bond energy of approximately 350 kJ/mol. The performance index of devulcanization processes depends on the selectivity coefficient (SC) which shows the ratio of crosslink bond breaks to chain bond breaks.

The following approaches enable researchers to control reaction outcomes (Martinez et al., 2021):

- The process requires temperature elevation between 180 and 250 °C and simultaneous application of mechanical forces through twin-screw extrusion or open-mill operations.
- The process uses coagents including disulfides, thiols, and amines as chemical reagents for control purposes.
- The process uses physical methods including microwave treatment, ultrasound application, and supercritical CO_2 processing to enhance results.

6.2 Mechanisms, Kinetics, and Thermodynamics

The elastic network in sulfur-cured rubbers develops through sulfur bridges which connect polymer chains at unsaturated sites of diene monomers. The crosslinks in these materials consist of poly-sulfidic (—S_x—) and di-sulfidic (—S—S—) and mono-sulfidic (C—S—C) bonds as mentioned which show rising bond strength. The elastic network allows for controlled bond breaking through thermal or chemical treatment because S—S and C—S bonds have lower strength than C—C bonds.

The process of devulcanization focuses on breaking crosslinks instead of completely degrading the molecular chains. The system experiences two opposing chemical processes which involve crosslink bond breaking (—C—S—S—C— → 2 —C—S•) and chain bond breaking (—C—C— → 2 —C•). The process achieves effective selective devulcanization when SC values exceed 2.

The following mechanisms may be applicable in devulcanization:

- The application of frictional shear and heat forces breaks down S—S bonds in the rubber network (R—S—S—R → R—S• + •S—R). The stabilization of radicals occurs through hydrogen donor reactions which produce thiols.
- Using disulfide agents, namely, diphenyl disulfide (DPDS) functions as a sulfur-exchange reagent which transforms R—S—S—R into R—S—S—Ph and Ph—S—S—R.
- The application of microwave radiation or ultrasonic cavitation creates specific areas for S—S bond breakdown (—C—S—S—C— → 2 —C—S•) while preserving the C—C bonds.

The system includes radical scavengers such as phosphines and phenolic antioxidants which stop uncontrolled oxidation and also, devulcanizing oils (aromatic or aliphatic) that enhance heat distribution. The system contains ZnO and accelerators which help perform specific sulfur exchange reactions.

The structural and analytical evidence of devulcanization process may be evaluated by FTIR/Raman spectra which demonstrate decreased S—S band intensity between 510 and 540 1/cm, while XPS data shows sulfur chemical-state modifications. The swelling tests indicate reduced crosslink density. The GPC results show no significant alteration in molecular-weight distribution which indicates that chain scission occurred to a minimal extent.

The devulcanization process of crosslinked rubber requires simultaneous breakdown of chemical bonds, radical reactions, and physical diffusion events. The kinetic and thermal properties of this process control both the degree of crosslink breakdown and the preservation of polymer chain integrity.

The devulcanization process follows pseudo-first-order kinetics because it depends on the amounts of crosslinks present in the material:

$$\mathrm{d}\left[C_x\right]/\mathrm{d}t = -k\left[C_x\right]$$

where $\left[C_x\right]$ is the concentration of sulfur crosslinks and k the apparent rate constant. The rate constant follows the Arrhenius equation:

$$k = Ae^{-E_a/RT}$$

The equation shows that E_a represents activation energy while R stands for gas constant and T represents absolute temperature. The experimental data shows that S—S scission requires 90–120 kJ/mol activation energy, while C—C cleavage needs 150–200 kJ/mol to occur which supports thermal devulcanization at 180–250 °C without damaging the main polymer chain.

The addition of mechanical shear, ultrasound, and microwave activation brings nonthermal energy elements which decrease the measured activation energy by 20–30%. The process becomes more efficient because these inputs create localized frictional heat and radical initiation which speeds up crosslink rupture.

The process of S—S bond breaking requires endothermic energy input of 270 kJ/mol but it needs less energy than C—C bond breaking which requires 350 kJ/mol. The process becomes favorable when the mechanical shear or microwaves provide enough energy (ΔH) to equal the bond-dissociation energy. Also, the process of de-crosslinking leads to increased chain mobility which results in positive entropy gain ($\Delta S > 0$). Accordingly, the Gibbs free energy change ($\Delta G = \Delta H - T\Delta S$) becomes negative at temperatures above 200 °C, which indicates that bond cleavage will occur spontaneously during processing. The addition of solvents or swelling agents decreases interchain bonding strength which results in decreased ΔH values that enable sulfur bond breaking at lower temperatures.

In practical implications, the process of optimizing devulcanization kinetics and thermodynamics needs to find the right balance between proper temperature and shear rate management while using the right amount of reagent for radical stabilization and keeping the residence time brief to prevent backbone damage and re-crosslinking to maintain high selectivity (Formela et al., 2022).

Industrial extruder-based systems achieve their best results through temperature control between 190 and 240 °C and brief processing times under five minutes to optimize energy consumption and product quality (Ostad Movahed et al., 2016a). The knowledge of devulcanization kinetic and thermodynamic properties allows scientists to create efficient recycling methods which produce high-value rubber products while using minimal energy and reducing carbon emissions.

6.3 Devulcanization Techniques

6.3.1 Mechanical

Mechanical devulcanization stands as a practical solution for transforming vulcanized rubber waste including end-of-life tires (ELTs) into usable materials. The process depends on mechanical power which creates high shear forces, compression, friction, and impact energies to break sulfur crosslinks (S—S, C—S) in rubber networks. The mechanical process employs mechano-chemical activation through shear stress to break molecular bonds while protecting the polymer backbone from thermal breakdown. The method provides environmental benefits because it eliminates dangerous chemicals and reduces the need for additional chemical substances. The process operates through open two-roll mills, internal mixers (Banbury or Brabender), and corotating twin-screw extruders (TSEs) (Figure 6.1) which provide scalable and reproducible continuous operation.

The traditional two-roll rubber mill represents the basic setup which operates as the first and simplest configuration. The process involves feeding ground rubber particles smaller than 1 mm between two heated rolls that operate at different speed settings. The process of frictional shear and compression work together to create progressive softening and de-crosslinking of the material. The process offers two main benefits because it requires minimal investment and allows users

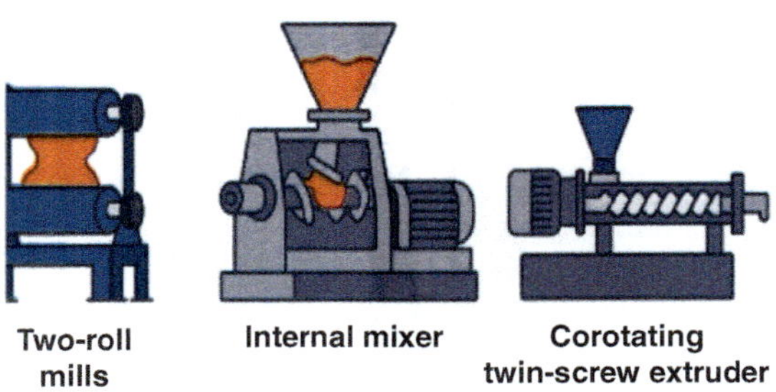

Figure 6.1 Schematic of open two-roll mills, internal mixers (Banbury or Brabender), and corotating TSEs.

to monitor the operation directly. The process temperature management restrictions result in re-crosslinking of materials when the shearing process exceeds recommended limits.

Internal mixer (Banbury or Brabender) is another equipment used for rubber devulcanizing process. The closed-chamber design enables better control of temperature and torque during the mixing process. The rotor operates at 40–80 rpm while maintaining chamber pressure between 3 and 5 bar to achieve complete mechanical energy distribution. The addition of DPDS or devulcanization oils serves as additives to stabilize radicals during the process. The process achieves 60–70% crosslink density reduction while using approximately 800 kWh of energy per ton of waste rubber.

TSE is the most advanced mechanical system performs continuous devulcanization through precise control of residence time, temperature profile, and screw shear intensity. The design elements affect the system's operational performance included the screw design with kneading blocks. The mixing zones produces different shear zones throughout the system. The operation speed of the screw should range between 100 and 200 rpm because faster speeds help break S—S bonds but may damage polymer chains. The barrel temperature follows a sequence from 160 to 240 °C throughout the screw path. The pilot-scale operation uses feed rates between 50 and 150 kg/h, but industrial production lines operate at 500 kg/h. The extruder torque measurement shows direct relationship with crosslink bond destruction while process control becomes possible through real-time torque and melt pressure monitoring.

The mechanical devulcanization process depends on mechanical work conversion into thermal and frictional energy which breaks specific S—S bonds as demonstrated below:

$$E_{\mathrm{m}} = \tau \cdot \dot{\gamma} \cdot t$$

where E_{m}, τ, $\dot{\gamma}$, and t denote as mechanical energy input per unit volume (J/m^3), shear stress (Pa), shear rate (1/s), and processing time (s), respectively.

Above equation expresses that the mechanical energy density which rubber receives during devulcanization depends directly on the applied shear stress, shear rate, and the length of processing time.

The process operates under typical conditions which include shear rates between 200 and 500 1/s, temperature range from 180 to 250 °C, and residence times between two and five minutes. The friction ratio between rolls or screw flights which ranges from 1.15 to 1.30 creates enhanced shear forces that stretch polymer chains until they break through mechanical activation. The energy distribution across the rubber particles becomes uneven during processing because the outer parts experience intense shear and heat while the inner section stays at lower temperatures. The temperature difference between surface and core areas enables molecular-weight preservation while crosslinks in outer layers undergo breakdown.

Friction ratio in two-roll mills refers to shear undergoes by rubber in two-roll mills. The rubber sheet undergoes shear and frictional forces when it passes through an open two-roll mill because the two rolls move toward each other at different speeds. Friction ratio measures as surface speed of the faster roll divided by surface speed of the slower roll. The typical values range between 1.1: 1 and 1.3: 1. The rolls operate at different speeds because one roll moves 10–30% faster than the other. For example, the fast roll operates at 22–26 rpm when the slow roll runs at 20 rpm. The rubber sheet undergoes high shear deformation because the two rolls operate at different speeds. The process of tearing and elongating polymer chains occurs through this method which leads to better filler dispersion and devulcanization sulfur crosslink breakdown. The compound temperature and mixing performance depend on the selected friction ratio value. The process requires a friction ratio below 1.1 to achieve proper shear and breakdown results. The process of chain scission (degradation) occurs when the friction ratio exceeds 1.3 while the compound temperature becomes too high. The optimal devulcanization range for friction ratio exists between 1.15 and 1.25 to achieve proper selectivity and molecular stability.

The Banbury (internal) mixer maintains friction ratio principles but operates through distinct mechanisms than an open two-roll mill. A Banbury mixer (or Brabender-type internal mixer) contains three main components which include, a mixing chamber contains two rotors which operate inside its closed design. The ram operates from above to generate pressure while the rotor drive system enables speed control of the rotors. The material stays confined between rotor tips and chamber walls while undergoing strong shear forces and compression that suits both compounding and mechanical devulcanization processes.

The surface speed ratio between two rotors defines the friction ratio in this context. Normally, Banbury mixers operate at friction ratios which fall between 1.05:1 and 1.25:1. This means the speed difference between the two rotors reaches 5–25% in operation. The rotors in Banbury mixers operate at different speeds to generate strong shear zones which produce internal friction that enables effective mixing or devulcanization. The Banbury operates through three-dimensional motion instead of using planar mechanics. The distribution of shear force becomes more intense when the friction ratio increases. The rubber material experiences continuous movement between high-shear and low-shear zones because of differential rotor speeds. The generation of frictional heat increases when differential speeds become higher which helps devulcanization but enhances risks to chain scission and over-curing.

In devulcanization context, the Banbury mixer performs mechanical devulcanization through its rotor differential which generates enough shear force to break S—S and C—S bonds. The mixer operates at speeds between 40 and 80 rpm for this process. The optimal friction ratio for this process should be between 1.1 and 1.2

because it enables crosslink bond breakdown while keeping temperatures below 250 °C. The extent of devulcanization becomes visible through torque measurements because the rubber becomes softer when crosslinks break. As mentioned, the friction ratio affects both shear intensity and heat generation speed in the process. The mixing process becomes mild when using a lower friction ratio which produces less shear energy. The ratio needs optimization to perform targeted crosslink bond breaking instead of causing random material breakdown. The devulcanization process depends on this parameter (friction ratio) to achieve proper selectivity between S—S and C—C bond scission and to produce high-quality DVR.

The waste rubber experiences selective de-crosslinking through chamber wall and rotor tip compression and shear forces in Banbury mixer even when it operates at 1:1 rotor speed (no friction ratio).

As a conclusion note, the design of early Banbury internal mixers used differential rotor speeds to enhance shear and mixing performance during rubber compounding operations. However, modern Banbury mixers including tangential mixer may operate with differential rotor speeds which create a friction ratio. The rotors of another modern Banbury mixer known as intermeshing mixer, operate at matching speeds.

The relative tangential velocity ($\vartheta = \pi DN$), where D and N represent screw diameter and rotational speed (rpm), respectively, between intermeshing screw flights in corotating TSEs functions as a substitute for friction ratio in two-roll mills and Banbury mixer. Engineers use three parameters to control shear gradient and energy input in extruders. They are screw speed (N), flight pitch and clearance (distance between flight edge and extruder wall), and mixing-zone geometry. The process works similarly to mill friction ratio control. The local shear rate increases when engineers operate at higher screw speeds or use tighter intermeshing clearances.

Shear forces cause polymer chains to stretch while their molecular vibrations create free radical species (R•) at points where crosslinks exist. The main chemical processes that occur during this process include:

$$\text{R—S—S—R} \rightarrow \text{2R—S}\bullet$$

$$\text{R—S}\bullet + \text{H—R}' \rightarrow \text{R—SH} + \text{R}'\bullet$$

The radical fragments either merge together or they interact with hydrogen donor agents, i.e., devulcanizing agents, which contain hydrogen to create stable end products. The final product quality depends on how well radicals form during the process and how well they get stabilized. The formation of excessive radicals through high shear forces results in chain breakage but insufficient shear leads to insufficient devulcanization. The system requires specific parameter adjustments to reach its maximum selection capability (Table 6.1).

Table 6.1 Optimum processing parameters for maximum selection capability.

Parameter	Typical range	Effect
Temperature	180–250 °C	Enhances S—S bond rupture; excess causes re-crosslinking
Shear rate	200–500 1/s	Higher rates improve de-crosslinking efficiency
Residence time	2–5 min	Balances completeness and molecular stability
Additive ratio	2–5 phr	Promotes selective sulfur cleavage
Feed particle size	<500 μm	Ensures uniform energy distribution

Mechanically devulcanized rubber (MDR) shows the following characteristics:

- The crosslink density of MDR decreases to 60–80% below original levels. The processing characteristics of MDR show some recovery because Mooney viscosity drops between 40% and 60%. The blending of MDR with virgin NR, SBR or EPDM rubber at 10–40 wt.% results in better material compatibility.
- The mechanical properties of MDR include tensile strength represents 70% of virgin compound strength and elongation at break ranges from 200% to 300%.

The mechanical devulcanization process requires less energy than chemical or pyrolytic recycling methods (0.6–0.9 MWh/MT) while producing fewer volatile organic compounds (VOCs). The process operates without solvent recovery or effluent treatment which enables its use as a clean and scalable method. The process achieves a GHG reduction of 0.8–1 MT CO_2-eq for every ton of rubber that replaces incineration. The recovery of waste heat from extruder barrels enables predrying feedstock which results in a 15% improvement of energy efficiency.

As industrial implementations, European commercial facilities at Kraton and Rubber Jet Valley and Japanese and Chinese operations run continuous twin-screw production lines to create 3,000–10,000 MT of DVR granules annually. The production output serves three main applications:

- The production meets requirements for tire retread and repair compounds which use up to 20% blend ratio.
- The product serves as a material for creating rubber–plastic blends which get used in automotive component manufacturing.
- The product functions as an asphalt modifier and sound-insulating mat material.

It was concluded that the mechanical devulcanization serves as a fundamental element of circular-rubber technology because it provides simple engineering solutions with safe environmental practices and large-scale production capabilities.

The combination of hybrid processes which include microwave and chemical posttreatments with mechanical devulcanization produces high-quality DVR for tire compound manufacturing. The development of future improvements will focus on three main areas which include in-line torque control systems, AI-based shear optimization, and reactive extrusion to achieve maximum process selectivity and sustainability.

6.3.2 Microwave-assisted

The technique of microwave-assisted devulcanization (MWD) has become popular for selective rubber recycling because it provides efficient volumetric heating, selective energy absorption, and potential large-scale production capabilities (Jourabchi and Ostad Movahed, 2024). The microwave energy penetrates directly into rubber molecular dipoles and conductive fillers which creates uniform and controlled heating that breaks sulfur-based crosslinks without damaging the rubber backbone. The two standard microwave frequencies for industrial use operate at 915 MHz and 2.45 GHz which produce wavelengths of 33 and 12 cm, respectively. The frequencies operate under standardized rules for industrial scientific and medical (ISM) applications. The polymer network interacts with microwave energy based on its dielectric properties which are measured through complex permittivity values (ε^*):

$$\varepsilon^* = \varepsilon' - j\varepsilon''$$

where ε' is the dielectric constant (energy storage capability) and ε'' is the dielectric loss factor (energy dissipation as heat). The loss tangent, defined as:

$$\tan\delta = \varepsilon''/\varepsilon'$$

determines the microwave absorbability of the rubber compound. The majority of uncured rubbers including NR, SBR, and EPDM do not absorb microwaves because they have nonpolar structures and low tan δ values below 0.01. However, the absorption properties of carbon black, metal oxides, and polar degradation products enable targeted internal heating which results in effective devulcanization.

Microwave energy operates as nonionizing electromagnetic radiation which causes materials to experience dipolar rotation and ionic conduction. The absorbed energy from microwaves transforms into thermal energy through molecular friction which creates localized temperature spikes at the filler–polymer interface. The microwave power dissipation (P) equation describes the devulcanization process as:

$$P = 2\pi f \varepsilon_0 \varepsilon'' E^2$$

where

P	is	the power dissipated per unit volume $\left(W/m^3\right)$,
f	is	frequency (Hz),
ε_0	is	the permittivity of free space $\left(8.854 \times 10^{-12}\ F/m\right)$,
ε''	is	dielectric loss factor,
E	is	the electric field strength (V/m)

The rubber achieves 180–250 °C temperature range within seconds because high field strengths between 10^3 and 10^4 V/m create fast and consistent energy absorption. The temperature range between 180 and 250 °C enables the breakdown of S—S (270 kJ/mol) and C—S (290 kJ/mol) bonds without causing excessive damage to the C—C backbone structure (350 kJ/mol). The short processing time reduces oxidative degradation, leading to DVR with superior mechanical and rheological properties compared to those produced by conventional heating and mechanical techniques.

A standard microwave devulcanization unit contains the following components:

1. The system accepts crumb or granulated rubber through its feeding system which handles particles between 1 and 3 mm in size.
2. The system uses a conveyor or rotating drum to distribute the material evenly for microwave exposure.
3. The magnetron produces 915 MHz or 2.45 GHz microwave energy through its operation.
4. The waveguide system directs microwave energy to specific areas of the treatment zone.
5. The system includes temperature sensors, infrared feedback control systems, and gas atmosphere control functionality which uses nitrogen or CO_2 to prevent oxidation.

The rubber feed undergoes microwave cavity treatment which uses electromagnetic waves to create equal internal temperature throughout the material. The devulcanization level depends on both the processing time between 10 and 120 seconds and the power density range from 0.5 to 2 kW/kg. The microwave devulcanized rubber (MWR) product undergoes cooling followed by pelletize before it gets reprocessed with virgin polymer or additional materials. The system achieves high energy efficiency between 60% and 75% because volumetric heating reduces heat loss. The selective coupling rubber to conductive fillers such as carbon black and silica directs heat to areas with highest crosslink density which improves the SC between crosslink and chain scission.

The experimental results demonstrate that MWD achieves 60–80% reduction in crosslink density while maintaining 70–85% of the original tensile strength of virgin rubber based on the specific formulation (Ostad Movahed et al., 2016b).

Table 6.2 The MWD technical and environmental benefits.

Feature	Advantage
Rapid, uniform heating	Avoids temperature gradients and overheating of surface layers
Selective crosslink scission	Preferential breakage of S—S and C—S bonds due to localized energy absorption
Short processing time	Reduces oxidative degradation
Lower energy consumption	0.4–0.7 kWh/kg, lower than conventional thermal devulcanization
Compatibility with continuous operation	Suitable for conveyor and twin-screw configurations
Cleaner process	In most cases, no solvent or chemical reagents required

The gel fraction shows significant decrease while molecular weight remains mostly unchanged which confirms selective de-crosslinking occurs. Table 6.2 summarizes the MWD technical and environmental benefits.

The MWD process provides multiple benefits, yet it encounters multiple operational difficulties. The uneven distribution of fillers throughout the material structure together with irregular particle shapes creates hot spots which result in chain breakdown. The microwave penetration depth of 3–6 cm restricts batch thickness so researchers need to design specific reactor shapes. The process requires continuous monitoring of both temperature and dielectric values to achieve selective results. Scientists solve these problems through the implementation of combined systems which combine microwave technology with mechanical and mechanical shearing extruders and chemical devulcanization using devulcanization agents, namely, DPDS for tire wastes, and subsequent chemical devulcanization. Researchers solve these problems through combining microwave technology with other devulcanization techniques.

Figure 6.2 represents a typical microwave-assisted rubber devulcanization production line.

The process of microwave devulcanization supports circular economy principles because it enables the sustainable transformation of waste tires and rubber materials. The process requires 0.4–0.7 kWh of energy to treat each kg of material while reducing greenhouse gas emissions by 0.7–1 tons of CO_2 equivalent per ton of virgin rubber production. The initial investment for 1 MT/h throughput ranges between \$0.6 and \$1.2 million while industrial pilot plants achieve two to three years payback periods.

The life-cycle assessments (LCAs) show that microwave devulcanization methods achieve 30–45% lower environmental impact through reduced global

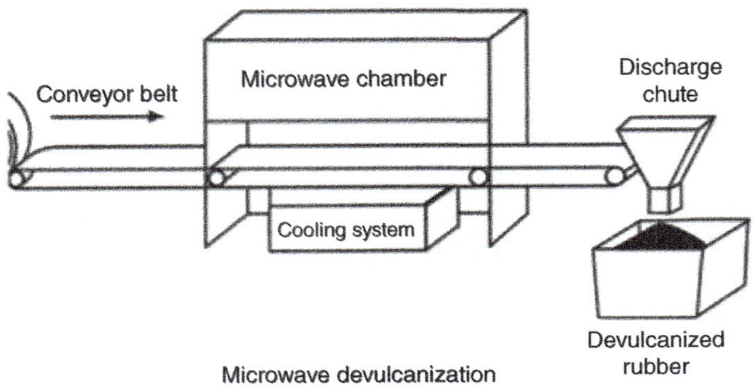

Figure 6.2 A typical microwave-assisted rubber devulcanization production line.

warming potential and cumulative energy demand compared to traditional reclaiming and thermal pyrolysis methods. The combination of sustainability metrics with enhanced process selectivity makes microwave devulcanization an advanced tertiary recycling method which shows great potential for the coming years.

In brief, MWD offers a method to extract valuable rubber materials through a clean process that operates with high selectivity and uses minimal energy. The technology achieves success through its ability to heat materials volumetrically while it couples energy between fillers and matrices and its fast controllable temperature system. The upcoming advancements in reactor engineering, real-time monitoring systems, and hybrid microwave-mechanical processing will establish this method as an industrial standard for treating massive amounts of tire and industrial rubber waste during the following 10 years.

6.3.3 Thermal

The process of thermal devulcanization stands as one of the earliest methods which remains useful for rubber crosslink material recovery. The steam-assisted devulcanization process in autoclaves stands out as a vital thermal method because it operates with ease and works with multiple rubber materials and allows production at different scales (Ostad Movahed et al., 2015b). The process uses heated saturated or superheated steam to break down vulcanized rubber crosslinks through temperature-controlled pressurized treatment. The steam autoclave system creates a moist heat environment which breaks down sulfur crosslinks through hydrolysis and performs minimal oxidation while protecting the polymer backbone from excessive damage. The steam in the process maintains uniform heat distribution which protects against hot spots that would otherwise occur during the crosslink scission process. Consequently, the crosslink scission-to-chain scission ratio (CSR) as a key quality indicator in rubber reclamation will be controlled with ease and efficiency.

As mentioned, the vulcanized rubber network contains sulfur bridges which exist as poly-sulfidic (S_x), di-sulfidic (S_2), and mono-sulfidic (S) structures. The S—S and C—S bonds in this temperature range (180–250 °C) under autoclave pressure (1–2.5 MPa) break before the C—C backbone bonds. The combination of water and high pressure enables hydrolysis and radical recombination reactions which break crosslinks and enable some recovery of material workability.

The devulcanization reaction operates according to a pseudo-first-order kinetic model:

$$\mathrm{d}\left[v\right]/\mathrm{d}t = -k\left[v\right]$$

where $\left[v\right]$ represents the crosslink density, and k is the temperature-dependent rate constant following the Arrhenius equation:

$$k = Ae^{-E_\mathrm{a}/RT}$$

The activation energy (E_a) for thermal devulcanization ranges between 80 and 110 kJ/mol which shows that this process needs less energy than pyrolysis which is >200 kJ/mol.

A steam autoclave functions as a heavy-duty pressure container which includes temperature and pressure management systems. The process follows these sequential steps for completion. The process starts with basket or tray placement of crumbed or granulated rubber waste particles that measure between 2 and 5 mm in size. The system performs an air evacuation process to remove all air particles which helps stop rubber degradation from oxidation. The system receives saturated or superheated steam at temperatures between 180 and 250 °C while operating under pressures from 1 to 2.5 MPa. The reaction period extends from 20 to 90 minutes based on the specific rubber material being processed (NR, SBR, and/or EPDM). The autoclave undergoes steam venting followed by a controlled temperature decrease through water or compressed air cooling. The post-treatment process involves mechanical refinement of the devulcanized material through milling or pelletizing operations. Figure 6.3 illustrates a typical steam autoclave-devulcanization system.

The internal water vapor functions as a thermal conductor and mild reducing agent which helps desulfurization and network relaxation during the process. Experimental evidence shows that the gel fraction amount reduces between 40% and 70%. The crosslink density shows a 60–80% decrease. The Mooney viscosity values and sol fraction measurements demonstrate that the material can be

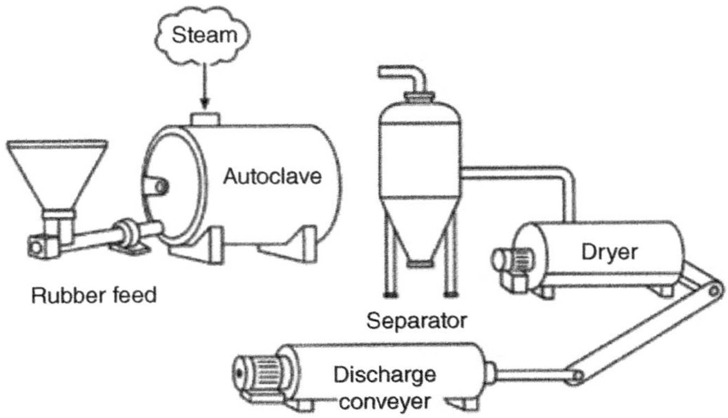

Figure 6.3 A typical steam autoclave-devulcanization system.

successfully reprocessed. The combination of extended heating periods and excessive temperature exposure results in chain breakdown which produces materials that become either sticky or brittle. The optimal temperature–time profiles at 210 °C for 30 minutes help developers reach peak devulcanization selectivity while preserving tensile strength and elongation performance similar to the original compound.

Steam performs multiple essential tasks which extend beyond its role as a heat transfer agent. The water molecules in steam act as catalysts for hydrolysis by breaking sulfur bonds through nucleophilic attacks which produce hydrogen sulfide and polysulfide residues. The matrix becomes temporarily flexible because of moisture absorption which enables stress relaxation and energy transmission. The process of condensing steam enables fast heat transfer between the material and its surroundings. The presence of small amounts of oxygen helps sulfur oxidation to produce sulfoxide and sulfone intermediates which make crosslinks more fragile. The process of devulcanization using dry air becomes less effective when compared with superheated steam with temperatures above 230 °C because it provides faster heat transfer and produces cleaner reaction products. Table 6.3 summarizes the advantages and limitations of steam-autoclave devulcanization

Industrial operations use hybrid systems which unite steam autoclaves with mechanical shearing or chemical additives to achieve better selectivity and improved energy efficiency. The addition of zinc stearate or disulfide oils at low concentrations during steam treatment helps to speed up desulfurization reactions while making the devulcanized product more workable.

The thermal autoclave devulcanization process operates at commercial scale in multiple facilities across Asia and Europe. A 2 MT/batch autoclave requires 0.6–0.9 GJ/MT of steam power to operate while processing rubber at costs

Table 6.3 Advantages and limitations of steam-autoclave devulcanization.

Advantages	Limitations
• Uniform heat distribution via steam condensation • No external chemical reagents required • High throughput and scalability • Compatible with mixed rubber wastes • Energy-efficient and environment-friendly	• Limited control over bond selectivity • Partial oxidation possible if air leaks • Moderate mechanical properties compared with chemical or microwave methods • Moisture removal step required before reprocessing • Requires high-pressure equipment and strict safety systems

between \$80 and \$120 per MT. The process generates 0.5–0.8 t CO_2-eq/t of GHG emission reductions while achieving more than 70% material recovery rates compared to incineration. The thermo-reclaimed rubber (TRR) obtained from this process can be used to manufacture tire carcasses, mats, conveyor belts, and general rubber products through the addition of fresh curing agents and plasticizers.

It concluded that the steam-assisted autoclave devulcanization method has proven itself as a fully developed environment-friendly process which offers better efficiency than chemical and mechanical devulcanization methods. The process offers three main benefits through its ability to distribute heat evenly and its lack of chemical solvents and its potential for large-scale production. The autoclave method provides essential devulcanization capabilities through its selectivity that is lower than microwave and supercritical-fluid-based processes.

6.3.4 Chemical

The process of chemical devulcanization stands as a top method for breaking sulfur crosslinks in vulcanized rubber because it provides both high selectivity and protects the polymer structure. The process of chemical devulcanization uses devulcanizing agents as reagents to break sulfur–sulfur (S—S) and carbon–sulfur (C—S) bonds through reduction, oxidation, and nucleophilic substitution reactions (Ostad Movahed et al., 2015a). The method produces rubber that maintains its molecular weight while breaking crosslinks with high precision which enables its use for recycling and blending operations that result in materials with virgin-like performance. The increasing need to recycle ELTs and other cross-linked elastomeric waste has led scientists to study chemical devulcanization methods which they combine with mechanical shear and ultrasonic energy and microwave irradiation to boost reaction rates and agent penetration.

The chemical devulcanization process breaks S—S and C—S bonds through chemical reactions which include reduction, oxidation, and substitution depending on the applied reagent. The reactions follow this simplified chemical structure:

$$R{-}S{-}S{-}R' + 2H \rightarrow 2R{-}SH$$

$$R{-}S{-}C + H \rightarrow R{-}SH + C{-}H$$

The S—S and C—S bonds in the network become accessible to reduction agents which include disulfides, mercaptans, thiols, and phosphines. The following common reduction agents included DPDS, triphenyl phosphine (TPP), triaryl disulfides (DADS), tetramethyl thiuram disulfide (TMTD), and zinc oxide and stearic acid synergistic systems. The reduction process starts when the reagent nucleophile attacks the sulfur bridge to create thiols (R—SH) while breaking down the sulfur network.

Also, the crosslink-breaking process of hydrogen peroxide, nitric acid, and peracids occurs through sulfoxide and sulfone intermediate formation which leads to hydrolysis. The oxidative treatment method proves successful, but it creates a risk of polymer backbone degradation through chain scission which results in embrittlement.

The C—S bond undergoes nucleophilic substitution when exposed to specific reagents which include amines, alcohols, and ionic liquids. The process of desulfurization using alcoholysis with ethylene glycol, aminolysis with ethanolamine, and ionic-liquid-based systems operates at 150–200 °C while producing minimal emissions. The research on modern devulcanization uses the chemical agents listed in Table 6.4.

The process of chemical devulcanization takes place in internal mixers (Banbury or Brabender), TSEs, and high-shear batch reactors.

The process involves:

- The process starts by adding ground rubber (20–40 mesh) to the chosen reagent and a minimal amount of process oil or solvent.
- The mixture undergoes controlled shear (60–120 rpm) at temperatures between 160 and 220 °C for 10–20 minutes.
- The process ends with volatiles removal followed by rubber drying after filtration.

In continuous operations, the devulcanizing agent needs to be injected through the TSE barrel to create specific reaction areas which maximize the interaction between the reagent and rubber.

The chemical process selectivity depends on three factors which include temperature levels, reagent concentrations, and mixing shear rates. The main objective involves maximizing the selectivity index (SI) through the measurement of S—S/C—S bond scission divided by total bond scission. The SI for disulfide-based

Table 6.4 The common chemical agents for chemical devulcanization of different rubbers.

Agent	Chemical type	Reaction type	Effective temperature (°C)	Remarks
DPDS	Aromatic disulfide	Reductive cleavage of S—S bonds	180–220	Highly selective; used in industrial reclaimers
TMTD	Sulfur donor accelerator	Reductive–nucleophilic	160–200	Efficient in NR and SBR devulcanization
TPP	Organophosphorus	Reductive	180–200	Produces clean cleavage without oxidation
2-Mercaptobenzothiazole (MBT)	Mercaptan	Reductive	150–190	Enhances re-crosslinking upon revulcanization
Ethanolamine	Aminolysis agent	Nucleophilic substitution	180–220	Eco-friendly and recyclable medium
Ionic liquids (e.g., [BMIM]Cl)	Salt-based solvents	Solvolysis	150–190	Green chemistry alternative; tunable selectivity
Hydrogen peroxide	Oxidative	Oxidation	140–160	High reactivity; risk of backbone degradation

systems usually falls between 0.7 and 0.9, while phosphine-based systems range from 0.5 to 0.7.

The optimization strategies consist of using the aromatic disulfides, i.e., DPDS, which serve as specific reagents to target sulfur bonds for crosslink cleavage. The addition of zinc stearate or stearic acid serves as catalysts to boost reaction performance. Also, mechanical shear application helps to increase reagent penetration into crosslinked areas.

A typical mechanism using DPDS proceeds as follows:

1. DPDS decomposes thermally into two radicals:

$$(C_6H_5S)_2 \rightarrow 2C_6H_5S\cdot$$

2. The radicals attack S—S and C—S crosslinks in the rubber matrix:

$$R{-}S{-}S{-}R' + C_6H_5S\cdot \rightarrow R{-}S\cdot + R'{-}S{-}C_6H_5$$

3. The recombination reactions produce mercaptan and thiophenol compounds through sulfur fragment release.

The process reduces poly-sulfidic crosslinks into di-sulfidic or mono-sulfidic bonds which eventually transform into thiol-terminated chain ends to enhance flow and reprocessability.

The reclaimed rubber undergoes substantial changes in its structural and mechanical properties through chemical devulcanization: the crosslink density of the material decreases by 60–85%, the sol fraction increases from its initial 10% to reach 40–50% of the total material. The tensile strength of DVR maintains 60–80% of its original value when compared to virgin rubber. The molecular mobility of the material increases which results in better elongation at break performance.

The gel–sol analysis and swelling index measurements serve as standard methods to evaluate devulcanization efficiency and selectivity. The FTIR and XPS analyses show the presence of thiol (—SH) and sulfone ($—SO_2—$) groups which prove that crosslink bonds break without causing substantial oxidation.

The process of chemical devulcanization provides better environmental benefits and economic value than pyrolysis and incineration methods. The process operates at temperatures between 160 and 220 °C which results in lower energy requirements of approximately 0.5 GJ/MT. The process achieves material recovery rates between 70% and 85%. The process produces a 1 t CO_2-eq/t reduction in greenhouse gas emissions when compared to incineration.

The process of using solvents and hazardous chemicals including amines, disulfides, and peroxides needs proper waste disposal systems and emission reduction measures. The development of green devulcanization agents using vegetable oil-derived disulfides, ionic liquids, and supercritical CO_2-compatible reagents has become popular because these agents produce less toxic waste and enable reagent recovery. The DVR material serves as a cost-effective substitute for virgin rubber in new compounds because it reduces material expenses by $400–800 per MT based on formulation and application requirements.

The use of hybrid systems which combine chemical processes with mechanical or microwave energy has gained significant popularity in recent times (Ostad Movahed et al., 2016b). The process of mixing reagents under shear conditions inside open two-roll mills or TSEs speeds up the breakdown of crosslink bonds. The application of microwaves for sulfur site heating produces faster reaction rates and better reaction uniformity. The combination of chemical processes with ultrasonic energy creates better reagent penetration and improved access to cross-links. The hybrid methods boost total conversion yields and selectivity while preserving operational efficiency and large-scale production capabilities.

In brief, the process of chemical devulcanization stands as a leading method for precise rubber waste recycling. The process requires exact control between chemical reactions, heat management, and mechanical processing. The development of new reagents through ionic liquids and organophosphorus catalysts and disulfide donors enables researchers to perform selective devulcanization at high levels without substantial polymer breakdown. The process enables sustainable material recycling through its high-value production method which supports circular economy standards and sustainable manufacturing objectives.

From this section, it can be summarized that the process of devulcanization involves two steps which include breaking specific chemical bonds and stabilizing free radicals. The process of crosslink removal becomes possible through proper management of energy levels, chemical reactions, and active techniques which protect the polymer structure. Hybrid systems that combine mechanical and chemical processes show potential for effective recycling of vulcanized rubber materials.

References

Formela, K., Cysewska, M., and Haponiuk, J. T. Recent progress in devulcanization technologies of waste rubber. *Polymers* 2022, *14*, 1003.

Jourabchi, S. and Ostad Movahed, S. Effect of microwave irradiation on the molecular diffusion of ethyl acetate into SBR/EPDM rubber blends. *Prog. Rubber Plast. Recycl. Technol.* 2024, *40*, 201–217.

Martinez, J., Formela, K., Kosmalska, A., and Haponiuk, J. Devulcanization of tire rubber: a review. *Polymers* 2021, *13*, 488.

Ostad Movahed, S., Ansarifar, A., and Estaji, S. Review of the reclaiming of rubber waste and recent work on the recycling of ethylene–propylene–diene rubber waste. *Rubber Chem. Technol.* 2016a, *89*(3), 54–78.

Ostad Movahed, S., Ansarifar, A., Ghanei, R., and Kermani, Y. Devulcanization of ethylene–propylene–diene waste rubber by microwaves and chemical agents. *J. Elastomers Plast.* 2016b, *48*(1), 122–144.

Ostad Movahed, S., Ansarifar, A., Karbalaee, S., and Atharifar, S. Devulcanization and recycling of waste automotive EPDM rubber powder using shearing action and chemical additive. *Prog. Rubber Plast. Recycl. Technol.* 2015a, *31*, 87–116.

Ostad Movahed, S., Ansarifar, A., Karbalaei Nezhad, S., and Atharifar, S. A novel industrial technique for recycling ethylene–propylene–diene waste rubber. *Polym. Degrad. Stab.* 2015b, *111*, 114–123.

7

Case Studies in Waste Rubber Devulcanization Techniques and Industrial Processes

7.1 Reclamation of EPDM Rubber

The **first case study** appeared in *Progress in Rubber, Plastics and Recycling Technology* when authors published "Devulcanization and recycling of waste automotive EPDM rubber powder by using shearing action and chemical additive" in 2015.

As mentioned, the 21st century brings about waste disposal management as a major environmental challenge for human societies. The rubber recycling continues to rise because of rising raw material expenses, decreasing natural resources, and expanding environmental consciousness, and nondegradable nature of polymeric materials makes waste polymer disposal a critical environmental issue. Due to the worldwide demand for sustainability needs, the current main challenge in rubber recycling grows from the restricted application of recycled rubber to actual recycling systems that produce new rubber products. The development of advanced breakdown methods for recycled rubber stands as a critical problem which affects worldwide sustainability efforts. The ASTM STP 184 A defines devulcanization through its description of "depolymerization, oxidation, and plasticity enhancement" because these processes happen together during reclamation operations. The process of devulcanization operates as the opposite of vulcanization. Sulfur vulcanization creates both C—S and S—S bonds which should break down during devulcanization through C—S and S—S bond cleavage. The perfect devulcanization process should break crosslinks without damaging the main polymer chain.

The first commercial introduction of ethylene–propylene–diene monomer (EPDM) rubber occurred in the United States during 1962. The manufacturing process of EPDM involves combining ethylene and propylene with a diene monomer to create the polymer. The diene monomer creates unsaturated sites and double bonds throughout the polymer chain. The market for general purpose rubber shows the fastest growth through EPDM rubber products. The material exhibits outstanding ozone resistance and demonstrates high solid filler tolerance. The total weight of vehicles contains 3% nontire rubber products which include weather strips, hose, vibration insulators, and miscellaneous parts. The disposal of scrap rubber parts from vulcanized filled EPDM rubber products has been a persistent challenge since their first production and use. The destruction of rubber networks during devulcanization processes falls into five distinct categories: chemical processes, thermochemical processes, mechanical processes, irradiation processes, and biological processes. The chemical process of rubber scrap powder requires disulfides, thiophenols, and their zinc salts, and mercaptans as reclaiming agents. The rubber network structure undergoes breakdown when reclaiming agents interact with the rubber material. The combination of heat with reclaiming agents serves as the method to break down rubber crosslinking points in thermochemical processes. The rubber network structure breaks down through mechanical stress which occurs when shearing forces are applied to the material. The shearing process for rubber network breakdown occurs through operations with two-roll mills, batch mixers, and single- or twin-screw extruders. The devulcanization process through irradiation uses microwave and ultrasonic wave technology. The three-dimensional rubber network structure becomes vulnerable to breakdown when exposed to microwave and ultrasonic wave treatment. The process of biological devulcanization through microbial metabolism serves as an effective method for environmental protection. The sulfur crosslinks in rubber become vulnerable to biological breakdown when *Thiobacillus* species and other microbes use their biological properties to oxidize sulfur into sulfate. The biological devulcanization process requires extended periods of time to achieve limited conversion rates. The devulcanization process using supercritical materials exists as a nonindustrial method among other miscellaneous approaches.

The recycling process for EPDM rubber allows manufacturers to restore its original state through devulcanization of polymer chain crosslinks or transform waste materials into functional products. The recycling process for EPDM faces multiple obstacles because devulcanizing agents fail to dissolve in rubber, and the rubber contains more than 50% of unbreakable monosulfidic crosslinks. The three bond types in EPDM rubber require 270 kJ/mol to break monosulfidic C—S bonds and 240 kJ/mol to break polysulfidic S—S bonds and 345 kJ/mol to break peroxide C—C bonds.

This research established a new technique to transform automotive part waste EPDM rubber powder into usable material through devulcanization followed by compound development for automotive part manufacturing. The research introduced a new approach by varying feed screw speeds and main rotor speeds of a twin-screw extruder during waste rubber devulcanization at fixed temperature conditions. The research employed a semi-industrial twin-screw extruder as shearing action device and chemical reactor. 2-Mercaptobenzothiazole-disulfide (MBTS) chemical used to assist to devulcanize waste power at 220 °C. The devulcanization process operated at 6 and 8 rpm feed screw speeds and 180 and 220 rpm main screw speeds while maintaining a constant temperature at 220 °C. The research combined different proportions of devulcanized waste powder with a virgin commercial EPDM-based automotive rubber to create eight blends. The devulcanized blends together with reference compound (RC) underwent semiefficient vulcanization (SEV) system to revulcanize. The research evaluated how devulcanized powder addition affected blend viscosity, cure properties, and mechanical strength through comparison with the RC.

The Part Lastic Company in Iran provided waste EPDM rubber powder for this study. The powder consisted of multiple aged and new rubber automotive materials from different sources with particles smaller than 1 mm in size. The first step involved using a Soxhlet apparatus to remove oil from the material. The extraction process used 300 mL of acetone to extract oil from 100 g of rubber sample at 70 °C for 16 hours according to ASTM D297-93. The rubber sample underwent oven drying with circulating air for 12 hours after extraction. The waste EPDM rubber powder contained 17.42 wt.% oil according to the average results from five extraction tests. The waste rubber powder underwent thermogravimetric analysis (TGA) using STA 1500 from Scinco Co. Ltd. to determine its chemical composition. The TGA test ran under nitrogen and air conditions at temperatures below and above 550 °C. The TGA results along with their corresponding compositions appear in Figure 7.3. The research employed EPDM (KEP 270 from KUMHO POLYCHEM in Korea with 57% ethylene and 4.5% diene monomer) and carbon black (N330 from Pars Carbon Ltd. in Saveh) as well as MBTS (Perkacit MBTS from Flexsys) and tetramethylthiuram disulfide (TMTD; Perkacit TMTD from Flexsys) as devulcanizing agents and accelerators. The research used Perkacit ZDBC from Flexsys as an accelerator and Harcros Durham Chemicals in the UK provided zinc oxide as an activator and Anchor Chemicals Ltd. in the UK supplied stearic acid as an activator. The research employed elemental sulfur from Solvay Barium Strontium in Hannover, Germany, as a curing agent and aromatic and aliphatic oils from Mehran Tyre 290 and Fariman in Iran as processing aids.

The waste rubber powder devulcanization process involves adding MBTS and aromatic oil and or paraffinic oil to a basic mixer with an agitator (Table 7.1). The waste rubber powder needed 24 hours of oil immersion to achieve complete penetration of the rubber material. The semi-industrial twin-screw extruder (CTE 65-52D from Coperion Keya Nanjing Machinery Co. Ltd.) with 100–250 kg (about 551.15 lb.)/h capacity performed the devulcanization process (Figure 7.1). The machine featured a feeder with adjustable feed rate screw and cooling and heating systems and a main screw that operated at different speed settings. The extruder hopper received 40 kg compounds which started the screw operation. The waste

Figure 7.1 The twin-screw extruder serves as a device to perform shearing operations during devulcanization. *Source*: Ostad Movahed et al. (2015) / with permission of Sage Publications.

Table 7.1 Formulation of the compound used in the devulcanization process by the extruder.

Material	Content (phr)
EPDM waste rubber powder	100
MBTS (reclaiming agent)	6
Aromatic oil	5
Paraffinic oil	10

powder underwent devulcanization to create compounds A and B through different feed screw speed and main screw speed combinations as shown in Table 7.2.

The process output produced compound A which appears in Figure 7.2. The laboratory Banbury mixer (1.5 L capacity, Werner Pfleiderer) with counter-rotating tangential rotors prepared different compound A and compound B ratios with virgin EPDM rubber, carbon black, and oil for 4.5 minutes. The mixing chamber operated at 40 rpm with 80% capacity while the initial temperature was 40 °C and the final temperature reached 85 °C. The required amounts of carbon black

Table 7.2 The devulcanization process of waste powder in the extruder, rotor speed, and temperature data.

Devulcanized compound	Main screw speed (rpm)	Temperature (°C)	Feed screw speed (rpm)
A	180	220	6
B	220	220	8

Figure 7.2 Devulcanized waste rubber after extrusions (compound A). *Source:* Ostad Movahed et al. (2015) / with permission of Sage Publications.

and oil were determined based on compound A and compound B contents in the blends before adding them in one stage. The chemical ingredients carbon black and oil originated from a RC with standard automotive rubber strip manufacturing formulation (Part Lastic Company, Iran) as shown in Tables 7.3 and 7.4. The mixing conditions were controlled by computer software, which also recorded all data.

The rubber blends received curing chemicals through two-roll mills with 14 cm diameter and 32 cm length rolls which operated for five minutes. The Banbury mixer and two-roll mills operate with built-in cooling systems. The required chemical amounts for rubber blend curing emerged from devulcanized compound A and compound B weight percentages in the blends. The final blend formulations received labels A-1–A-4 for compound A blends with virgin rubber and chemical ingredients and labels B-1–B-4 for compound B blends with virgin rubber and

Table 7.3 Formulations of the RC and blends with different amounts of compound A.

		A-1		A-2		A-3		A-4		
Material	**RC**	**Ref.**[a]	**A**	**Ref.**	**A**	**Ref.**	**A**	**Ref.**	**A**	**Compound A**
EPDM	100	90	10	80	20	60	40	40	60	—
N330	112	101	0	90	0	67	0	45	0	—
Oil	40	36	0	32	0	24	0	16	0	—
Sulfur	2.5	2.25	0.07[b]	2	0.14	1.5	0.28	1	0.42	0.7
ZnO	5	4.5	0.14	4	0.28	3	0.56	2	0.84	1.4
Stearic acid	1	0.9	0.03	0.8	0.06	0.6	0.12	0.4	0.18	0.3
MBTS	0.5	0.45	0.015	0.4	0.03	0.3	0.06	0.2	0.09	0.15
ZDBC	1.8	1.6	0.054	1.4	0.108	1.08	0.216	0.72	0.324	0.54
TMTD	0.7	0.63	0.02	0.56	0.04	0.42	0.08	0.28	0.126	0.2

[a] All amounts in phr or g.
[b] Typical calculation: $0.3888 \times 82.58 \times 0.9219 \times 0.1 \times 0.025 = 0.07$.

chemical ingredients. The devulcanized waste powder received chemical ingredient additions for property evaluation (Tables 7.3 and 7.4).

The cure traces were obtained through an oscillating disk rheometer cure meter (ODR, SANTAM SRT-200B, Santam Company) at 180 ± 2 °C with a ±3° angular displacement and 1.7 Hz test frequency. The rheometer operated for a maximum duration of three minutes. The cure rate index (CRI) served as a measurement to determine the rubber curing speed.

The rubber compounds underwent compression molding at 180 °C under 160 atm pressure to achieve their optimum cure times. The milled sheets received rubber pieces which weighed about 57 g each. The molding process allowed the rubber pieces to expand in every direction because they received equal pressure distribution. The rubber pieces received equal pressure distribution during molding to stop anisotropy from developing in the final product. The $15 \times 15\ cm^2$ rubber sheets with 2.3 mm thickness served to determine tensile strength, elongation at break, modulus, hardness, tear strength, and compression set of the cured rubber compounds.

The LEO 1530 VP field emission gun scanning electron microscope (SEM) with field emission gun was used to evaluate how filler particles and other impurities spread throughout the rubber. The rubber pieces received three minutes of liquid nitrogen treatment before they underwent fracturing to obtain two new surfaces.

Table 7.4 Formulations of the RC and blends with different amounts of compound B.

		B-1		B-2		B-3		B-4		
Material	**RC**	**Ref.**[a]	**B**	**Ref.**	**B**	**Ref.**	**B**	**Ref.**	**B**	**Compound B**
EPDM	100	90	10	80	20	60	40	40	60	—
N330	112	101	0	90	0	67	0	45	0	—
Oil	40	36	0	32	0	24	0	16	0	—
Sulfur	2.5	2.25	0.07[b]	2	0.14	1.5	0.28	1	0.42	0.7
ZnO	5	4.5	0.14	4	0.28	3	0.56	2	0.84	1.4
Stearic acid	1	0.9	0.03	0.8	0.06	0.6	0.12	0.4	0.18	0.3
MBTS	0.5	0.45	0.015	0.4	0.03	0.3	0.06	0.2	0.09	0.15
ZDBC	1.8	1.6	0.054	1.4	0.108	1.08	0.216	0.72	0.324	0.54
TMTD	0.7	0.63	0.02	0.56	0.04	0.42	0.08	0.28	0.125	0.2

[a] All amounts in phr or g.
[b] Typical calculation: $0.3888 \times 82.58 \times 0.9185 \times 0.1 \times 0.025 = 0.07$.

The SEM analysis of gold-coated 60 mm^2 samples with 5 mm thickness produced images for examination. The researchers analyzed filler particles and impurity distribution in rubber through visual inspection of SEM images.

The TGA results (Figure 7.3) indicated that EPDM started degrading at 287.5 °C and completed its degradation process at 547.8 °C. The oil presence in the sample caused mass loss to occur between 26.2 and 287.6 °C. The sample underwent a mass loss transition when the atmosphere switched to air at 550 °C during the 550–613 °C temperature range. The combustion of carbon black in the sample occurred when the atmosphere switched to air at 550 °C. The unburned residue consisted of minerals and metal components which remained in the sample. The rubber content in the residue reached 32.1% of the original waste powder weight before acetone extraction while the total residue weight reached 38.88%.

The waste powder required an effective reclaiming agent to achieve proper devulcanization. The authors demonstrated that TMTD did not function as an effective devulcanizing agent. The authors conducted devulcanization tests with TMTD but found no significant results, so they excluded all TMTD test data from their report. The twin-screw extruder (devulcanization reactor) appears in Figure 7.1 together with the water bath used for cooling devulcanized profiles from 220 °C reactor temperature to room temperature. The devulcanization of rubber through the extruder required specific screw design elements and material feeding rates

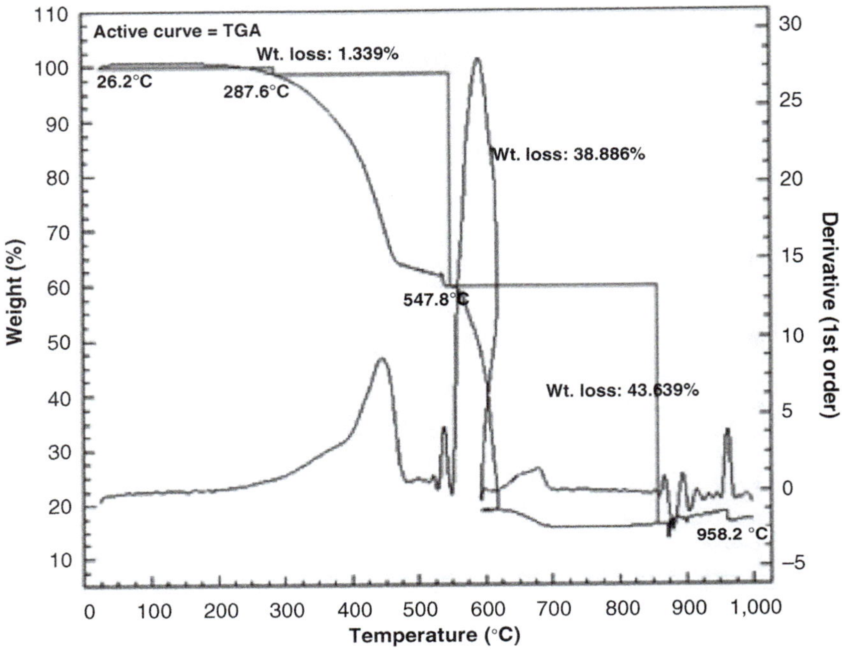

Figure 7.3 The TGA traces show the EPDM rubber waste powder after acetone extraction of oil.

to achieve continuous profile output. The screw channel contained two distinct operational areas. The first section of the screw channel used shearing forces to break down rubber into small pieces while heating it to 220 °C for devulcanization reactions. The second section of the screw channel enabled rubber devulcanization through shearing forces and MBTS devulcanizing agent. The screw shearing action in the first zone transformed large rubber pieces into small particles while simultaneously heating them to processing temperature before moving them into the devulcanization reaction area. The rubber experienced restricted flow during devulcanization while the reaction zone fill factor reached maximum levels. The rubber experienced extreme compression which led to increased hydrostatic pressure throughout the material. The reactor produced devulcanized rubber as a continuous strand which emerged from its head (Figure 7.2). The devulcanized rubber surface quality depended heavily on the devulcanization process parameters which included screw speed and screw channel temperature and material feed rate and raw material composition. The strand surface showed roughness when the operating parameters were not set correctly.

The research team performed multiple experiments under specific conditions which included 180 and 220 rpm main screw speeds and 220 °C temperature and 6 and 8 rpm feed screw speeds to achieve the best devulcanized rubber surface quality. The feed composition consisted of 100 phr EPDM waste rubber powder and 6 phr MBTS and 15 phr of aromatic and aliphatic oils (Table 7.1). The specific mixing conditions produced the highest devulcanization percentage from the waste powder when using this formulation. The devulcanization process benefited from the oil's dual function of enhancing rubber plasticity and its ability to speed up rubber oxidation and stop gel formation through radical acceptance. The oils expanded the polymer structure while making polymer chains more mobile, which made crosslink breaking during devulcanization more accessible. The formulations of compounds A and B used reclaiming oils with high compatibility. The devulcanization results for compounds A and B showed 92.19% and 91.85%, respectively, which proved MBTS effectiveness as a devulcanizing agent. The results demonstrated that the crosslink density (CLD) decreased from 186 to 14.5 and 15.16 mol/m^3, while the sol–gel content increased from 0.5% to 17.6% and 13.1% for the initial waste rubber powder and devulcanized compounds A and B, respectively. The polymer chains and chemical crosslinks underwent breakdown during devulcanization when MBTS was present in a twin-screw extruder because of its large shearing forces. The shearing process in the extruder broke down MBTS molecules into their corresponding radical forms (Figure 7.4).

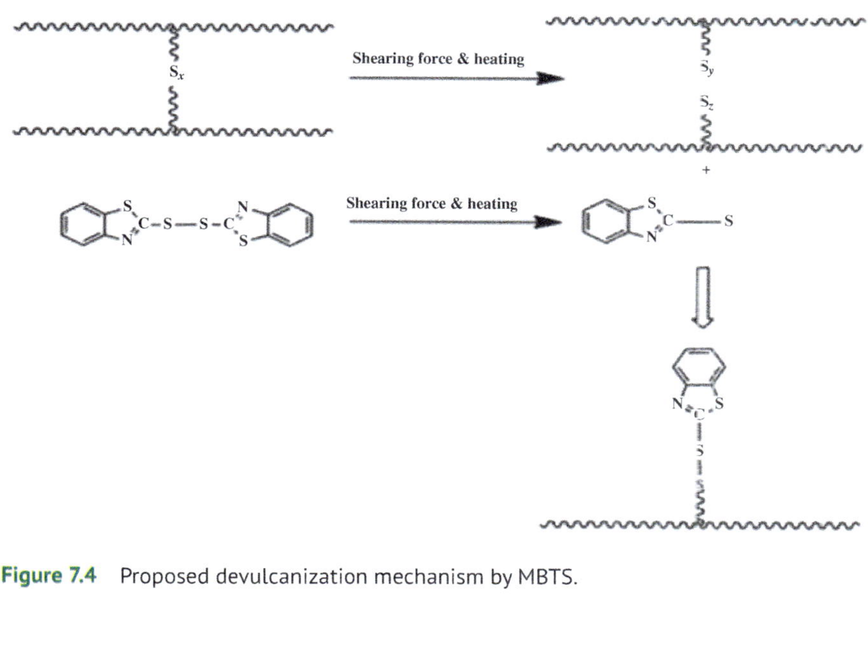

Figure 7.4 Proposed devulcanization mechanism by MBTS.

The MBTS radicals showed better stability than TMTD radicals because their potential resonance stabilization weakened the central S—S bond in the devulcanizing agent structure, which helped create radicals. The broken polymer chain radicals combined with these radicals create accelerator-terminated polysulfidic pendant groups on polymer chains that became ready for revulcanization (Figure 7.4). The compound A devulcanization percentage reached 92.19%, while its CLD reached 14.5 mol/m^3, and its sol–gel percentage reached 17.6%. The devulcanization results of compound A exceeded those of compound B. The devulcanization process benefited from extended screw channel residence time which occurred when the screw speed reached 180 rpm. The longer compound residence time in the screw channel decreases industrial production speed. The Banbury mixer prepared blends compounds A and B with virgin EPDM rubber, carbon black, and oil through temperature and mixing time measurements. The experimental results demonstrated that compound temperature rose steadily with increasing mixing duration for all blend combinations, RC and compounds A and B. The RC experienced the most significant temperature increase from 40 to 83 °C throughout the mixing process. The initial temperatures of the compound blends exceeded the RC temperature before starting the mixing process. The A-1 blend compound temperature increased from 53 to 83 °C throughout mixing time while for compound A-4 with maximum compound A content, the temperature started at 68 °C before rising to 81 °C after the mixing process finished. The B-1–B-4 blends showed identical temperature patterns. The addition of compound A or compound B to the blend resulted in elevated starting temperatures, which reached their peak values more quickly than the RC during mixing. The tested blends demonstrated identical temperature patterns throughout their evaluation.

The RC showed viscosity at 57 MU. The viscosity of the blend reached 62 MU when compound A reached 50 wt.%, but it dropped to 30 MU when compound A reached 100 wt.%. The viscosity of the blend followed a similar pattern when compound B reached 30 wt.% as 60 MU before decreasing to 24 MU for 100 wt.% of compound B. The rubber chains in compounds A and B experienced major chain breaks during extruder shearing which made them softer. The blends became more fluid when compound A and compound B percentages increased because their softening effect reduced viscosity.

The rubber compounds underwent cure property evaluation based on compound A and compound B percentages in the blends, and the results plotted as a graph. The correspondent graph demonstrates how values change when compound A reaches different weight percentages in the blend. The ΔTorque values stayed constant until compound A reached 20 wt.% in the blend. The ΔTorque value dropped significantly when compound A reached its maximum concentration in the blend. The addition of compound A beyond 20 wt.% in the blend did not enhance the CLD of the blend. The addition of compound A to the blend affected cure times through different mechanisms. The addition of compound A to the blend resulted in shorter cure times, which improved the overall cure process. The blends showed identical patterns when compound B levels were raised in the mixture.

A graph demonstrates how ΔTorque changes when compound B weight percentage increases in the blend. The CLD remained unaffected by compound B additions up to 20 wt.%, but ΔTorque values plummeted when compound B reached 100 wt.%. The addition of compound B to the blend resulted in longer cure times because the property showed significant reduction. The CRI showed substantial growth when compounds A and B reached 50 wt.% or higher in the blends. The rubber curing process became faster, which resulted in a shorter overall curing time. The CRI values increased because the blends contained more active crosslinking sites than the RC.

The rubber compounds' mechanical properties depended heavily on the percentage of compounds A and B present in the blends. The addition of compounds A and B to their maximum levels resulted in an 87% decrease of tensile strength and an 82% decrease of elongation at break. The modulus values decreased when the blends contained their maximum amounts of compounds A and B. The RC showed a modulus of 11 MPa, but the addition of compounds A and B resulted in a modulus decrease to 2 MPa. The rubber showed higher modulus values when it experienced extended elongation. The rubber modulus reached 6.5 MPa when stretched to 100%, but it dropped to 3 MPa when stretched to 20%. The addition of compound A or compound B to the RC did not improve its modulus values.

The hardness values reached their maximum when compounds A and B reached 40 wt.%, but the values dropped significantly after that point. The tear strength values decreased dramatically by 80% when compounds A and B reached their optimal blend concentrations.

The gasket industry relies on compression sets as a critical property which represents the ratio between elastic and viscous rubber responses under specific deformation conditions. The compression set results showed interesting patterns during the study. The compression set value started at 68% (without compounds A and B) before increasing to 69% when the blend contained 10% compounds A and B. The measurement reached its minimum point at 59% before returning to its value of 70% when the blends contained the maximum amount of compounds A and B. The blend achieved its minimum compression set at 20% compound A content.

It seemed that the mechanical properties of blends showed no improvement when using high amounts of devulcanized waste powder (compound A or compound B), but smaller additions proved beneficial. The addition of 40 wt.% devulcanized waste powder did not affect viscosity, while 20 wt.% additions did not change CLD, the tensile strength remained unchanged at 10 wt.%, elongation at break stayed constant at 10 wt.%, and hardness remained unaffected at 50 wt.% of devulcanized waste powder in the blends. The addition of devulcanized waste powder to the RC to make correspondent blends resulted in the shortest possible cure cycle because it optimized scorch time, optimum cure time, and CRI.

The SEM images revealed the prepared rubber compounds and blends. The final blend morphology determined the mechanical performance of all prepared rubber compounds. The complete distribution of filler particles throughout the rubber matrix produced the highest possible reinforcement impact on rubber mechanical properties after curing. The RC contained carbon black particles that measured 20 μm on average while dispersed throughout the rubber matrix. The rubber contained both filler particles and additional contaminants during its composition process.

The TGA results in Figure 7.3 show that non-rubbery residue made up 15.5% of the total sample. The grinding process that turned rubber waste into powder might have introduced metal contaminants into the system. The final product contained impurities and contaminants which appeared as white spots during micrograph examination (SEM). The elimination of contamination remains impossible during all manufacturing operations. The rubber contained cavities (Figure 7.5) which created areas of weakness that reduced the tensile strength of the cured rubber. The rubber matrix contained undesirable cavities which needed to be avoided. The internal rubber structures appear identical in all compound A and

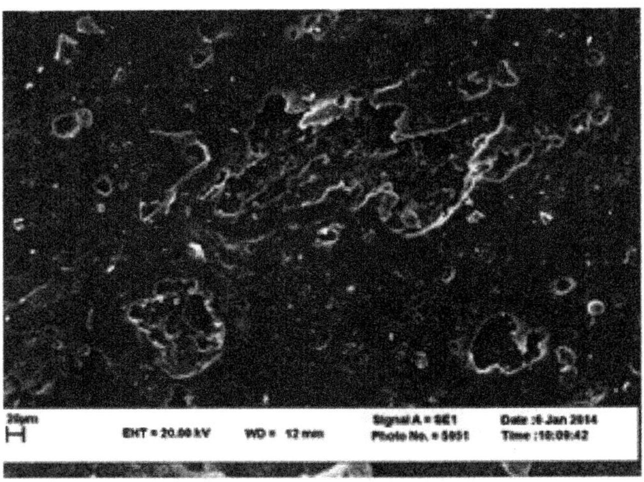

Figure 7.5 SEM micrograph showing the microstructure of compound A. *Source*: Ostad Movahed et al. (2015) / with permission of Sage Publications.

compound B ratios in the blends with no discernible effect. The results showed that different feed screw speeds and main rotor speeds enhanced waste rubber powder devulcanization efficiency when the temperature remained constant. The authors discovered that main screw speeds of 180 and 220 rpm combined with 220 °C temperature and feed screw speeds of 6 and 8 rpm produced the best conditions for devulcanizing rubber powder into a continuous smooth surface. The discovery proved to be quite fascinating.

In summary, the research findings showed that the semi-industrial twin-screw extruder with shearing action and reactor using MBTS created an effective method to devulcanize waste automotive EPDM rubber powder for commercial EPDM-based automotive rubber strips compound reuse or recycling. The optimal results for specific properties required precise determination of devulcanized waste powder amounts in commercial rubber strips compounds. The research results matched previous studies which demonstrated that EPDM rubber vulcanizate properties including processing, mechanical strength, and cure characteristics depended on the amount of devulcanized waste rubber added

The **second case study** appeared in *Polymer Degradation and Stability* when authors published “A novel industrial technique for recycling ethylene–propylene–diene waste rubber” in 2015.

As mentioned, the waste rubber recycling process through autoclave operation with the aid of a chemical agent falls under thermo-chemical processing. The process uses high-pressure steam, i.e., at 280 °C temperatures and 5.6–6.9 MPa pressures together with a devulcanizing agent. The devulcanization method was employed by specific researchers to treat natural rubber (NR), butyl, and silicone rubber materials. The main goal of this research involves using high-pressure steam industrial autoclave to devulcanize waste EPDM rubber powder which originated from discarded automotive parts. The devulcanization process received assistance from MBTS and TMTD organic disulfide devulcanizing agents. The waste powder required soaking in aromatic and aliphatic oils before undergoing devulcanization. The devulcanized powder received two different weight ratio combinations with virgin EPDM and carbon black and oil before undergoing semiefficient revulcanization in the last stage. The research evaluated how devulcanized waste powder addition affects blend properties through measurements of viscosity, cure, and mechanical tests.

The Part Lastic Company provided waste EPDM rubber powder which appears in Figure 7.6. The powder consisted of different aged and new automotive rubber materials which had particles smaller than 1 mm in size. The Soxhlet apparatus extracted oil from the sample during the first stage of processing. The extraction process used 300 mL of acetone to extract oil from 100 g of sample during 16 hours at 70 °C according to ASTM D297-93. The sample underwent oven drying with circulating air for 12 hours after extraction. The waste EPDM rubber powder

Figure 7.6 EPDM waste rubber powder appearance. *Source*: Ostad Movahed et al. (2015) / with permission of Elsevier.

Table 7.5 The physical characteristics together with chemical makeup of EPDM rubber waste powder following acetone extraction for oil removal.

Property	Amount
Sol content (wt.%)	0.5
Density (g/cm^3)	1.4072
CLD (mol/m^3)	186
TGA results	
Carbon black and calcium carbonate (wt.%)	15.45
Carbon black and calcium carbonate (wt.%)	43.60
Oil in the sample which could not be extracted in the first extraction stage EPDM (wt.%)	2.07

contained 17.42 wt.% oil according to five extraction tests that were conducted. The waste rubber powder underwent TGA analysis (STA 1500 from Scinco Co. Ltd.) to identify its chemical makeup after acetone extraction. The TGA test operated under nitrogen and synthetic air conditions to measure sample compositions at temperatures below and above 550 °C. The TGA results along with their corresponding chemical compositions appear in Table 7.5.

The other used materials included EPDM (KEP 270 with 57 wt.% ethylene and 4.5% diene monomer from virgin rubber of KUMHO POLYCHEM in Korea) and carbon black (N330 as filler from Pars Carbon Ltd. in Saveh) and MBTS (Perkacit MBTS served as devulcanizing agent and accelerator from Flexsys) and TMTD (Perkacit TMTD functioned as accelerator and devulcanizing agent from Flexsys) and zinc dibutyl dithiocarbamate (Perkacit ZDBC acted as an accelerator from Flexsys) and zinc oxide (activator from Harcros Durham Chemicals in the UK) and stearic acid (activator from Anchor Chemicals Ltd. in the UK) and elemental sulfur (curing agent from Solvay Barium Strontium in Hannover, Germany), and aromatic and aliphatic oils used as processing aids (Mehran Tyre 290 and Fariman).

The waste powder underwent devulcanization through two separate procedures. The researchers combined waste powder with various oil combinations and TMTD quantities to create compounds A-1–A-4. The compounds A-1 and A-4 lacked TMTD as an ingredient. The researchers combined waste powder with different oil ratios and MBTS amounts to create compounds B-1–B-6. The compounds B-1 and B-2 lacked MBTS according to Table 7.6. The compounds underwent preparation through a basic mixer system that included an agitator. The compounds received 24 hours of oil penetration into rubber before undergoing industrial autoclave devulcanization at Savalan Sanat Co. Ltd. The industrial autoclave measured 4.5 m in length and 1.5 m in diameter while maintaining an 8,000 L nominal capacity. The autoclave received 250 kilograms (about 551.15 lb.) of oil-treated waste powder before the door closed for operation. The compounds

Table 7.6 Devulcanization formulations of the compounds in phr.

Compound no.	EPDM	Aromatic oil	Aliphatic oil	TMTD	MBTS
A-1[a]	100	15	5	—	—
A-2	100	15	5	6	—
A-3	100	5	15	6	—
A-4	100	5	15	—	—
B-1[b]	100	15	5	6	—
B-2	100	5	15	6	—
B-3	100	—	20	—	6
B-4	100	5	15	—	6
B-5	100	7	13	—	6
B-6	100	10	10	—	6

[a] Devulcanization conditions for A compounds: steam pressure was 6 bar, steam temperature 165 °C, and devulcanization time eight hours.

[b] Devulcanization conditions for B compounds: steam pressure was 8 bar, steam temperature 175 °C, and devulcanization time eight hours.

underwent devulcanization through steam treatment at 165 °C under 6 bar pressure to create A-1–A-4 compounds and at 175 °C under 8 bar pressure to create B-1–B-6 compounds. The entire process required eight hours to finish. The research investigated how different steam pressures between 6 and 8 bar affected waste powder devulcanization because these pressures match current industrial standards for automotive radiator hose vulcanization at Part Lastic Co. in Iran. The temperature increase stemmed from selecting specific steam pressure levels.

The devulcanized waste powder showed spongy texture with an inconsistent appearance (Figure 7.7) before researchers dried it at room temperature for 24 hours to achieve processing readiness. The laboratory Banbury mixer (1.5 L capacity, Werner Pfleiderer) with counter-rotating tangential rotors processed compounds B-3–B-6 for 4.5 minutes to create blends with virgin EPDM, carbon black, and oil. The experiments used a 40 rpm rotor speed with 80% chamber filling and temperature settings from 40 to 85 °C. The mixing chamber operated at 40 °C during its initial phase before reaching 85 °C at the end of the process. The required amounts of carbon black and oil were determined based on the basis of devulcanized powder weight in each blend before adding them simultaneously to the mixture. The RC contained chemical ingredients amount as recommended by Part Lastic Company (Iran) which used a standard formulation to produce automotive rubber strips. The mixing system operated under computer control which recorded all processing data.

The research used two different weight ratios to create blends. The weight ratio of virgin EPDM rubber to devulcanized powder used in the study was

Figure 7.7 Devulcanized compound B-3 after removal from the autoclave and subsequent drying in air for 24 hours. *Source*: Ostad Movahed et al. (2015) / with permission of Elsevier.

80/20 and 40/60. The weight percentages of carbon black, oil, and chemical curatives were determined through calculations based on the established weight ratios. The two blends from compound B-3 received labels B-3-1 (80/20) and B-3-2 (40/60), while compound B-4 received labels B-4-1 (80/20) and B-4-2 (40/60), compound B-5 received labels B-5-1 (80/20) and B-5-2 (40/60), and compound B-6 received labels B-6-1 (80/20) and B-6-2 (40/60).

The final stage involved adding curing chemicals including elemental sulfur, ZnO, stearic acid, MBTS, ZDBC, and TMTD to the blends through a two-roll mill operation that lasted five minutes. The two-roll mills operated with 14 cm diameter and 32 cm long rolls which used a 3–4 mm gap for mixing. The Banbury mixer and two-roll mills operate with built-in cooling systems. The chemical curatives needed for rubber curing were determined through calculations based on the blend compositions. The blends underwent viscosity and cure property assessment followed by semiefficient revulcanization to evaluate their mechanical characteristics. The research included compounds B-3–B-6 as additional test materials.

The LEO 1530 VP field emission gun SEM with field emission gun was used to evaluate carbon black particle distribution and rubber impurities. The rubber pieces underwent three minutes of liquid nitrogen exposure before they were fractured to obtain two new surfaces. The SEM analysis of gold-coated 60 mm^2 samples with 5 mm thickness produced images for examination and photography. The researchers analyzed filler particle distribution and rubber impurities through SEM image analysis.

The solvent used for determining sol % and CLD was toluene. The rubber sample of 5 g underwent 16 days of swelling in 300 mL of solvent at 21 °C until equilibrium was reached. The sample weight received daily measurements until equilibrium was achieved before the solvent extraction process. The samples underwent nine hours of air drying followed by 24 hours of oven drying at 85 °C before they received an additional 24 hours of drying at 23 °C before their final weight measurement. The CLD calculation used the Flory–Rehner equation to determine the value:

$$\rho_c = -\frac{1}{2V_s}\frac{\ln(1-\upsilon_r^0) + \upsilon_r^0 + x(\upsilon_r^0)^2}{(\upsilon_r^0)^{1/3} - \upsilon_r^0/2}$$

where ρ_c is the CLD (mol/m^3), V_s is the molar volume of toluene (1.069×10^{-4} m^3/mol at 25 °C), υ_r^0 is the volume fraction of rubber in the swollen gel, and χ is the interaction parameter, which was calculated using the following equations:

$$\chi = 0.429 + 0.218\upsilon_r^0$$

$$\upsilon_r^0 = \frac{1}{1 + \frac{d_r}{d_s}\left(\frac{1-f_{sol}}{1-f_{sol}-f_{fil}}\right)\left(\frac{w_s}{w_0} - 1\right)}$$

where d_r and d_s are the densities of the rubber and solvent, respectively; f_{sol} is the weight fraction of soluble material in the initial sample (sol fraction); f_{fil} is the initial weight fraction of filler in the sample; w_s is the weight of the swollen gel; and w_0 is the weight of the dried sample. The Kraus correction was omitted because the study focused on CLD comparisons between tested compounds. The rubber compounds underwent compression molding at 180 °C under 160 atm pressure after finishing the ODR tests. The milled sheet received 57 g pieces which were used for the experiment. The mold center position received each rubber piece to allow complete directional flow under pressure application. The rubber pieces received uniform pressure during molding to stop anisotropic properties from developing in the final product. The rubber vulcanizates underwent mechanical property testing using 15 cm × 15 cm sheets which measured 2.3 mm in thickness.

The TGA results for acetone-extracted waste rubber powder appear in Table 7.5. The TGA results indicate that EPDM started degrading at 287.6 °C until it reached complete degradation at 547.8 °C. The oil presence in the sample caused mass loss to occur between 26.2 and 287.6 °C. The TGA curve showed a transition between 550 and 613 °C which scientists linked to carbon black combustion after they switched the atmosphere to air at 550 °C. The TGA curve displayed a distinct peak at 700 °C which scientists linked to the calcium carbonate content. The waste powder originated from multiple rubber items which contained both carbon black and calcium carbonate. The TGA results showed that calcium carbonate was decomposed into calcium oxide while losing weight at temperatures above 700 °C. The unburned residue consisted of minerals and metals which remained in the sample. The waste rubber contained 38.88 wt.% EPDM rubber which represented 32.1 wt.% of the original waste rubber before acetone extraction. The waste powder required an efficient devulcanizing agent to achieve proper devulcanization results. The compounds displayed spongy textures after devulcanization as shown in Figure 7.7. The compounds became smooth in appearance after two minutes of processing on the two-roll mills while their consistency improved.

The devulcanization percentage of compounds A-1 and A-4 showed the lowest values while their CLD measurements reached the highest points at 28.6% and 132.8 mol/m^3 and 10.1% and 167.2 mol/m^3, respectively. The devulcanization process benefited more from compound A-1's high aromatic to aliphatic oil weight ratio of 3 than from compound A-4's lower weight ratio of 0.3. The waste powder expands at a lower rate when submerged in aromatic oil compared to aliphatic oil. The use of aromatic oil requires toxic substances which makes any process that reduces its usage more desirable. The addition of TMTD led to a 33% increase in devulcanization percentage for compound A-1 and a 500% increase for compound A-4. The devulcanization percentage of compound A-3 became higher because TMTD contains aliphatic structures which dissolve better in aliphatic oils. The devulcanization process benefits from the oil's multiple functions which include waste powder plasticity enhancement and rubber oxidation acceleration and radical acceptor properties for gel prevention. The oil serves two main functions in devulcanization by speeding up rubber oxidation and blocking gel formation through radical acceptance. The oil swells the polymer network which enables better chain movement that makes crosslink breaking easier during devulcanization.

The addition of TMTD proved advantageous for waste powder devulcanization when compounds A-1 and A-2 contained identical aromatic/aliphatic ratios of 3. The devulcanization results showed 28.6% without TMTD and 38.1% with TMTD for compounds A-1 and A-2 (Table 7.6). The devulcanization results for compounds A-3 and A-4 with aromatic/aliphatic ratios of 0.3 showed a similar pattern. The devulcanization results for these compounds reached 64.9% with TMTD addition but only 10.1% without TMTD. The same pattern of results appeared in compounds B. The devulcanization process of waste powder received benefits from MBTS addition when compounds B-2 and B-4 contained the same **aromatic/aliphatic ratio of 0.3**. The devulcanization results for these compounds showed 76% without MBTS and **93.9% with MBTS**. The thermal properties of MBTS follow typical accelerator behavior since it melts above 100 °C but undergoes decomposition at 200 °C. The autoclave process uses steam pressure and high temperature to break down both polymer chains and crosslink bonds in the presence of MBTS during devulcanization. The MBTS molecules underwent decomposition to produce free radicals during the process. The MBTS radicals showed greater stability than TMTD radicals because their molecular structure allowed resonance stabilization, which reduced the strength of their central S—S bond. The broken polymer chain radicals joined with MBTS radicals to create accelerator-terminated polysulfidic pendant groups on polymer chains which became ready for revulcanization (Figure 7.4).

The freeze-fracture images, in Figure 7.8, demonstrate the filler distribution and internal arrangement of the RC and unvulcanized and vulcanized compounds. The mechanical properties of the blends received influence from their resulting structural patterns as stated earlier. The rubber matrix achieves its maximum reinforcing potential through complete filler particle dispersion. The rubber matrix contains well-dispersed filler particles which measure 20 μm on average. The micrographs of unvulcanized rubbers show white particles that appear as impurities together with the filler material. Also, the white particles in mentioned compounds appear to be metallic contaminants. The microstructural patterns between the blends showed no distinguishable variations.

The rubber blends contained different levels of impurities because the devulcanized powder weight ratio varied between them. The mechanical properties of rubber blends suffered from impurity presence because compound B-3-1 contained fewer devulcanized powder impurities than compound B-3-2 which

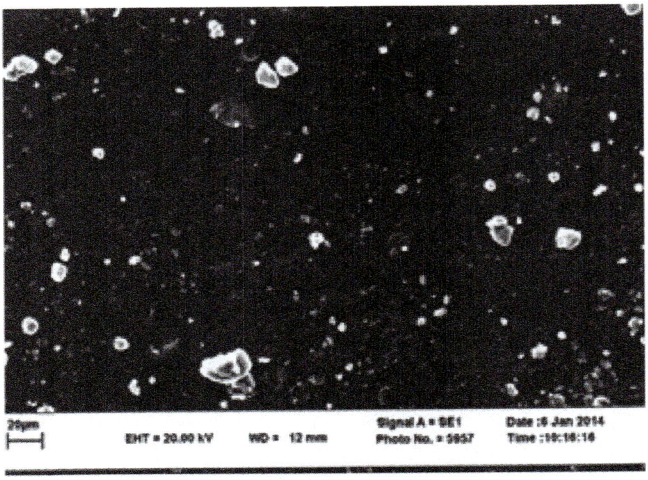

Figure 7.8 SEM micrograph showing the interior of the RC after freeze-fracture. *Source*: Ostad Movahed et al. (2015) / with permission of Elsevier.

had more devulcanized powder. The rubber blends showed different mechanical behavior because of their varying impurity content. The mechanical properties of compound B-3-1 exceeded those of compound B-3-2 because B-3-1 contained fewer impurities and demonstrated better tensile strength, elongation at break, tear strength and lower compression set. The rubber blends showed deteriorating mechanical properties because of impurity contamination. The mechanical properties of rubber blends would improve significantly when impurities are minimized during devulcanization. As mentioned, the TGA results in Table 7.5 show that non-rubbery residue makes up about 15.5% of the total residue. The grinding process that turned rubber waste into powder might have introduced metal contaminants into the system. The contamination during the grinding process most likely caused impurities to enter the final product which appeared as white spots under microscopic examination. The elimination of contamination remains impossible during such manufacturing operations. The RC showed a Mooney viscosity measurement of 57 MU. The compound viscosity decreased when researchers added more devulcanized rubber powder to the mixture. The weight ratio of virgin rubber to devulcanized powder in compound B-3-1 was 80/20, but it decreased to 40/60 in compound B-3-2. The viscosity measurements decreased by approximately 4% when the weight ratio of virgin rubber to devulcanized powder in the blend changed from 80/20 to 40/60. The viscosity measurements of compounds B-4–B-6 showed similar results. The viscosity of compound B-4 decreased by 13%, while compound B-5 showed minimal change and compound

B-6 experienced a 16% decrease in viscosity when the virgin rubber to devulcanized powder weight ratio in the blend changed from 80/20 to 40/60. The devulcanized powders (compounds B-3–B-6) showed viscosities of 32.2, 40.3, 35.2, and 37.5 MU, respectively, which were lower than all other compounds. The viscosity reduction in the blends became explainable because of the devulcanization process which caused chain scission and degradation.

The scorch time of all tested compounds obtained from the ODR test showed a decrease when researchers changed their blend compositions. The scorch time of compound B-3-2 decreased by 24% when researchers used less virgin rubber and more devulcanized powder in their blend (RC versus compound B-3-2). The other compounds showed identical patterns in their test results. The blend of compound B-4-2 showed a 20% scorch time reduction when virgin rubber content decreased against RC. Similarly, the scorch time of compound B-5-2 decreased by 32% when virgin rubber content decreased against RC. The scorch time of compound B-6-2 decreased by 16% when virgin rubber content decreased against RC. The addition of unvulcanized powder to blends also resulted in shorter optimum cure times which decreased by 46% for all tested compounds. The shortened cure cycle became more noticeable because of this improvement.

The results indicated that ΔTorque followed a distinct pattern because it measured rubber CLD variations. The results from the study showed that rubber CLD decreased when devulcanized powder amounts increased in the blends. The ΔTorque value decreased by 40% for compound B-3-2 and by 44% for compound B-4-2 and by 33% for compound B-5-2 and by 46% for compound B-6-2. The substantial decrease in CLD would negatively impact rubber vulcanizate mechanical properties.

The CRI measurement, which shows rubber cure speed, produced the most significant improvement. The RC started with a CRI of 0.99 1/min, while this parameter increased by 98% for compound B-3-2, for compound B-4-2 by 94%, for compound B-5-2 by 77%, and for compound B-6-2 by 106%. The industrial process will benefit from the fast cure rate that these compounds provide. It seemed that the addition of up to 60 wt.% devulcanized powder to virgin rubber in the RC formulation produced beneficial processing results through viscosity reduction. The substitution of up to 20 wt.% virgin EPDM rubber with devulcanized powder in the formulation did not cause any negative impact on rubber scorch time or optimum cure time and CLD. The cure rate improved significantly when the virgin rubber was completely replaced by devulcanized powder.

The mechanical properties of the blends showed different responses when devulcanized powder amounts increased in the blends. The data showed distinct patterns for certain properties, but other properties displayed unclear results. Compound B-5-2 achieved the highest hardness increase at 19% among all the tested compounds. The 40/60 weight ratio of virgin rubber to devulcanized powder in compound B-5-2 produced the highest results. The RC showed lower hardness than all blends which indicated that substituting 60% of virgin rubber with devulcanized powder did not harm the rubber's hardness. The blends demonstrated higher modulus values at 20% elongation except for compound B-4-2. The moduli values decreased when elongation reached 50% and continued to decrease at 100% and 200% elongations. The devulcanized powder addition to rubber blends resulted in no positive effect on rubber modulus when stretching beyond 20% elongation. The substitution of 60% virgin rubber with devulcanized powder in blends did not result in any negative impact on the 20% elongation modulus. The elongation at break measurements showed no specific pattern between different blends because they produced different results. Compound B-3-1 achieved the most significant improvement by increasing its elongation at break by 20% compared to the RC, while compound B-5-2 demonstrated the worst result with a 36% decrease in elongation at break. The tear strength, compression set, and tensile strength values decreased for all compounds that underwent testing. The tear strength of compound B-4-2 experienced the most significant decrease at 38%. The RC demonstrated better tear strength than all other compounds in this study. The gasket industry relies on a compression set as a critical property which measures how rubber responds to deformation through elastic and viscous components, as stated earlier in this text. The compression set results show that most compounds experienced deterioration in their compression set values. The compression set of compound B-3-1 showed the best result at 75% lower than the RC, while compounds B-3-2 and B-4-2 achieved only 55% lower compression set than the RC. The authors need to highlight that the test results show significant variation because the waste powder used might contain low-quality materials.

The results showed that adding small amounts of devulcanized powder did not significantly harm the mechanical strength of the blend. The mechanical properties of the blend deteriorated when researchers added excessive amounts of devulcanized powder to the mixture. The high amount of contaminants in the devulcanized powder caused this negative effect. The research shows that partial substitution of virgin rubber with devulcanized waste rubber in automotive rubber strip common formulations will not harm viscosity, and cure properties or selected mechanical characteristics. The discovered method enables waste powder recycling through substitution of virgin rubber in rubber formulations.

The **third case study** appeared in the *Journal of Elastomers and Plastics* when authors published "De-vulcanization of ethylene–propylene–diene waste rubber by microwaves and chemical agents" in 2016.

In this study, the microwave oven served as a heating device which combined with chemical substances and aromatic and aliphatic oils to devulcanize cured EPDM waste powder obtained from automotive part residues between three temperatures, 200, 230, and 260 °C. The devulcanized waste powder underwent revulcanization through sulfur-cure processing followed by property evaluation of both cure characteristics and mechanical strength.

The electromagnetic radiation spectrum of microwaves spans from 1 mm to 1 m in wavelength and operates between 300 MHz and 300 GHz frequencies. The two primary microwave frequencies used in applications are 0.915 and 2.45 GHz. The energy from microwaves functions to break down crosslinks through its ability to target C—S and S—S linkages. The method provides volumetric heating that starts from the material core and moves toward the surface without needing extended heating times to achieve temperature changes and produces uniform heating of rubber. The heating process of materials through microwaves requires materials to possess natural polarity as their fundamental property. The microwave absorption and dispersion properties of EPDM rubber materials show poor performance. The addition of polar fillers including silica and carbon black to rubber materials could solve the absorption problems.

The EPDM waste powder originated from Part Lastic Company based in Iran. The powder consisted of multiple rubber components from automotive waste which included new and aged materials with particles smaller than 1 mm in size. The original powder consisted of 1 cm sized particles which proved unsuitable for devulcanization because big particles fail to devulcanize effectively. The powder required grinding to achieve 1 mm particle size before starting the devulcanization process. The researchers used a sieve to analyze 1 kg of powder for determining its particle size distribution. The analysis revealed that 2% of the particles measured below 0.599 mm and 5% below 0.710 mm and 8% below 0.853 mm, while the rest of the particles were smaller than 1 mm. The first stage of oil extraction from the powder used a Soxhlet apparatus. The extraction process used 300 mL of acetone to extract oil from 100 g of rubber during 16 hours at 70 °C according to ASTM D297-93. The samples underwent oven drying with circulating air for 12 hours after extraction. The waste powder contained 17.42 wt.% oil according to the average results from five extraction tests. The waste powder composition after acetone extraction was determined through TGA using STA 1500 from Scinco Co. Ltd. The TGA measurements operated under nitrogen gas at temperatures below 550 °C and air gas at temperatures above 550 °C. The remaining ingredients included *N*-cyclohexyl-2-benzothiazole sulfenamide (Perkacit CBS from Flexsys), dipentamethylenethiuram tetrasulfide (DPTT; Perkacit DPTT from Flexsys), hexadecylamine (HDA from Cayman Chemical Co.), 2-mercaptobenzothiazole (MBT; Perkacit MBT from Flexsys), MBTS (Perkacit MBTS from Flexsys), TMTD (Perkacit TMTD from Flexsys), zinc dibutyl dithiocarbamate (Perkacit ZDBC from Flexsys), zinc oxide (activator from Harcros Durham Chemicals, UK), stearic acid (activator from Anchor Chemicals Ltd., UK), elemental sulfur (curing agent from Solvay Barium Strontium, Hannover, Germany), and aromatic and aliphatic oils with boiling points of 270 and 280 °C, respectively, as processing aids (Mehran Tyre 290 and Fariman).

The first step involved combining CBS, DPTT, HDA, MBT, MBTS, and TMTD chemicals which served as devulcanizing agents with aromatic oil before adding them to 500 mL glass beaker containing waste powder. The waste powder received 24 hours of oil soaking to achieve complete rubber penetration. The process resulted in six different compounds which researchers named compounds A–F. The RC received aromatic oil mixing without any devulcanizing agent. The 300 g weight of compounds in the beaker underwent microwave treatment inside the GMO-530 Gosonic oven which operated at 2,000 MHz and 900 W power output and 30 L internal volume. The microwave oven processed three different temperature settings for three minutes to create separate compounds for each temperature. The microwave oven processed compound A at three different temperatures which resulted in compounds A-1–A-3. The researchers performed this process for all compounds in Table 7.7 to generate 18 compounds which they listed in Table 7.8.

Also, the research investigated how various amounts of aromatic and aliphatic oils affect waste powder devulcanization when using a devulcanizing agent.

Formulation of the RC: devulcanized waste powder, 100 phr; elemental sulfur, 2.5 phr; ZnO, 5 phr; stearic acid, 1 phr; MBTS, 0.5 phr; ZDBC, 1.8 phr; and TMTD, 0.7 phr. Note that the curing chemicals were from a common formulation for rubber automotive strips. phr: parts per hundred rubber by weight.

The preparation method for these compounds followed the same procedure as before. The microwave oven extraction of devulcanized rubber compounds produced large bulky materials, which are shown in Figure 7.9. The laboratory two-roll mills with 14 cm diameter and 32 cm length cylinders processed the

Table 7.7 Formulations for devulcanizing the waste rubber powder.

Compound	Waste rubber (phr)[a]	CBS[b]	DPTT[b]	HDA[b]	MBT[b]	MBTS[b]	TMTD[b]	Aromatic oil (phr)
A	100	6.7						26.7
B	100		6.7					26.7
C	100			6.7				26.7
D	100				6.7			26.7
E	100					6.7		26.7
F	100						6.7	26.7
Reference	100							26.7

[a] After acetone extraction.
[b] Devulcanizing agent.

Table 7.8 Devulcanization conditions, sol content, CLD, and devulcanization % of the waste rubber powder.

Compound	Devulcanization chemical	Temperature (°C)	Sol content (%)	CLD (mol/m^3)	Devulcanization (%)
RC-1	—	200	4.5	51	72
RC-2	—	230	5.6	49	73
RC-3	—	260	5.8	45	75
A-1	CBS	200	4.3	17	90
A-2	CBS	230	5.9	15	91
A-3	CBS	260	9.6	12	93
B-1	DPTT	200	2.8	43	76
B-2	DPTT	230	4.3	41	77
B-3	DPTT	260	4.3	38	79
C-1	HDA	200	5.5	16	91
C-2	HDA	230	7.2	12	93
C-3	HDA	260	13.6	10	94
D-1	MBT	200	6.2	17	90
D-2	MBT	230	7	14	91
D-3	MBT	260	10.1	12	93
E-1	MBTS	200	5	43	76
E-2	MBTS	230	5.2	40	77
E-3	MBTS	260	5.8	36	80
F-1	TMTD	200	3	46	75
F-2	TMTD	230	4	44	75
F-3	TMTD	260	5.5	35	80

devulcanized rubber compounds for two minutes to create workable sheets. The researchers studied devulcanized rubber sheet surface appearance to determine how microwave-induced temperature increase affects rubber surface sheet.

The rubber compounds in Tables 7.7 and 7.8 required sulfur cure system for revulcanization. The cure system has operated in rubber automotive strip production at Part Lastic Company in Iran for multiple years and the researchers chose it for this research. The formulation consisted of EPDM rubber at 100 phr, N330 carbon black at 112 phr, oil at 40 phr, elemental sulfur at 2.5 phr, ZnO at 5 phr, stearic acid at 1 phr, MBTS at 0.5 phr, ZDBC at 1.8 phr, and TMTD at 0.7 phr.

Figure 7.9 A typical devulcanized waste powder after removing from the glass beaker and after heating in the microwave oven. *Source*: Ostad Movahed et al. (2015) / with permission of Sage Publications.

The chemical curatives from this formulation were applied to devulcanized compounds for curing purposes. The correct amounts of curatives for these compounds were determined through analysis of their individual compositions. The compounds underwent mixing before researchers evaluated their curing characteristics. The rubber compounds underwent compression molding at 180 °C under 160 atmosphere pressure after measuring their cure properties. The mechanical property evaluation of revulcanized rubbers used 15 cm × 15 cm sheets which were 2.3 mm thick to obtain test samples.

The TGA results for acetone-extracted waste rubber powder appear that the EPDM material started degrading at 287.6 °C before reaching its final degradation point at 547.8 °C. The oil content in the sample caused mass loss to occur between 26.2 and 287.6 °C. The sample underwent a transition when the atmosphere changed to air at 550 °C which resulted in carbon black combustion. The unburned residue consisted of minerals and metals which remained in the sample. The rubber content in this waste rubber material reached 38.88 wt.%, which equates to 32.1 wt.% of the original waste rubber before acetone extraction. The waste powder required an effective chemical agent to perform devulcanization.

The waste powder particle size needed to be controlled because it determined the devulcanization temperature which affected the devulcanization process efficiency. The sample underwent complete degradation when hot spots formed from 1 cm diameter particles during microwave exposure. The devulcanization process worked efficiently when the waste rubber particles measured 1 mm in size. The inner-to-surface material heating process of microwave technology provides volumetric heating that requires minimal time to produce temperature changes. The devulcanization process became more efficient when the waste powder particle size decreased because less rubber volume needed to be heated by microwaves. The waste powder required longer devulcanization times because oil addition slowed down temperature rise rates. The waste powder in the reactor experienced reduced heat transfer rates because oil diluted the material and microwaves only interacted with carbon black particles that had polar properties. The waste powder temperature rose at a slow rate. The waste powder contained metal contaminants created processing challenges. The microwave exposure caused metal impurities to heat up quickly, which produced hot spots that damaged surrounding rubber material.

As stated, in this study, researchers used six chemicals and aromatic oil to devulcanize waste powder in a microwave oven at RC containing only aromatic oil without any chemicals (Table 7.7). The RC led to lower devulcanization percentages, while CLD decreased with increasing devulcanization temperature. The sol content and devulcanization percentages of compounds A-1–A-3 increased from 4.3% to 9.6% and from 90% to 93%, respectively, when the temperature rose from 200 to 260 °C. The devulcanization percentage increased while CLD decreased from 17 to 12 mol/m^3 when temperature rose according to the recorded data. The devulcanization process showed identical patterns for all compounds in Table 7.8. The maximum devulcanization temperatures in the microwave oven stayed below the boiling points of aromatic and aliphatic oils which reached 270 and 280 °C, respectively. The devulcanization process did not result in any oil evaporation because the boiling points of the oils exceeded the maximum oven temperature. The results showed that temperature increase enhanced waste powder devulcanization and made the process more efficient. The results in Table 7.8 demonstrate that devulcanization effectiveness placed in order of HDA > MBT = CBS > MBTS = DPTT > TMTD.

TMTD achieved the lowest devulcanization rate at 75–80%, while the compound containing HDA reached the highest devulcanization rate at 91–94% when the temperature increased from 200 to 260 °C. Research studies have examined the application of different amines and their derivatives for rubber devulcanization processes. The crosslinks in vulcanized rubber get broken down through nucleophilic reactions when amines are used for devulcanization. The devulcanization process of rubber becomes more efficient when researchers use amine nucleophiles with strong reactivity. The devulcanization reaction depends on three main factors which include amine structure type, basicity, steric hindrance, and presence of an α-H atom. The devulcanization reaction depends most heavily on the presence of an α-H atom. The devulcanization process of EPDM rubber shows high effectiveness when using α-H-containing aliphatic amines such as HDA. The research findings from this study match the results that have been documented in previous studies. The devulcanization process of HDA occurs through nucleophilic reactions, but the other five chemicals require radical-based devulcanization mechanisms. The devulcanization efficiency of HDA, CBS, and MBT exceeded those of MBTS, DPTT, and TMTD. The aromatic structures of CBS and MBT allowed them to dissolve better in aromatic oil, which provided improved access to cleaved rubber radicals during devulcanization. The devulcanization efficiency of MBTS was lower than CBS and MBT because its large molecular structure created steric barriers that blocked access to cleaved rubber radicals. The compounds containing TMTD showed the lowest devulcanization because TMTD consisted of only aliphatic structures.

The investigation of waste powder devulcanization through aromatic and aliphatic oil quantity variations required the preparation of nine additional compounds (E-1–E-9). The compounds contained MBTS as their devulcanizing agent and underwent devulcanization at 200, 230, and 260 °C for three minutes in the microwave oven. The devulcanization process received multiple benefits from the oil addition. The oil served two purposes during devulcanization by enhancing rubber plasticity and speeding up rubber oxidation while blocking sol prevention through radical acceptor functionality. The rubber chains became more distant from each other because the oil caused rubber swelling which resulted in better chemical agent access to crosslinks. The devulcanization process achieved better results because the oil addition made crosslinks more accessible to devulcanizing chemical agents. The devulcanization results for compounds E-1–E-3 containing aromatic oil showed a slight increase from 76% to 80% when the temperature rose from 200 to 260 °C. The devulcanization percentage reached its lowest point when using aliphatic oil but reached its highest point when using aromatic oil. The devulcanization behavior of waste powder became less affected by oil composition when the temperature reached its highest point at 260 °C. The compounds with equal oil proportions showed results that were positioned between the two sets of data. The aromatic structure of MBTS allowed it to dissolve quickly in aromatic oil. The aromatic oil entered the rubber matrix to enable the microwaves-generated rubber chain radicals to combine heat-generated radicals during devulcanization. The rubber chain radicals formed accelerator-terminated polysulfidic pendant groups through this process which became available for revulcanization. The devulcanization level increased when researchers used higher aromatic to aliphatic oil ratios in the compound.

The rubber surface appearance in Figure 7.10 shows the results of mixing the RC with oil followed by milling and then microwave oven devulcanization at 230 °C for three minutes. The compound surfaces displayed an uneven texture during this test. The rubber contained significant amounts of contaminants, and the devulcanization process achieved only a low percentage of success (Table 7.8). The rubber surfaces maintained their rough appearance with visible impurities after using chemicals to enhance waste powder devulcanization. The rubber surface appearance did not show any noticeable change from devulcanization levels.

The ODR test results presented the scorch and optimum cure times along with CRI and ΔTorque values for the revulcanized rubber compounds. The ΔTorque

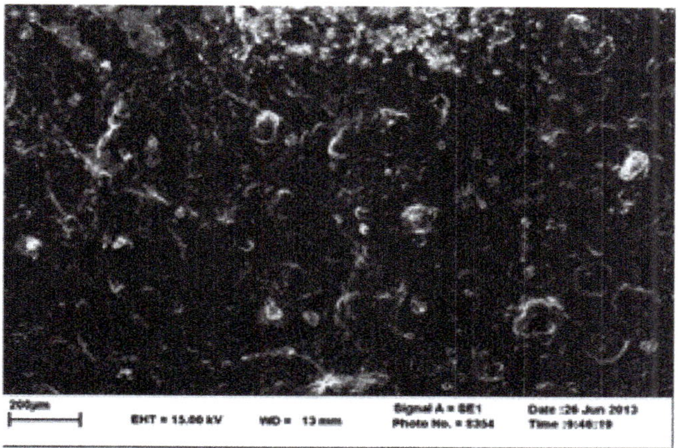

Figure 7.10 Appearance of the devulcanized rubber surfaces (RC) after two-minute processing on two-roll mills. Devulcanization temperature at 230 °C. *Source*: Ostad Movahed et al. (2015) / with permission of Sage Publications.

values of the RC and compounds A-1–A-3 and D-1–D-3 and E-1–E-3 all decreased. The compounds contained CBS, MBT, and MBTS chemicals and underwent devulcanization at temperatures of 200, 230, and 260 °C showed increasing in ΔTorque values (compounds C-1–C-3, B-1–B-3, and F-1–F-3). The ΔTorque values for compounds C-1–C-3 showed the lowest readings at 3.3–6.3 dN·m, which indicated these compounds needed the least amount of curing. The ionic nature of cleaved rubber chains prevented them from combining with accelerator radicals that formed during revulcanization but this problem minimized when the devulcanization temperature increased. The compounds E-1–E-3 showed the highest ΔTorque values which reached 12.5 to 10.1 dN·m to indicate their dense cross-link structure and excellent curing performance compared to other compounds. The devulcanization process of waste powder affected both scorch, optimum cure times, and CRI results. The entire process required brief time periods. The scorch and optimum cure times for compounds A-1–A-3 ranged from 0.10–0.11 and 0.30–0.35 min:s, respectively. The RC required the longest scorch and optimum cure times which spanned from 0.21–0.25 and 0.51–1.08 min:s. The CRI showed different values based on the specific waste powder composition and devulcanization process. The results showed that all compounds underwent curing processes.

The mechanical properties of the revulcanized rubber compounds shoed that the Shore A hardness values of compounds C-1–C-3 reached their lowest points at 49–55 Shore A. The compounds received HDA devulcanization treatment at 200, 230, and 260 °C before the revulcanization process. The RC showed higher Shore A values of 70–71 compared to the devulcanized compounds with HDA which had values between 49 and 55 Shore A. The devulcanization process with MBTS produced the hardest compound which became known as compounds E-1–E-3. The Shore A hardness of this compound reached 79. The devulcanization temperature increase from 200 to 260 °C did not increase noticeable change in the revulcanized compound hardness. The testing results showed no effect from temperature increases on all evaluated compounds. The revulcanized compounds showed comparable tensile strength values. The tensile strength measurements reached their peak at 3.7–2.7 MPa for compounds E-1–E-3, while the Reference compound showed the lowest strength at 1.5–2.3 MPa. The RC tensile strength values matched those of compounds F-1–F-3.

The elongation at break measurements showed wide fluctuations based on the waste powder composition and devulcanization treatment methods. The compounds C-1–C-3 produced the highest elongation at break values, which reached 211–186%, while compounds F-1–F-3 produced the lowest values at 34–44%. The available data for 100% elongation modulus showed values between 1.2 and 1.6 MPa for compounds C-1–C-3 and between 3.3 and 2.7 MPa for compounds D-2 and D-3. The modulus at 200% elongation was measured for compounds C-1 and C-2 at 2 and 2.2 MPa, respectively. The results showed that compounds E-1–E-3 achieved the highest CLD through their ΔTorque values which ranged from 12.5 to 10 dN·m. The compounds achieved the highest hardness rating of 79 Shore A while demonstrating the best tensile strength between 3.7 and 2.7 MPa. The compounds underwent devulcanization with MBTS chemical at three different temperatures.

It was concluded that using 100% revulcanized rubber in rubber compounds showed insufficient mechanical strength for applications that require heavy duty performance. The poor mechanical properties stem from waste powder contaminants and substandard material quality, which remains difficult to manage. Compound E demonstrates the most promising mechanical performance which makes it appropriate for producing automotive carpet and mat materials. The discovery of this new recycling method for automotive EPDM waste powder presents a potential alternative for its reuse. Also, the research demonstrated that microwave heating at 200, 230, and 260 °C combined with CBS, HDA, and MBT chemicals and aromatic and aliphatic oils enables successful devulcanization of automotive EPDM waste. The devulcanization process achieved better results when the temperature reached higher levels. The sulfur cure system enabled the devulcanized waste powder to become recyclable for automotive mat and carpet production.

7.2 Reclamation of Waste Tire Rubbers

This **case study** appeared in *Polymer Degradation and Stability* when authors published "Reclaiming waste tire rubber by an irradiation technique" in 2016.

The recycling process known as tire rubber recycling or reclaiming handles vehicle tires which become unusable because of wear or permanent damage such as punctures. The tire market will experience annual growth of 4.3% until it reached 2.9 billion units in 2017. The worldwide generation of used tires amounts to 17 million tons annually while the EU produces 3.3 million tons of used tires each year. The environmental threat from end-of-life tires requires immediate recycling solutions.

The waste rubber recycling process includes rubber reclaiming as one of its methods. The process of rubber devulcanization occurs through thermal, mechanical, and irradiation of energy applications with or without chemical devulcanizing agents to break S—S bonds in vulcanized rubber networks. The resulting product from this process becomes a partially devulcanized rubber which serves as a substitute for virgin rubber in certain applications and allows manufacturers to produce rubber products through revulcanization. The standard rubber composition for tire production includes NR for heavy vehicle tires and blend of styrene–butadiene rubber (SBR) and butadiene rubber (BR) in passenger vehicles. Multiple approaches have been documented for tire rubber recycling operations. The researchers employed different methods to devulcanize tire rubbers through their work. The researchers employed two-roll mill mechanical shear and chemical agents for their devulcanization process. They employed internal mixers with devulcanization aids for their devulcanization process. Also, they employed single- and twin-screw extruders for their devulcanization process. Other researchers employed supercritical carbon dioxide and chemical agents for their devulcanization process. The devulcanized materials showed minimal devulcanization levels which led to unsatisfactory mechanical and dynamical performance during revulcanization. The three-dimensional rubber network structure becomes vulnerable to breakdown through irradiation (microwave). The researchers examined the physical and chemical transformations which occur in tire rubber waste after multiple microwave treatment durations. They conducted multiple tests on tire rubber containing NR and SBR polymers as polymeric materials to study its properties. The microwave treatment caused sulfur crosslinks to break down, which resulted in higher rubber fluidity. The exposure duration beyond a certain point caused both degradation and changes in specific properties of the material. The research showed that temperature elevation triggers devulcanization which depends on both carbon black concentration and microwave exposure duration. The authors examined how waste tire materials absorb microwaves when heated at different temperatures. The researchers identified the specific conditions which enable microwave heating of waste tires.

Most published studies have used microwave irradiation as a standalone method to devulcanize waste rubber without adding chemical agents. The literature lacks any study which evaluates different chemical agents for microwave waste rubber devulcanization. The authors have already investigated how different operational factors and multiple devulcanizing agents affect microwave devulcanization of EPDM as stated in previous sections. This research investigates tire rubber devulcanization by adding new aspects to the investigation. The waste tire rubber from passenger cars was collected from local city areas before being processed into particles with 319 μm median size (Figure 7.11).

The waste powder composition was determined through TGA using STA 1500 equipment from Scinco Co. Ltd. The TGA analysis operated under nitrogen gas at temperatures below 550 °C and air gas at temperatures above 550 °C. The TGA results together with waste rubber waste powder composition data appear in Figure 7.12. The processing aids consisted of aromatic and paraffinic oils which started boiling at 270 and 280 °C, respectively (Mehran Tyre 290 and Fariman). The following chemicals were used as ingredients: 2-propanethiol, hexylamine, cyclohexanethiol (Sigma Aldrich), diphenyl disulfide (DPDS from Sigma-Aldrich), *N*-cyclohexyl-2-benzothiazole sulfenamide (Perkacit CBS from Flexsys), HDA (Cayman Chemical Co.), MBTS (Perkacit MBTS from Flexsys), *N*-cyclohexyl-2-benzothiazole sulfenamide (SANTACURE CBS from Flexsys), zinc oxide (activator from Harcros Durham Chemicals, UK), stearic acid (activator from Anchor Chemicals Ltd., UK), elemental sulfur (curing agent from Solvay Barium Strontium, Hannover, Germany), and *N*-*tert*-butyl benzothiazole-2-sulfenamide (TBBS; SANTACURE TBBS from Flexsys).

The first step involved using acetone to extract waste powder samples which removed oils and curing chemical residues and unlinked rubber materials through a Soxhlet apparatus with 300 mL of acetone and 100 g of rubber at 70 °C for 24 hours according to ASTM D297-93. The samples required oven drying

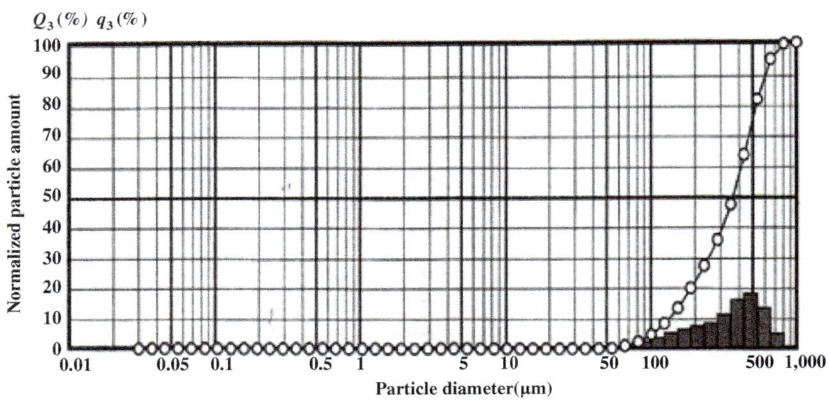

Figure 7.11 Waste tire rubber particle size distribution.

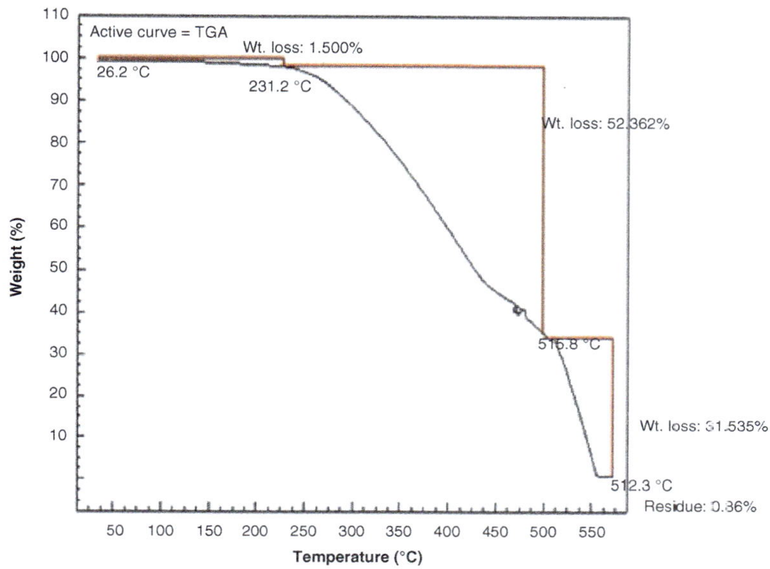

Figure 7.12 TGA test results for waste tire rubber powder.

with circulating air for 12 hours after extraction. The extracted material served as the starting material for devulcanization. The devulcanizing agents CBS, DPDS, TMTD, and HDA received appropriate amounts before being combined with 30 g of oil which was then added to 100 g of waste powder in a 500 mL glass beaker. The waste powder required 12 hours of soaking at 80 °C in an oven to achieve complete penetration of the oil into the rubber material. The process resulted in four different compounds which are listed in Table 7.9. The RC received aromatic oil treatment but lacked any chemical additives. The formulation of compounds contained substantial amounts of oil relative to other ingredients. The devulcanizing agent dissolves in oil while the rubber experiences enhanced swelling properties. The laboratory microwave oven GMO-530 from Gosonic operated at 900 W with 2,000 MHz frequency and 30 liters of internal capacity to reach different final temperatures. The microwave oven produced devulcanized rubber compounds that became large and bulky after removal from the microwave oven. The laboratory two-roll mills processed the devulcanized rubber compounds into workable sheets through a two-minute ambient temperature operation with 14 cm diameter and 32 cm length cylinders and 4 mm nip width.

The researchers studied waste tire powder devulcanization through seven additional compounds which used different DPDS amounts and aromatic and paraffinic oil combinations (compounds E–H in Table 7.10 and I–K in Table 7.11).

Table 7.9 Devulcanizing formulations for the waste tire rubber powder (effect of the devulcanizing agent, devulcanization temperature, 200 °C).

Compound	Waste rubber (phr)	HDA[a]	DPDS[a]	CBS[a]	MBTS[a]	Aromatic oil (phr)
A	100	5				30
B	100		5			30
C	100			5		30
D	100				5	30
Reference (RC)	100					30

[a] Devulcanizing agent (phr).

Table 7.10 Devulcanizing formulations for the waste tire rubber powder (effect of the devulcanizing agent amount, devulcanization temperature, 240 °C).

Compound	Waste rubber (phr)	DPDS[a]	Aromatic oil (phr)
E	100	2	30
F	100	4	30
G	100	6	30
H	100	8	30

[a] Devulcanizing agent (phr).

Table 7.11 Devulcanizing formulations for the waste tire rubber powder (effect of oil content, devulcanization temperature, 240 °C).

Compound	Waste rubber (phr)	DPDS[a]	Aromatic oil (phr)	Paraffinic oil (phr)
I	100	2	15	—
E	100	2	30	—
J	100	2	45	—
K	100	2	—	30

[a] Devulcanizing agent (phr).

The devulcanization process for all compounds in Tables 7.10 and 7.11 occurred at 240 °C in the microwave oven using the method described above. The radiation duration reached three minutes.

After microwave irradiation and devulcanization process, the first stage involved extracting polar oils, accelerator, and curing chemical residues from samples using acetone followed by toluene extraction of nonpolar components including nonpolar oils and non-crosslinked rubber residues at 110 °C for 48 hours through a Soxhlet apparatus. The extraction process used 300 mL of acetone and/or toluene to extract 100 g of rubber. The samples underwent oven drying with circulating air for 12 hours after extraction. The sol fraction calculation combined the acetone and toluene soluble fractions while subtracting the oils that were added to the samples before devulcanization. The determination of CLD required toluene as the solvent. The extraction process involved placing 5 g of extracted rubber into 300 mL of solvent within labeled bottles which then swelled for 16 hours at 21 °C. The weight measurements of the samples occurred daily until the samples reached their equilibrium point. The samples underwent nine hours of air drying followed by 24 hours of oven drying at 85 °C before they received an additional 24 hours of storage at 23 °C before their final weight measurement. The Flory–Rehner equation as stated earlier enabled researchers to determine the CLD after completing the measurement process. The interaction parameter in the mentioned equation (χ) was used as 0.405.

The following calculations determined the sol content and devulcanization percentage:

$$\text{Sol content} = \text{Sol fraction} \times 100 = \frac{w_\text{i} - w_0}{w_\text{i}} \times 100$$

where w_i is the initial weight of the sample before double extractions, and w_0 is the weight of the dried sample;

$$\text{Devulcanization } (\%) = 100 \times (\text{Initial waste rubber CLD} \\ -\text{Final rubber CLD})/\text{Initial waste rubber CLD}$$

The initial waste rubber CLD value amounts to 165 mol/m^3. The results from CLD measurements did not require Krause correction because the study focused on comparing different compound CLD values. The devulcanization temperature and devulcanization percentage and sol content and CLD values of devulcanized waste rubbers appear in Table 7.12.

The Horikx analysis served to evaluate how sol content affects CLD reduction when a network degrades through devulcanization. The relative decrease in CLD occurs through main chain scission during devulcanization according to:

$$1 - \frac{\vartheta_\text{f}}{\vartheta_\text{i}} = 1 - \left[\frac{(1 - s_\text{f}^{0.5})^2}{(1 - s_\text{i}^{0.5})^2}\right]$$

Table 7.12 Devulcanization temperature, sol content, CLD, and devulcanization % of the rubbers.

Compound	Devulcanization chemical	Final temperature (°C)	Sol content (%)	CLD (mol/m^3)	Devulcanization (%)
RC	—	200	7.5	165	0
A	HDA	200	12	71	57
B	DPDS	200	16.5	85	48
C	CBS	200	12	96	42
D	MBTS	200	10	93	43
B-1	DPDS	180	10	44	73
B-2	DPDS	220	14	71	57
B-3	DPDS	240	15	81	51
B-4	DPDS	260	12	78	53
B-5	DPDS	280	18	79	52
E	DPDS	240	12	107	35
F	DPDS	240	23	87	47
G	DPDS	240	14	89	46
H	DPDS	240	18	114	31
I	DPDS	240	11	107	35
J	DPDS	240	10	99	40
K	DPDS	240	12	160	3

and when pure crosslink scission takes place during devulcanization, the relative decrease in CLD is given by:

$$1-\frac{\vartheta_{\mathrm{f}}}{\vartheta_{\mathrm{i}}}=1-\left[\frac{\gamma_{\mathrm{f}}\left(1-s_{\mathrm{f}}^{0.5}\right)^{2}}{\gamma_{\mathrm{i}}\left(1-s_{\mathrm{i}}^{0.5}\right)^{2}}\right]$$

where s_{i}, ϑ_{i} and s_{f}, ϑ_{f} are sol fractions and CLDs before and after devulcanization, respectively. In these experiments, s_{i} and ϑ_{i} are 0.075 and 165 mol/m^3, respectively. γ_{i} and γ_{f} are the average numbers of crosslinks per chain in the remaining insoluble network before and after devulcanization, respectively. The relationship between γ and s is as follows:

$$s=\frac{\left[(2+\gamma)-(\gamma^{2}+4\gamma)^{0.5}\right]}{2\gamma}$$

The value of γ_i was calculated by replacing s in above equation by the value of 0.075 (initial sol content).

By fixing the values of γ_i and s_i in two former equations and with cooperation of the latter equation, Horikx curves can be drawn.

The measurement of rubber compound crosslinks between mono-, di-, and polysulfidic structures used 2-propanethiol and hexylamine and cyclohexanethiol according to published methods (Figure 7.13). The researchers used

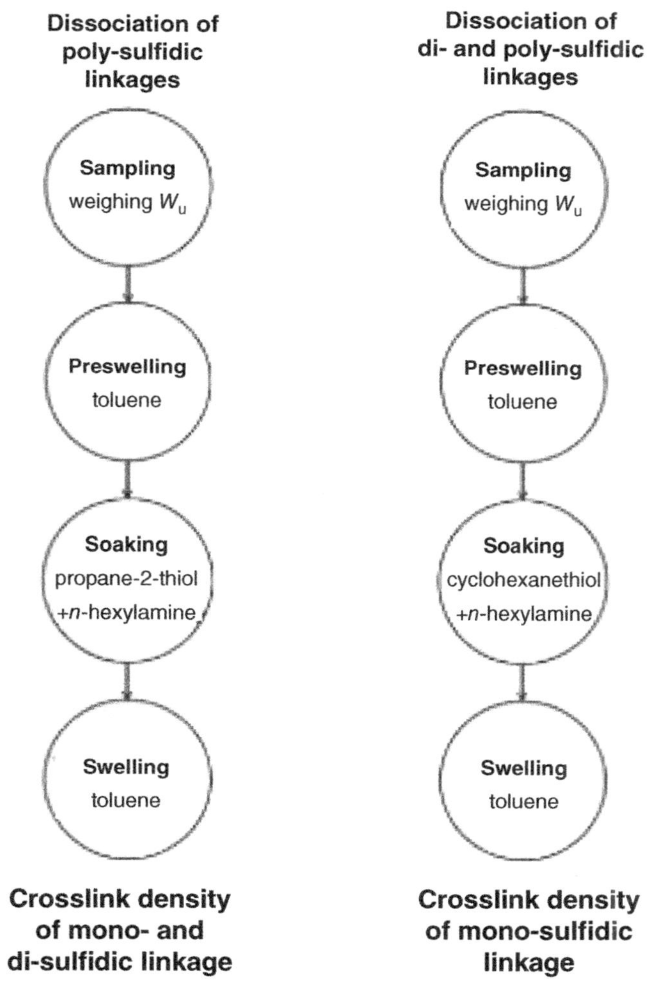

Figure 7.13 The experimental process for sulfur crosslink type and density measurement has been established. The research used cyclohexanethiol instead of hexanethiol to enhance the efficiency of S—S di-bond cleavage.

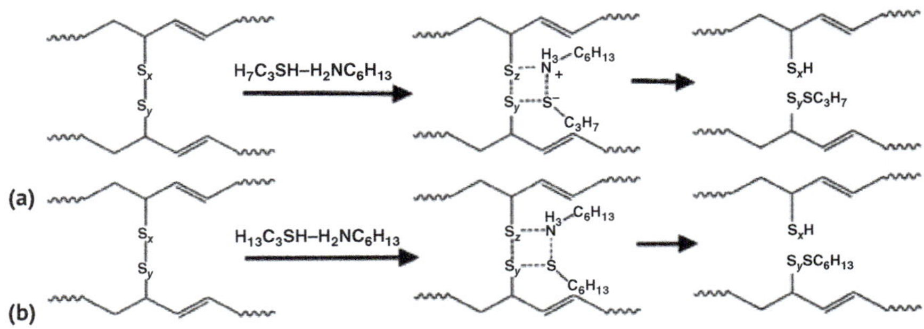

Figure 7.14 Mechanism of sulfur crosslinks breakage by (a) 2-propanethiol + *n*-hexylamine and (b) cyclohexanethiol + *n*-hexylamine. *Source*: Mollanorouzi and Ostad Movahed (2016).

Table 7.13 Mono-, di-, and polysulfide fractions in sulfur bridges for selected rubber compounds.

Compound no.	Total CLD (mol/m^3)	Monosulfide (%)	Disulfide (%)	Polysulfide (%)
Initial waste rubber powder	165	60	18	22
I	107	80	18	2
B-1	44	88	12	—

cyclohexanethiol instead of hexanethiol to enhance the efficiency of S—S bond cleavage in this study. The breakdown process of sulfur crosslinks through 2-propanethiol and *n*-hexylamine and cyclohexanethiol and *n*-hexylamine reactions is shown in Figure 7.14a and b, respectively. Table 7.13 shows the percentage of mono-, di-, and polysulfide bonds in sulfur bridges for selected rubber compounds.

The waste powder particle size determines devulcanization temperature as researchers discovered this relationship in previous sections. The sample develops hot spots when microwave irradiation occurs with large particles measuring between 0.5 and 1 cm in diameter. This causes the degradation of one side of the sample while the other side remains at room temperature. The image in Figure 7.11 displays waste rubber powder with 319 μm median particle size. The waste rubber particle size distribution shows that 10% of particles are smaller than 133 μm, while 50% are below 367 μm and 90% are less than 614 μm. The used waste rubber powder underwent efficient devulcanization according to the results presented in Table 7.12.

The TGA curve (Figure 7.12) indicates that rubber degradation started at 231 °C and ended at 516 °C. The initial rubber waste contained 1.8% oil which caused mass loss between 33 and 231 °C. The mass loss took place between 516 and 573 °C which related to the carbon black content of rubber sample due to combustion when the atmosphere switched to air during this process. The remaining unburned material consisted of minerals and metals which were present in the sample. The proposed tire rubber devulcanization process through radical and ionic mechanisms appears in Figures 7.15 and 7.16. The radical mechanism of DPDS, CBS, and MBTS requires microwave radiation to break polymer chains into two radical segments through sulfur bridges. The radicals that form through microwave radiation react with reagent radicals to create stable compounds. The devulcanization process of HDA occurs through ionic reactions. The polymer chains split into two ions through their sulfur crosslinks when HDA interacts with microwave radiation.

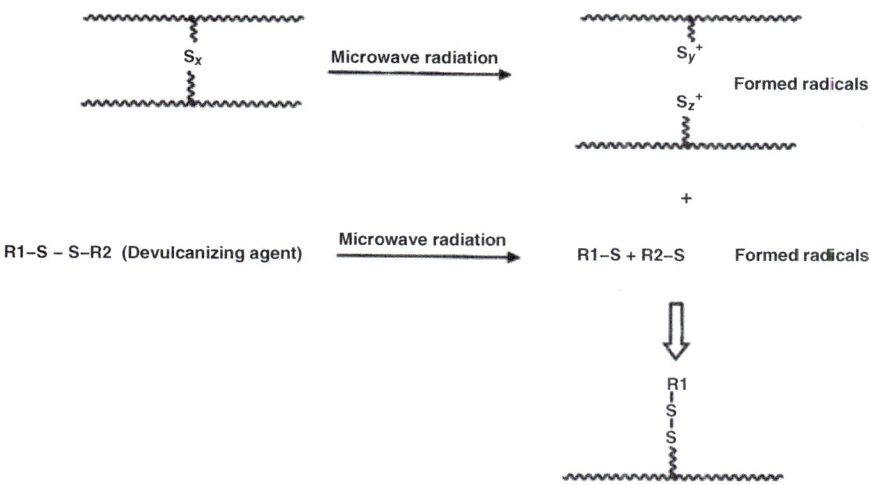

Figure 7.15 Proposed radical devulcanization mechanism of tire rubber.

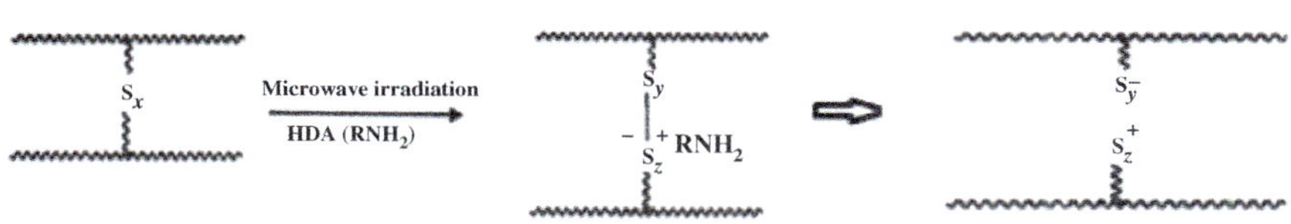

Figure 7.16 Proposed ionic devulcanization mechanism of tire rubber by HDA.

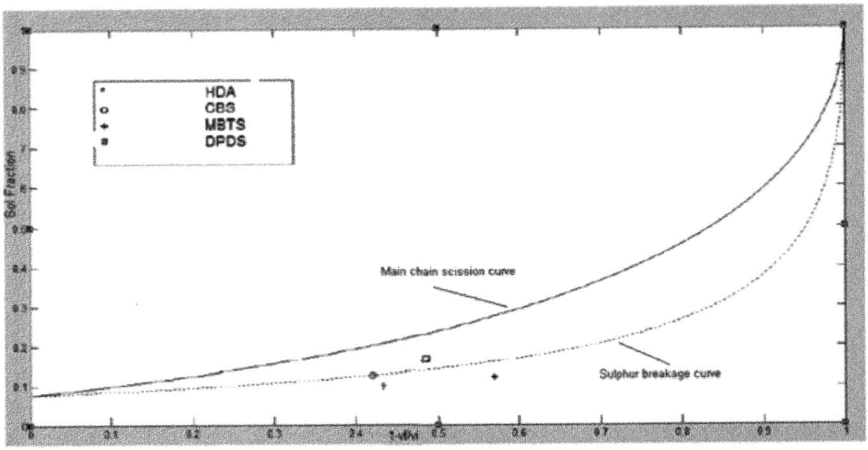

Figure 7.17 The sol fraction versus relative reduction in CLD $\left(1 - \frac{\vartheta_f}{\vartheta_i}\right)$ for compounds A–D and RC (final devulcanization temperature, 200 °C).

The sol fraction versus relative CLD reduction for irradiated compounds made with different devulcanizing agents HDA, DPDS, CBS, MBTS, and RC (compounds A–D and RC without devulcanizing agent, Table 7.9) at 200 °C appears in Figure 7.17. The CLD of RC maintained the same as the original waste rubber, so it did not appear in the figure.

As observed, the compound containing DPDS at 16.5% sol fraction and 85 mol/m^3 CLD achieved 48% devulcanization (Table 7.12, compound B). The experimental results from this compound matched exactly with the sulfur crosslink scission curve described in Figure 7.17. The devulcanization process would produce data points that match the dotted curve when sulfur crosslinks break down, but the points would match the solid curve when the main chain experiences complete scission. The area between these two curves indicates that devulcanization causes both sulfur crosslinking and main chain scission to happen. The data points that approach the dotted curve indicate higher sulfur crosslink breakdown rates than main chain scission which makes them more desirable.

The sol fraction and CLD values for RC, HDA, CBS, and MBTS (compounds RC, A, C, and D, Table 7.12) were 7.5% and 165 mol/m^3, 12% and 71 mol/m^3, 12% and 96.1 mol/m^3, and 10% and 93 mol/m^3, respectively. The CLD measurement for compound RC showed identical results to the original waste rubber powder before devulcanization (165 mol/m^3). The microwave treatment without devulcanizing agent failed to break down sulfur crosslinks in the material. The data points for compounds A and D fall below the sulfur crosslink scission curve

(Figure 7.17). The devulcanization percentage of compound A exceeded the values from compounds B, C, and D, but DPDS proved to be the better devulcanizing agent because of the above-mentioned reason. Another graph demonstrated how the sol fraction changes when CLD decreases for DPDS devulcanized compounds (compounds B and B-1–B-5, Table 7.12) at different final microwave temperatures from 180 to 280 °C. The devulcanization percentages of the compounds showed a downward trend as the temperature increased.

The reason may refer to a set of fresh main chain reactions which scientists call "main chain unsaturation volatile formation crosslinking" and or the development of new mono- and polysulfide sulfur bridges as shown in Table 7.13. Figure 7.18 demonstrates the radicals which form after microwave exposure followed by chain and sulfur crosslinks expansion.

The devulcanization percentage values reached their peak at 180 and 200 °C for compounds B-1 and B with 73% and 48%, respectively. The devulcanization percentages showed minimal variation between 51% and 53% when temperatures exceeded 240 °C. The three compounds B-1, B-2, and B-4 showed completely under the sulfur crosslink breakage curve at 180, 220, and 260 °C temperatures which makes them unacceptable. The points from compounds B, B-3, and B-5 at 200, 240, and 280 °C follow the sulfur crosslink breakage curve and/or exist between two curves which indicates sulfur breakage occurs independently or together with main chain scission during irradiation. The compound B-3 point shows minimal distance from the sulfur crosslink breakage curve so 240 °C should be used as the best end devulcanization temperature. The selected mechanical properties of the

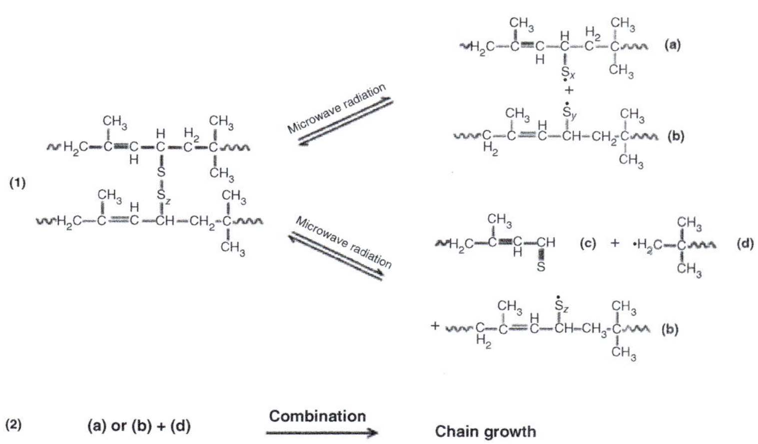

Figure 7.18 Polymer sulfur crosslink and chain growth mechanism. *Source:* Mollanorouzi and Ostad Movahed (2016).

revulcanized compounds versus a RC showed that most of the mechanical properties of the revulcanized rubber were inferior when compared with those of the reference rubber. As an illustration, the tensile strength, elongation at the break, and tear strength were reduced from 19 to 10 MPa, 540% to 226%, and 24 to 18 N/mm for RC and revulcanized B-3, respectively.

Another graph showed that the relationship between sol fraction and CLD reduction for compounds E–H with different DPDS amounts at 2, 4, 6, and 8 phr. The compounds contained 30 phr aromatic oil in their formulations before undergoing devulcanization at 240 °C. The devulcanization process did not improve with increasing DPDS amounts. The devulcanization percentage reached its lowest point at 31% for compound H with 8 phr DPDS, while compound F with 4 phr DPDS achieved the highest devulcanization at 47%. The data points from compounds E and G with 2 and 6 phr DPDS, respectively, match the sulfur crosslink breakage curve, but compound G shows higher devulcanization at 46% which indicates 6 phr DPDS provides sufficient devulcanization at 240 °C.

The aromatic oil content values for compounds I, E, and J were 15, 30, and 45 phr, respectively (Table 7.11). The production of compound K required 30 phr paraffinic oil. The formulations of all compounds contained 2 phr DPDS as their main ingredient. The devulcanization process received positive effects from aromatic oil content increases between 15 and 30 phr, but no changes occurred in devulcanization percentage between compounds I and E (Table 7.12). The data points for compound E followed the sulfur crosslink breakage curve when aromatic oil content reached 30 phr. The devulcanization process suffered from the 45 phr aromatic oil addition because the data point moved below the sulfur crosslink breakage curve while the devulcanization percentage rose to 40% from 35%. The devulcanization process suffered from paraffinic oil usage because compound E with 30 phr aromatic oil showed better results than compound K with the same amount of paraffinic oil (Table 7.12). The reason stems from DPDS having lower solubility in paraffinic oil compared to aromatic oil. The DPDS radicals that form cannot easily access the broken macromolecules.

Table 7.13 shows mono-, di-, and polysulfide percentages in sulfur bridges for three rubber compounds including I, B-1, and waste rubber powder. The sulfur bridges in the first waste rubber sample contain 22% polysulfide, 60% monosulfide, and 18% disulfide bonds. The waste rubber underwent curing through an efficient or semiefficient cure system with minimal elemental sulfur content. The microwave treatment of compounds I and B-1 resulted in polysulfide bond breakdown which decreased their CLD values while monosulfide crosslinks stayed intact. The waste rubber CLD value reached 165 mol/m^3 but microwave treatment decreased it to 107 mol/m^3 for compound I and 44 mol/m^3 for compound B-1.

The remaining sulfur crosslinks in the materials consisted mainly of monosulfide bonds.

The internal structures of waste rubber and compounds I and B-1 and revulcanized compound B-3 were examined by SEM after freeze-fracture treatment. The material displays complete heterogeneity through its coarse texture. The internal structure of compound I becomes visible because it contains 35% devulcanization. The non-devulcanized rubber shows a rough texture which devulcanized rubber covers. The microstructure of compound B-1 contains 73% devulcanized rubber as its main component and shows non-devulcanized rubber with fine structure inside a devulcanized rubber matrix. The border between matrix and vulcanized rubber in this compound appears to be a smooth surface. The high amount of oil in this compound formulation at 30 phr weight percentage might explain this observation. Also, the surface of revulcanized compound B-3 was examined by SEM. The main component of this material consists of devulcanized rubber. The border between devulcanized and revulcanized rubbers appears more refined but shows signs of rubber revulcanization.

The research findings lead to the following conclusions:

- Microwave irradiation successfully devulcanizes waste tire rubber. The successful devulcanization of waste rubber depends heavily on the distribution of particle sizes in the material. The devulcanization process required the waste rubber to reach a median particle size of 319 μm.
- The results demonstrated that waste rubber powder underwent successful devulcanization. DPDS proved to be the most effective devulcanizing agent among all the agents tested. The optimal devulcanization conditions occurred when the compound contained 30 phr aromatic oil and 6 phr DPDS at 240 °C.
- The microwave irradiation process resulted in a complete or partial breakdown of polysulfide crosslinks while maintaining mostly monosulfide structures in the remaining sulfur crosslinks.
- The most devulcanized compounds showed coarse morphology patterns where non-devulcanized rubber surfaces were covered by devulcanized rubber layers. The results showed that oil content levels had a significant impact on the process.

7.3 Reclamation of General Cured Butyl Rubber

This **case study** appeared in the *Journal of Applied Polymer Science* when authors published "Butyl rubber reclamation by combined microwave radiation and chemical reagents" in 2016.

The copolymer of isobutylene and 1–3% isoprene which forms butyl rubber undergoes cationic solution polymerization at −100 °C to produce this rubber. The development of butyl rubber occurred in 1937 when William J. Sparks and Robert M. Thomas worked at Standard Oil of New Jersey's Linden N.J. laboratory. The presence of isoprene in the polymer structure enables vulcanization through its unsaturated bonds. The main properties of butyl rubber stem from its fully saturated structure which blocks gas molecule penetration. The recycling process for butyl rubber becomes essential after this rubber material reaches its service life endpoint. So far, research studies about butyl rubber recycling and devulcanization processes remain limited.

The authors in previous sections investigated how different operating conditions with multiple devulcanizing chemicals affect microwave devulcanization of EPDM rubber. This research investigates additional aspects of general cured (sulfur or peroxide) butyl rubber microwave devulcanization (de-crosslinking in case of peroxide curing) while examining the same parameters from previous studies.

The waste tubes from bicycle and motorcycle butyl rubber were collected from local city areas before being reduced into particles with 279 μm median size. The waste powder composition was determined through TGA using STA 1500 equipment from Scinco Co. Ltd. The TGA test operated under nitrogen gas at temperatures below 550 °C and air gas at temperatures above 550 °C. The TGA results along with their corresponding chemical composition appear in Figure 7.19.

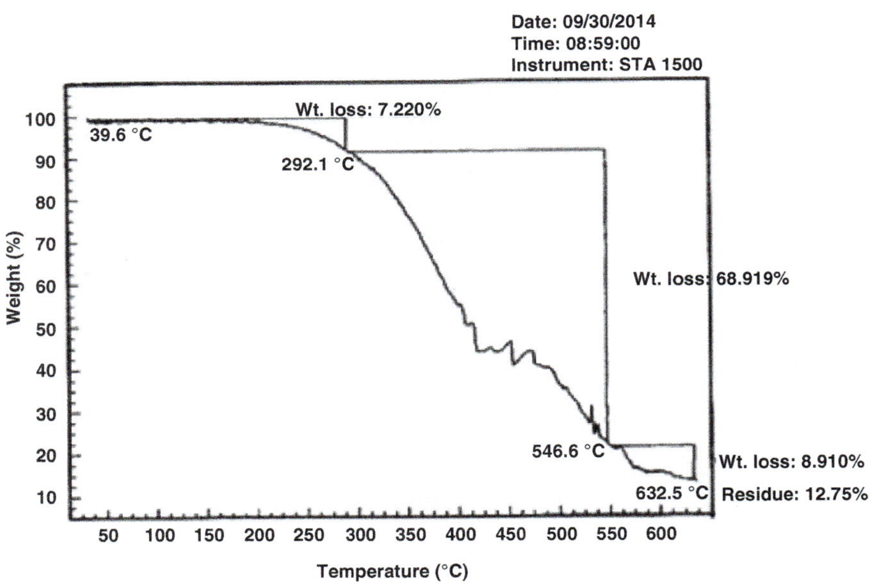

Figure 7.19 TGA test results for waste butyl rubber powder.

The following materials were used for the experiment: DPDS (Sigma-Aldrich), *N*-cyclohexyl-2-benzothiazole sulfenamide (SANTACURE CBS from Flexsys), HDA (Cayman Chemical Co.), TMTD (Perkacit TMTD from Flexsys), zinc oxide (activator from Harcros Durham Chemicals, UK), stearic acid (activator from Anchor Chemicals Ltd., UK), elemental sulfur (curing reagent from Solvay Barium Strontium, Hannover, Germany), and aromatic and paraffinic oils with 270 and 280 °C boiling points for processing aid purposes (Mehran Tyre 290 and Fariman). The researchers used 2-propanethiol and hexylamine and cyclohexanethiol from Sigma Aldrich to measure mono-, di-, and polysulfidic crosslinking.

The Soxhlet apparatus extracted waste powder samples with acetone to eliminate oils, un-crosslinked rubbers, and curing chemical residues through a 24-hour extraction process at 70 °C using 300 mL of acetone and 100 g of rubber according to ASTM D297-93. The extraction process required the samples to undergo oven drying with circulating air for 12 hours. The devulcanizing process required 4 g each of CBS and DPDS and TMTD and HDA which were combined with 30 g of oil before addition to 100 g of waste powder in a 500 mL glass beaker. The waste powder required 12 hours of soaking at 80 °C in an oven to achieve complete rubber penetration by the oil. The process resulted in four compounds which are listed in Table 7.14. The RC contained aromatic oil but no chemical additives. The beaker containing the mixture underwent microwave treatment inside the GMO-530 Gosonic oven which operated at 2,000 MHz and 900 W power output and had a 30 L internal volume. The devulcanized rubber compounds became large in size after they emerged from the microwave oven. The laboratory two-roll mill with 14 cm diameter and 32 cm length cylinders processed the devulcanized rubber compounds for two minutes to create workable sheets.

The research examined how various amounts of HDA devulcanizing reagent and aromatic and paraffinic oils affect waste rubber devulcanization through the preparation of six additional compounds (compounds E–J, Table 7.14). The devulcanization process for all compounds in Table 7.14 occurred at 180 °C except for compounds M and N which used 140 °C for devulcanization. The procedure of parameter measurement explained in the previous section.

The research by authors demonstrated that waste powder particle dimensions determine the required devulcanization temperature. As stated, the sample developed hot spots during microwave exposure when the diameter reached 0.5 cm. The research team encountered identical problems when working with butyl waste rubber powder that contained particles larger than 279 μm. The size distribution of waste rubber powder appears in Figure 7.20. The powder contained particles with diameters that measured 279 μm at their median point. The microwave radiation process revealed efficient devulcanization through changes in particle form and state. The swelling test results validated this observation by showing devulcanization percentages of devulcanizates. The diameter distribution of waste rubber

Table 7.14 Devulcanizing formulation for the waste rubber powders (devulcanization temperature, 180 °C).

Compound	Waste rubber (phr)[a]	HDA[a]	DPDS[a]	CBS[a]	TMTD[a]	Aromatic oil (phr)	Paraffinic oil (phr)
Reference (RC)	100					30	
A	100	4				30	
B	100		4			30	
C	100			4		30	
D	100				4	30	
E	100	2				30	
F	100	6				30	
G	100	8				30	
H	100	6					15
I	100	6					30
J	100	6					45
M[b]	100	6				30	
N[b]	100	8				30	

[a]Devulcanizing reagent (phr).
[b]Devulcanization temperature, 140 °C.

particles shows that 10% of particles are smaller than 130 μm, while 50% are below 279 μm and 90% are less than 500 μm.

The TGA results (Figure 7.19) indicate that IIR waste rubber started degrading at 292 °C and reached its completion point at 549 °C. The sample showed mass loss between 40 and 292 °C because it contained oil. The sample experienced two different mass loss events during heating from 549 to 633 °C. The carbon black combustion occurred when the atmosphere switched to air at 550 °C. The remaining unburned materials consisted of minerals and metals present in the sample. The devulcanization process of butyl rubber through DPDS and CBS and TMTD and HDA occurs through radical and ionic mechanisms as shown in previous sections (Figures 7.15 and 7.16). The first step of DPDS and CBS and TMTD devulcanization involves microwave radiation to split polymer chains into two radicals through sulfur bridges. The formed radicals react with reagent radicals which also form through microwave radiation to create stable compounds. The devulcanization process of HDA occurs through ionic reactions. The polymer chains split into two ions through their sulfur crosslinks when HDA interacts with

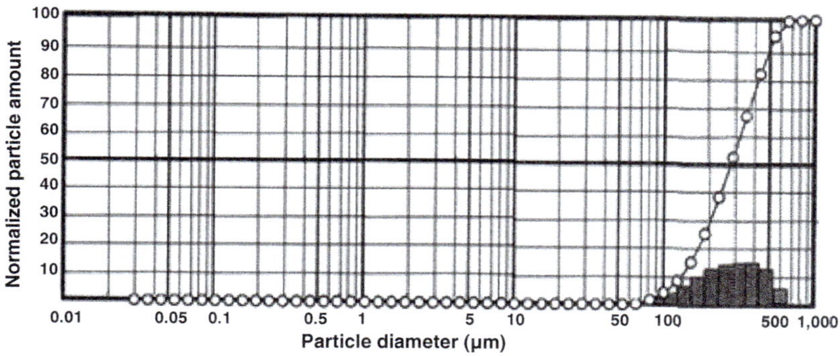

Figure 7.20 Butyl waste rubber particle size distribution.

them under microwave radiation. The sol fractions versus relative reduction in CLDs $\left(1-\frac{\vartheta_f}{\vartheta_i}\right)$ for irradiated compounds prepared with different devulcanizing reagents, HDA, DPDS, CBS, TMTD, and RC (compounds A–D and RC, without devulcanizing reagent, Table 7.14) at 180 °C are depicted in Figure 7.21 (CLDs of B and D were more than CLDs of the initial rubber waste and consequently, their data were not indicated in Figure 7.21).

The observation shows that HDA with 14% sol fraction and 25 mol/m^3 CLD achieves the highest devulcanization percentage at 64% while showing the closest match to the sulfur crosslink scission curve (dotted curve in Figure 7.21). The dotted curve in Figure 7.21 shows the expected sulfur crosslink breakdown pattern during devulcanization, while the solid curve represents complete main chain scission as described in earlier section. Lying data between these two curves means a mixing of sulfur crosslink and main chain scission occurred during devulcanization. The closer data points to the dotted curve show more sulfur crosslinks breakdown when compared with main chain scission. The sol fraction and CLD values for RC, CBS, DPDS, and TMTD (compounds RC, C, B, and D) were 19% and 40 mol/m^3, 17% and 35 mol/m^3, 25% and 139 mol/m^3, and 11% and 151 mol/m^3, respectively. Interestingly, for compounds B and D, the CLDs were more than CLD of the initial waste rubber powder before devulcanization (69 mol/m^3). Here, for these two compounds, with increasing sol fractions, CLDs were also increased.

Generally, increasing the rubber sol fraction is accompanied with CLD reduction, and it indicates the extent to which the rubber main chains and crosslinks are broken. The analysis shows that compounds B and D formed new crosslinks after radiation exposure. The formation of new crosslinks in compounds B and D after radiation exposure became unexpected. The main chain unsaturation volatile formation crosslinking reaction together with mono- and polysulfide sulfur

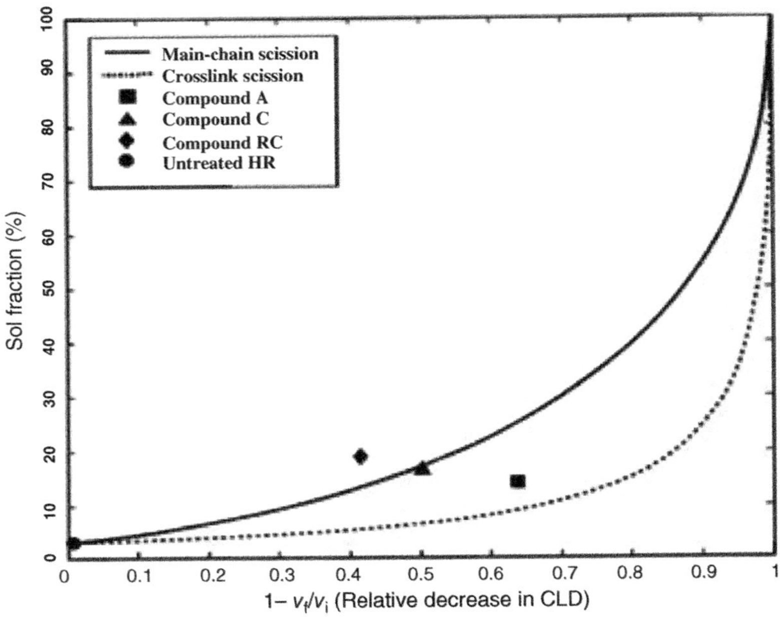

Figure 7.21 The relative decrease in CLD versus the sol fraction of compounds A–D and RC (devulcanization temperature, 180 °C).

bridge formation created these new crosslinks. Here, DPDS and TMTD function as vulcanizing agents instead of devulcanizing agents in this specific case.

The devulcanization process of compounds RC and C in Figure 7.21 resulted in an undesirable main chain breakdown. The compounds follow the main chain scission curve (solid curve) at an exact position. The devulcanization percentage of compound A reaches 64%, which makes it the highest among compounds C and RC. The results indicate that HDA functions as the best devulcanizing agent for this research. Another graph demonstrates how sol fractions relate to CLD reduction percentages for HDA-treated compounds A and A-1–A-6 at temperatures from 140 to 260 °C. The devulcanization percentage of compounds reached its maximum at 220 °C when it reached 75%. For 240 °C, the devulcanization percentage reduced to 72%, but for 260 °C, the situation completely changed, and in the same compounds B and D, sudden increase in CLD was observed (203 mol/m^3). The data for three compounds, A, A-1, and A-2 are completely close to the sulfur crosslink breakage curve (dotted curve). However, at 140 °C, the point lies a little under the curve. The highest devulcanization percentage among the compounds belongs to compound A which makes it the best selection for optimum compound.

Another graph shows the relationship between sol fraction and CLD reduction percentage for compounds A, E, F, G, and RC with HDA amounts of 4, 2, 6, and 8 phr and without HDA. The results demonstrate that higher HDA content leads to better results in devulcanization operations. The results from compounds E and RC exceed the main chain scission curve and produce unacceptable outcomes. The compound F with 6 phr HDA showed the closest match to sulfur crosslink breakage curve compared to the other compounds A and G which used 4 and 8 phr of HDA, respectively.

The research examined how different oil concentrations and types affect devulcanization operations. The paraffinic oil content in compounds H–J reached 15, 30, and 45 phr, respectively. The devulcanization percentage remained unchanged when paraffinic oil content increased from 15 to 30 phr according to compound H and I result. The data points from the sulfur crosslink breakage curve became more aligned with the devulcanization process when paraffinic oil content reached 30 phr (dotted curve). The devulcanization process suffered negative effects when paraffinic oil content exceeded a certain point. The process of devulcanization reagent penetration into macromolecules becomes challenging when excess oil is present. The devulcanization process shows negative effects from aromatic oil when compounds F and I receive equal amounts (30 phr) of aromatic and paraffinic oils The reason may refer to the lower solubility of HDA in aromatic oil when compared with paraffinic oil.

Figure 7.22 compares the devulcanization parameters of three compounds A, M, and N with HDA values of 4, 6, and 8 phr, at 140 °C, respectively. The study shows that HDA levels between 4–6 and 8 phr result in better devulcanization outcomes because they increase devulcanization percentages to 73% and 71% for compounds M and N. The devulcanization results for compound M with 6 phr HDA fall between the two curves, which makes it more suitable than compound N with 8 phr HDA.

Table 7.15 shows mono-, di-, and polysulfide percentages in sulfur bridges for rubber compounds H, N, B, D, and A-6 (devulcanization temperature, 260 °C) and waste rubber powder. The waste rubber structure contains 28% polysulfides and 72% disulfides without any monosulfide present. The waste rubber underwent conventional curing with excessive elemental sulfur based on its structure. The compounds B, D, and A-6 show disulfide bond breakdown after microwave treatment which produces new mono- and polysulfide bonds. The resulting CLD values became higher than the initial waste rubber CLD values. The waste rubber CLD value started at 69 mol/m^3 but increased to 139 mol/m^3 for compound B and 151 mol/m^3 for compound D and 203 mol/m^3 for compound A-6. The devulcanization process for compounds B and D used 4 phr DPDS and TMTD, respectively, at 180 °C. The initial disulfide bond breakdown in these compounds led to the

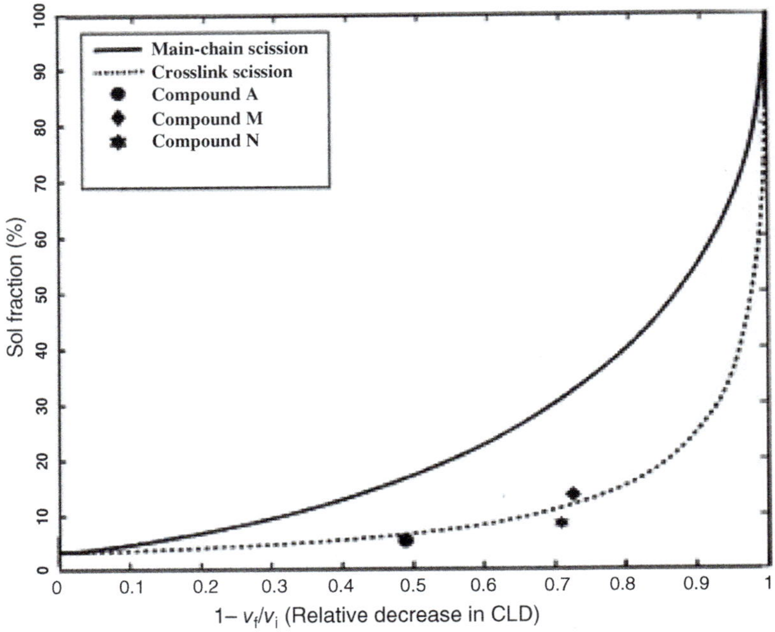

Figure 7.22 The relative decrease in CLD versus the sol fraction of compounds A, M, and N (devulcanization temperature, 140 °C).

Table 7.15 Mono-, di-, and polysulfide fractions in sulfur bridges for selected rubber compounds.

Compound no.	Total CLD (mol/m^3)	Monosulfide (%)	Disulfide (%)	Polysulfide (%)
Initial waste rubber powder	69	0	72	28
H	12	19	54	27
N	20	5	56	39
B	139	19	5	76
D	152	16	49	35
A-6	203	20	17	63

formation of new mono- and polysulfide crosslinks through released sulfur atoms. The study shows that DPDS outperformed TMTD in disulfide crosslink destruction during compound B processing. The devulcanization process suffered from DPDS and TMTD reagents which functioned as vulcanization agents. Compound A-6 stands alone because its behavior differs from the other compounds A and

Figure 7.23 SEM micrograph showing the interior of the waste rubber after freeze-fracture.

A-1–A-5. The breakdown of sulfur crosslinks occurred at temperatures between 240 °C and no additional crosslinks formed during this period. The process of devulcanization transformed into revulcanization when the temperature reached 260 °C because new sulfur and non-sulfur crosslinks formed. The CLD values of compounds H and N remained lower than the waste rubber values. The devulcanization process occurred for these compounds. The sulfur crosslink composition of these compounds transformed into mostly monosulfide structures with 5% and 19% di- and polysulfide contents for compounds N and H, respectively.

The freeze-fracture image in Figure 7.23 demonstrates how waste rubber fillers distribute throughout the material while showing its internal structure. The micrograph demonstrates poor carbon black distribution throughout the waste rubber material with particles measuring under 10 μm in size. The comparison between SEM microstructures of studied devulcanized rubbers shows more or less no distinguishable variations between each other. The compound E internal structure becomes visible because it contains the lowest devulcanization percentage at 22%. The micrograph shows devulcanized rubbers as distinct areas which exist within a vulcanized rubber matrix. The devulcanized rubbers form a smooth background that contains bulky vulcanized rubber agglomerates. The filler particles exist throughout both devulcanized and vulcanized areas of the material. The compound H microstructure shows devulcanized rubber as the main component with vulcanized rubber and filler present in small amounts. The high amount of devulcanized rubber in this compound makes it difficult to identify the boundaries between vulcanized and devulcanized rubber zones. The compound contains 15 phr of oil which results in a rough and nonsmooth border between the vulcanized rubber matrix and other components. The high oil content in compound J at

45 phr makes it challenging to identify devulcanized and vulcanized rubber zones. The border between these two zones became smooth because of the oil present in the material.

The research findings demonstrate the following results:

- Waste sulfur cured butyl rubber undergoes effective devulcanization through the combination of microwave radiation with chemical reagents. The successful devulcanization process depends heavily on the specific distribution of waste rubber particle sizes. The research findings indicate that waste rubber powder with 279 μm median particle size works best.
- HDA proved to be the most effective devulcanizing reagent among all the tested options. The compounds achieved their best devulcanization results when they contained paraffinic oils and 6 phr HDA at 180 °C during devulcanization.
- Microwave radiation caused the breakdown of disulfide crosslinks in most compounds which produced new mono- and polysulfide bridges from released sulfur atoms. The devulcanized compounds showed higher CLD values than their original waste rubber counterparts.
- The micrograph images of the most devulcanized compounds showed vulcanized rubber centers surrounded by devulcanized rubber matrices. The oil content levels produced noticeable effects in the study.

7.4 Reclamation of Phenolic Cured Butyl Rubber

This **first case study** in this section appeared in *Progress in Rubber, Plastics and Recycling Technology* when authors published "The phenolic cured butyl rubber reclamation using different de-crosslinking agents" in 2024.

Traditional cured rubbers include IIR which holds significant value for industrial applications and everyday rubber products. The material provides excellent protection against chemical substances, water, steam, and gas permeation. The phenolic-cured butyl rubber exhibits superior heat resistance when exposed to high temperatures. The butyl rubber's exceptional properties made it the perfect material for creating inner liners for tubeless tires. The synthetic rubber, known as butyl rubber, received its name after Thomas and Sparks developed it through synthesis in 1937. The production of butyl rubber through cold cationic copolymerization uses isoprene (*cis*-2-methyl-1,3-butadiene) and isobutylene (2-methylpropene) as its monomers. The polymerization process requires a Lewis catalyst to function in an organic solvent solution. The copolymer structure contains between 1% and 3% isoprene by weight. The rubber requires isoprene to create unsaturation sites which enable crosslinking reactions. The rubber exists in an unusable state before undergoing curing treatment. The three primary rubber crosslinking methods include sulfur vulcanization, phenolic resin curing, and peroxide crosslinking. The final properties of rubber articles depend on the specific curing method used during their production. The phenolic and peroxide curing methods produce rubbers with superior heat resistance because they create durable C—C bonds between rubber molecules during the crosslinking process.

The phenolic cured IIR-made bladder demonstrates two main characteristics which include thermal stability and minimal gas permeability. The rubber article serves as a curing material for tire production. The metal mold contains the bladder which receives the green tire assembly consisting of multiple uncured rubber parts including tread, carcass, and bead. The high-pressure steam curing process transforms the green tire assembly into its final form. The phenolic rubber curing process accepts various methods and materials including allyl phenol–formaldehyde Novolac, bisoxazoline–phenolics, polybenzoxazines, phenol–epoxy systems, phenolic resins with phenyl maleimide functions, pendant phenol functional linear polymers, propargyl ether functional phenolics, and phenolic resins with phenyl ethynyl groups. The traditional phenolic (phenol–formaldehyde) resins appear in Figure 7.24 for the rubber-made bladder curing process. The Novolac and Resole structures contain methylene groups that link phenolic units through di-methylene ether linkages.

Figure 7.25 demonstrates the established Chroman mechanism which is used for curing. The three-dimensional network formed through phenolic resin and rubber macromolecule bridging according to the described mechanism. The cationic reaction mechanism represents an alternative pathway which scientists

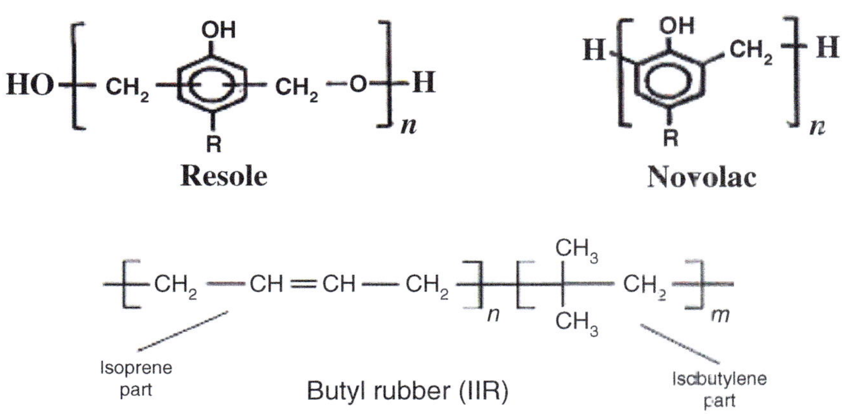

Figure 7.24 IIR and traditional phenolic (phenol–formaldehyde) resins, Resole, and Novolac. *Source*: Miranian et al. (2024).

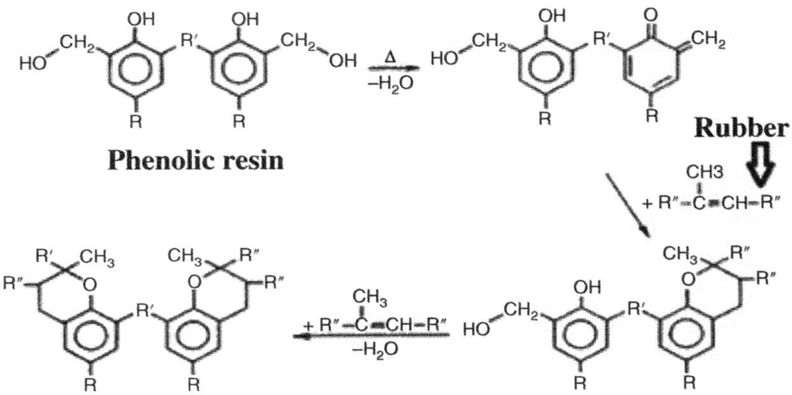

Figure 7.25 Chroman mechanism of rubber curing by phenolic resin. *Source*: Ostad Movahed et al. (2024).

have identified. The reaction sequence follows cationic pathways where benzylic cations act as reactive intermediates that attack the most accessible olefinic C-atom. The phenolic curing process shows distinct differences from traditional rubber curing methods which include sulfur and peroxide crosslinking. The phenolic resin attacks the unsaturated double bond in butyl rubber (isoprene part) during the curing process. The cured rubber contains fewer double bonds than the original uncured rubber.

The waste stream receives all polymeric articles after their life service ends. The three essential reasons for recycling this article include environmental protection, conservation of raw materials, and financial gain. The main approaches for polymeric waste recycling include reusing, mechanical processing, chemical treatment, and energy conversion which scientists classify as primary, secondary, tertiary, and quaternary, respectively. The chemical recycling process transforms polymeric waste into its original monomers or valuable light hydrocarbons. The resulting materials serve as feedstock for petrochemical operations and polymer production. The chemical recycling process in this study focuses on breaking the phenolic crosslinks which connect butyl rubber molecules. The de-crosslinking process of butyl rubber produces reclaimable material which can be used to create new rubber products either alone or blended with virgin rubber. The de-crosslinking process needs to break only crosslinks to achieve high-quality reclaiming rubber. The de-crosslinking process leads to main rubber chain scission, which decreases the quality of reclaimed rubber. Researchers have developed multiple methods to de-crosslink cured rubber as stated earlier in this text, which include microwave, ultrasonic irradiation, mechanical shear forcing, and thermal and chemical treatments. The chemical technique for de-crosslinking uses a specific chemical agent known as a de-crosslinking agent. The techniques for de-crosslinking rubber possess both beneficial aspects and detrimental characteristics.

The author has identified only a few published studies which focus on phenolic-cured butyl rubber recycling. The already research implemented restricted de-crosslinking methods with limited de-crosslinking agents, and no thermo-chemical approaches were tested. This research employed thermal and chemical methods to de-crosslink waste bladder butyl rubber powder. The de-crosslinking process occurred inside a heating oven which served as the thermal source while different de-crosslinking agents (chemicals) were applied at specific temperatures. The Yazd rubber industrial complex located in Yazd, Iran, provided the waste rubber powder. The phenolic-cured butyl rubber material served as the main component for producing waste rubber-made tire curing bladders. The laser diffraction particle size analyzer (Mastersizer 3000, Malvern Panalytical, UK) measured the powder size distribution (Figure 7.26).

The TGA instrument STA 1500 from Scinco Co. Ltd. (South Korea) performed chemical composition analysis of the rubber powder. The TGA test operated under nitrogen gas at temperatures below 400 °C. The atmosphere shifted to air when the temperature reached 400 °C to determine the carbon black content in the rubber powder. The results from the TGA test appear in Figure 7.27. The local market provided aromatic oil (Fariman Co., Iran) as a processing aid with a boiling point of 270 °C. The de-crosslinking agents used in this study included TMTD (Perkacit TMTD, Flexsys, Belgium), DPDS (Sigma-Aldrich, Germany), benzothiazyl-2-dicyclohexyl sulfenamide (DCBS, AcroChem, United States), TBBS (Western Reserve

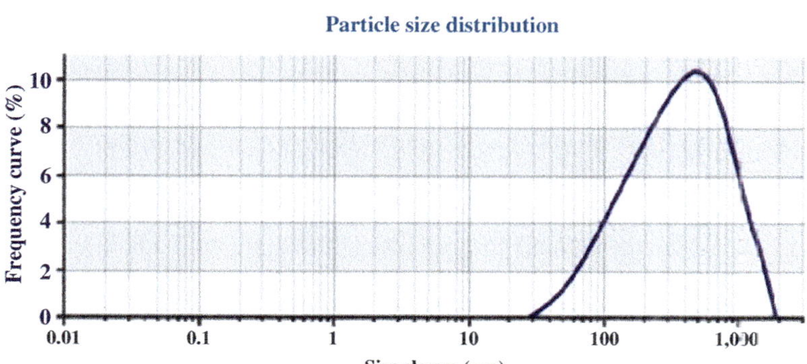

Figure 7.26 Cured rubber-made waste bladder powder size distribution. The particles were dispersed in the water before size distribution measurement.

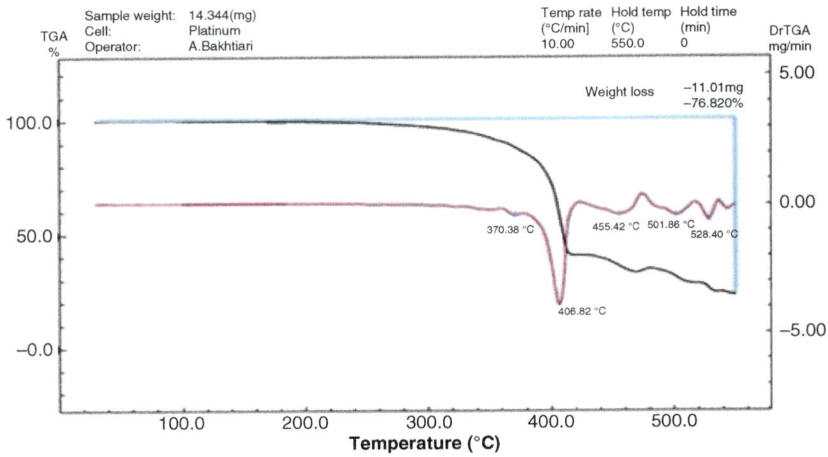

Figure 7.27 The TGA curves for the cured rubber made bladder waste powder.

Chemicals, United States), MBTS (NOCIL Limited, India), and MBT (NOCIL Limited, India).

The de-crosslinking agent required heating at 40 °C for 15 minutes to achieve uniform distribution throughout the aromatic oil. The treated aromatic oil soaked the waste rubber powder in a 1 L glass container for 72 hours (about 6 days) at 80 °C in an oven. The oil penetration reached its highest point through this method. The beaker containing the rubber compound underwent de-crosslinking treatment in a laboratory heating oven at 180, 200, 220, 240, and 250 °C for five minutes. The TEHRAN sterilize from Tehran Iran operated as a 30 L capacity heating oven. The temperature control system of the oven enabled users to set the internal oven temperature. The two-roll mill processed the treated compounds into 3-mm thick rubber sheets through sheeting at 45 °C for one minute. The used waste butyl rubber served as the waste rubber-made bladder. The tire underwent a curing process inside the mold using this material. The de-crosslinking process shows strong dependence on rubber powder particle size distribution and average particle dimensions. The rubber surface develops hot spots when oven heating occurs for particles that exceed 0.5 cm in diameter. The described events create temperature variations throughout the sample material which leads to uneven rubber de-crosslinking. The waste rubber required grinding to create a powder with 371 μm median size (Figure 7.26). The waste rubber particles consisted of 25%, 43%, 64%, and 86% volumes which had diameters smaller than 183, 293, 469, and 750 μm. The smallest particle measured 28 μm, while the largest reached 1,518 μm in diameter.

The waste butyl rubber powder composition appears in Figure 7.27. The light hydrocarbons and oil components in the rubber sample underwent degradation between 160 and 340 °C resulting in 6% mass reduction. The rubber started to decompose at 340 °C until the atmosphere switched from nitrogen to air at 455 °C. The rubber sample lost 54% of its mass during the process. The carbon black filler underwent 17% mass loss during the temperature range from 455 to 545 °C. The remaining mass at 545 °C consisted of unburned residues which included minerals that made up 23% of the total mass.

The Fourier transform infrared spectroscopy (FTIR) graphs analysis of waste bladder butyl rubber powder identified phenolic resin through its spectral peaks (Figure 7.28). The peaks at 3,412, 1,539, 1,621, 1,180, and 715 1/cm indicate the presence of phenolic hydroxyl (OH), C=C in aromatic ring, the original IR band of the aromatic ring, C—O—C (ether crosslink), and the C—H bending in the aromatic ring of the phenolic resin, respectively. However, the intensities of the latter two peaks were significantly lower than the former peaks.

The rubber industry uses de-crosslinking agents as standard curing accelerators. The agents function as de-crosslinking agents when temperatures exceed 160–180 °C. The agents function as de-crosslinking agents when temperatures rise above conventional levels to speed up reverse crosslinking reactions. The waste rubber powder underwent a transformation from its initial solid state to a bulky form when it passed through the heating oven. The observation of the rubber powder transformation into a bulky state proved that the rubber powder underwent de-crosslinking. The SEM images in Figure 7.29 and the de-crosslinking percentage values in Table 7.16 supported this observation.

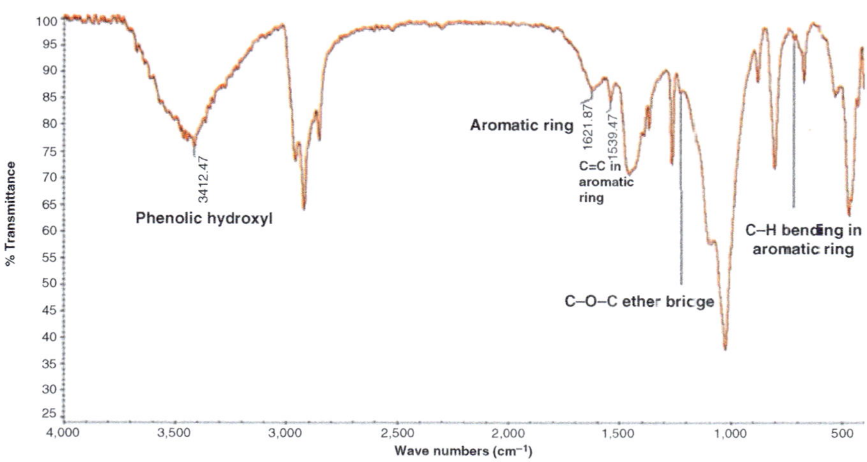

Figure 7.28 The FTIR graph for the used waste butyl rubber powder.

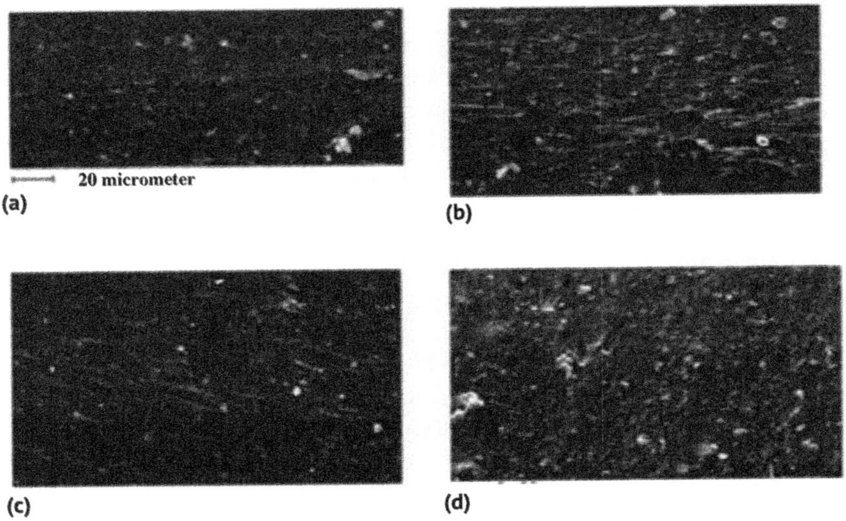

Figure 7.29 The SEM micrographs showing the interior of the (a) waste butyl rubber powder, (b) RC, (c) compound E with the lowest de-crosslinking %, and (d) compound G-2 with the highest de-crosslinking %. *Source*: Miranian and Ostad Movahed (2024) / with permission of Sage Publications.

The internal structure of waste rubber powder appears in Figure 7.29a. The rubber matrix contains poorly distributed carbon black filler particles which measure 7 μm in diameter. The filler particles formed agglomerates when the sample underwent oven heating without any de-crosslinking agent (Figure 7.29b) resulting in filler size growth to 12 μm. The rubber matrix surface transitioned from its original smooth appearance to a coarser texture after heating. The de-crosslinked rubbers exposed most filler particles through their surface just as the original waste rubber powder did. The SEM image of compound E (Figure 7.29c) showed a similar condition to the RC because it had the lowest de-crosslinking percentage at 56% (Table 7.16). The situation for compound G-2 with its maximum de-crosslinking percentage of 73% showed complete differences from the other treated compounds (Figure 7.29d). The rubber-de-crosslinked material shows distinct borders with the filler particles. The rubber macromolecules which already bonded together became more mobile. The macromolecules are separated from the filler surface.

The sol fractions versus relative reduction in CLDs or $\left(1-\frac{\vartheta_f}{\vartheta_i}\right)$ for oven-heated rubber compounds at 180 °C formulated with different de-crosslinking agents, DPDS, MBT, TMTD, MBTS, TBBS, DCBS, and RC without de-crosslinking agent (compounds G, A, C, B, D, E, and RC), are shown in Figure 7.30. The compounds

Table 7.16 De-crosslinking parameters, sol fraction, CLD, and de-crosslinking % of the heated compounds in heating oven.

Compound	De-crosslinking agent	De-crosslinking temperature (°C)	Sol fraction (%)	CLD (mol/m^3)	De-crosslinking (%)
RC	—	180	22	102	53
A	MBT	180	17	102	53
B	MBTS	180	16	95	66
C	TMTD	180	17	104	62
D	TBBS	180	19	103	62
E	DCBS	180	17	123	56
G	DPDS	180	18	84	70
H	DPDS	180	16	87	68
I	DPDS	180	19	105	62
J	DPDS	180	22	115	59
K	DPDS	180	16	112	60
G-1	DPDS	200	22	82	71
G-2	DPDS	220	22	77	73
G-3	DPDS	240	21	78	72
G-4	DPDS	250	19	81	71

required four phr of each de-crosslinking agent and 30 phr of aromatic oil as processing aid for their preparation.

The Horikx analysis shows that the solid and dotted curves in Figure 7.30 demonstrate complete main macromolecule chain scission and crosslink scission, respectively. The points between these curves show that de-crosslinking causes both crosslink degradation and macromolecule chain breakdown. The process of de-crosslinking produces two effects where the first one is favorable, but the second one is undesirable. The distance between points and the dotted curve determines the efficiency of de-crosslinking operations. The de-crosslinking process in heating ovens causes both crosslink degradation and macromolecule chain breakdown for all tested de-crosslinking agents, according to Figure 7.30. The studied compounds show no substantial variation in their behavior.

The de-crosslinking percentage of compounds G (DPDS) and E (DCBS) reached 70% and 56%, respectively, which made them the top and bottom performers among all studied de-crosslinking agents. The de-crosslinking percentage

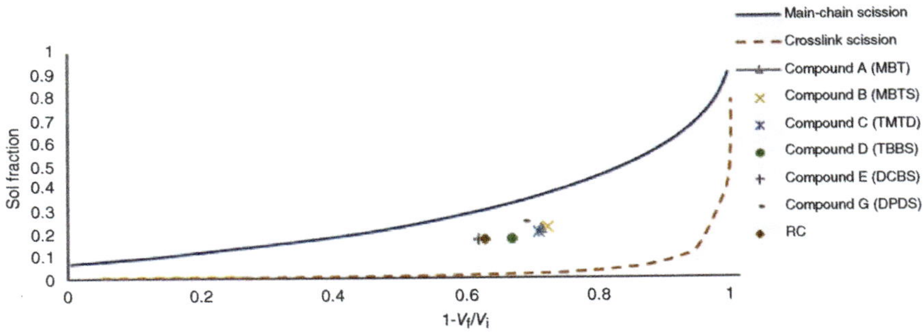

Figure 7.30 The relative decrease in CLD versus the sol fraction of compounds A–G and RC. De-crosslinking took place inside of heating oven at 180 °C.

of compound E proved to be lower than the RC at 63%. The delayed action of sulfenamide agent DCBS operated as a de-crosslinking process retardant. The de-crosslinking percentage of left compounds spanned from 62% for compound D with TBBS to 66% for compound B with MBTS. The relationship between CLD reduction and rubber sol fraction (%) increase depends on the ratio of main chain macromolecules to crosslinks that break down. The RC showed 102 mol/m^3 CLD and 22% sol fraction. The values decreased to 84 mol/m^3 and 18% for compound G according to Table 7.16. The results indicate that DPDS functions as an effective de-crosslinking agent for this kind of waste rubber treatment.

The CLD reduction pattern compared to sol fraction content of compounds with different DPDS amounts from 2 to 10 phr (compounds H, G, I, J, and K, Table 7.16) were also evaluated. The analysis showed that all compounds with varying DPDS levels experienced both crosslinking and main chain breakdown. The highest degree of crosslink scission occurred in compound H, while compound J showed the lowest degree of crosslink scission. The de-crosslinking percentage decreased from 68% to 70% in compounds H and G with 2 and 4 phr DPDS to 59–60% in compounds J and K with 8 and 10 phr DPDS in their formulations. The addition of more DPDS to compound formulations did not lead to any improvement in de-crosslinking performance.

Table 7.16 shows the de-crosslinking parameters for selected compounds (DPDS) with heated in oven at selected temperatures 180, 200, 220, 240, and 250 °C (compounds G and G-1–G-4). As observed, the oven temperature increase did not produce any significant improvement in de-crosslinking percentage results. The de-crosslinking percentage rose from 70% to 73% when the temperature range shifted from 180 to 220 °C. The parameter showed a minimal response when the oven temperature reached 220 °C. The crosslinking reactions began again when temperatures exceeded this point. The de-crosslinking agent forms an active complex through its reaction with available zinc oxide and fatty acid components in waste rubber formulations. The active complex from the next stage attacks rubber crosslinks to break phenolic bonds which results in two separate de-crosslinked rubber molecules. The de-crosslinking percentage of compounds I, J, and K decreased below the RC level when they contained 5, 8, and 10 phr DPDS in their formulations. The additional de-crosslinking agent (DPDS) present in the mixture functioned as an accelerator which caused the active complex to form new sulfur crosslinks with unreacted rubber in the waste rubber. The oven temperature increase resulted in a deteriorating effect on sol fraction values except for compound G-4 which showed a 250 °C temperature. The oven temperature increase led to a rise in sol fraction values from 18% for compound G to 22% and 21% for compounds G-1–G-3. The Horikx analysis results showed no significant variation between the tested compounds. The oven temperature increase produced equivalent effects on both crosslinking and macromolecular main chain scission.

From this study, it was concluded that the FTIR graphs served as evidence to confirm phenolic resin presence in used waste bladder butyl rubber powder structure. All studied de-crosslinking agents showed effective de-crosslinking results. The SEM images together with CLD measurements from the reclaimed rubbers confirmed the de-crosslinking process. The de-crosslinking percentage reached its peak when using 4 phr DPDS at 220 °C, and its lowest value occurred with 4 phr DCBS at 180 °C. The de-crosslinking parameters showed DPDS to be an effective de-crosslinking agent for the used waste bladder butyl rubber. The oven heating process from Horikx analysis demonstrated that it simultaneously destroyed crosslinks and primary rubber chains. The de-crosslinking process received relatively positive effects from increased oven heating temperatures, which caused equal damage to crosslinks and main chain structures.

This **second case study** in this section appeared in *Advance in Environmental Waste Management & Recycling (AEWMR)* when authors published "The microwave assisted reclamation of the phenolic cured butyl rubber" in 2023.

As stated, the cured rubber made bladder serves as a green tire curing tool because it maintains thermal stability while preventing gas from passing through. The uncured rubber compounds (green tire) received their different parts before being placed inside a hollow mold for heat and pressure-based curing. The green tire receives internal shaping and curing through a bladder that contains hot steam. The bladder consists of cured butyl rubber as its base rubber material which undergoes phenolic curing to achieve its final form. The lifespan of bladder products depends on the type of cured rubber material used for their manufacturing process. The recycling process of rubber materials into useful products serves two main purposes by saving resources and protecting the environment. The four main rubber recycling strategies include primary or reusing, secondary or mechanical, tertiary or chemical, and quaternary or energy recovery as described. The preferred rubber recycling method uses primary recycling because it produces uncured reclaim rubber which can be recured to create new rubber products.

The reusing strategy in rubber recycling involves devulcanization for sulfur curing and de-crosslinking for all other rubber curing methods including phenolic curing. The de-crosslinking process breaks down three-dimensional rubber networks which form during crosslinking (curing) reactions. The de-crosslinking process works to remove all crosslinks while maintaining the rubber chain structure.

The researchers proposed multiple methods to de-crosslink cured rubbers which included chemical approaches, mechanical methods, biological solutions, thermo-chemical treatments, and irradiation processes. The microwave irradiation technique stands as the most suitable method because microwaves can specifically disrupt chemical bonds within the cured rubber network. This research investigates how microwave irradiation affects waste bladder butyl rubber powder when de-crosslinking agents are added at different operational settings. The Yazd rubber industrial complex in Yazd, Iran, provided the cured rubber waste bladder for evaluation. The TGA results appear in Figure 7.27 and Table 7.17. The TGA operated under nitrogen gas at temperatures below 410 °C and used air as the gas at temperatures above 410 °C.

The de-crosslinking agents used in this study included DPDS (Sigma-Aldrich in Germany), TMTD (Perkacit TMTD from Flexsys in Belgium), DCBS (AcroChem in the United States), DPTT (Western Reserve Chemicals in the United States), TBBS (Western Reserve Chemicals in the United States), and MBT (NOCIL limited in India). The crosslinking agent was dissolved in aromatic oil through

Table 7.17 Physical properties and composition of the used cured rubber made waste bladder powder.

Property		Amount
Sol content (%)		7
Density (g/cm^3)		1.10
CLD[a] (mol/m^3)		281
TGA	Unburned minerals (%)	23
	Carbon black (%)	17
	Rubber (%)	54
	Light materials including oil (%)	6

[a]CLD was calculated using the Flory–Rehner equation.
Source: Dust and Ostad Movahed (2023) / OPAST GROUP LLC / CC BY 4.0.

the application of heat. Then, it introduced the waste rubber powder into a 1 L glass beaker and permitted the rubber powder to absorb the treated oil in an oven for 24 hours at 80 °C. This process ensures optimal oil penetration into the rubber powder. In the subsequent phase, to de-crosslink the rubber compound, the beaker is placed in a laboratory microwave oven and exposed to the chosen temperatures with maximum output power. The microwave oven utilized (GMO-530, Gosonic, Turkey) possessed an output power of 900 W, a frequency of 2,000 MHz, and a capacity of 30 L. It is also designed with a temperature regulator to manage the internal temperature of the microwave oven. The irradiated compounds were then processed on a two-roll mill with cylinder diameters and lengths of 14 and 32 cm, respectively, to create rubber sheets for additional testing. The hot spots in rubber samples become visible when particles grow to sizes exceeding 0.5 cm during microwave exposure. The waste rubber particles showed volume percentages of 25, 43, 64, and 86 for diameters smaller than 183, 293, 469, and 750 μm. The mentioned problem can be prevented through waste rubber powder production with 371 μm median size. The observed phenomena create temperature variations throughout the sample which leads to uneven de-crosslinking of the rubber material. The smallest particle measured 28 μm, while the largest reached 1,518 μm.

The rubber sample loses 6% of its mass when it experiences temperatures between room temperature and 340 °C because its oil content and light hydrocarbons start to decompose. The carbon black filler lost 17% of its mass between 455 and 545 °C. The rubber sample lost 54% of its mass during the process. The rubber started degrading at 340 °C until it reached 455 °C before the atmosphere switched from nitrogen to air. The remaining mass after this temperature point consisted of unburned residues containing minerals which accounted for 23% of the total mass loss.

The FTIR spectra show five distinct peaks which appear at 3,412, 1,539, 1,621, 1,180, and 715 1/cm that correspond to phenolic hydroxyl (OH), C=C in aromatic ring, the original IR band of the aromatic ring, C—O—C (ether crosslink), and C—H bending in the aromatic ring of the phenolic resin. The rubber powder transformed from solid particles into a compact adhesive state when exposed to microwave radiation. The visible transformation of rubber powder into an adhesive state proved that the rubber powder underwent de-crosslinking. The observed de-crosslinking percentages in treated compounds matched the results shown in SEM images.

The internal structure of waste rubber powder appears in Figure 7.31a. The rubber matrix contained poorly distributed carbon black filler particles which measured 7 μm in diameter. The filler particles formed agglomerates when subjected to microwave treatment without any de-crosslinking agent (Figure 7.31b and Table 7.18) which resulted in an average diameter of 15 μm. The rubber

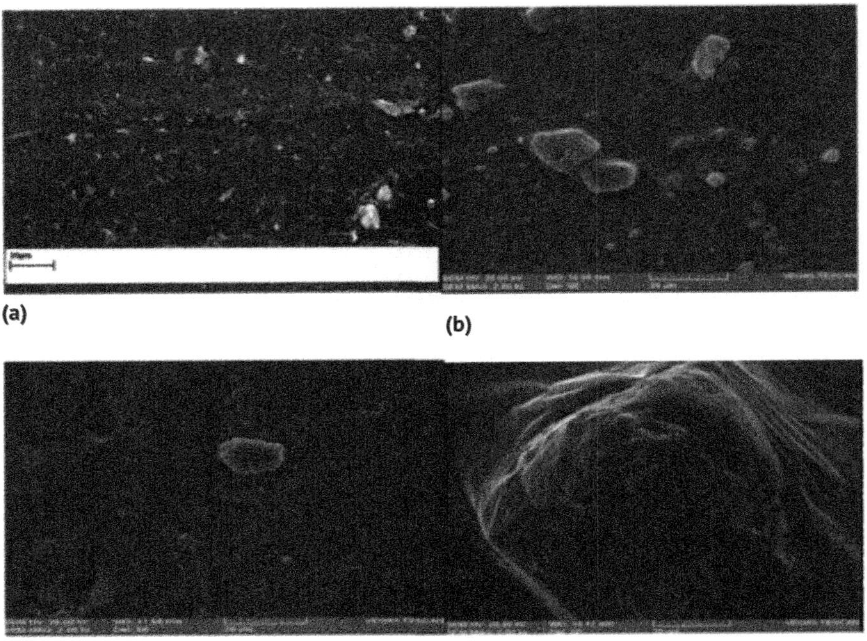

(a) (b) (c) (d)

Figure 7.31 The SEM micrographs showing the interior of the (a) waste butyl rubber powder, (b) RC, (c) compound E with the lowest de-crosslinking %, and (d) compound J-1 with the highest de-crosslinking %. *Source*: Dust and Ostad Movahed (2023) / OPAST GROUP LLC / CC BY 4.0.

matrix structure began in an unsmooth condition before it evolved into a more pliable form. The SEM micrograph (Figure 7.31c) of compound E shows a similar pattern to its low de-crosslinking percentage (Table 7.18). The de-crosslink rubber produced during processing covered most filler particles better than the crosslinked rubber found in the original waste rubber powder. The compound J-1 (Figure 7.31d) produced an entirely different result because its de-crosslinking percentage reached 83.2%, which became the highest value among all treated compounds (Table 7.18). The border between the newly formed de-crosslinked rubber and the filler particles disappeared from view. The rubber material formed a complete encasement around all filler particles. Note, compounds A–F have 4 phr de-crosslinking agents and 30 phr aromatic oil in their formulations. Also, compounds G–K had DPDS as de-crosslinking agent with the values of 2, 4, 6, 8, and 10 phr for compounds H, G, I, J, and K, respectively.

Horikx analysis results indicated that during the de-crosslinking process, crosslinks and the main chain scission are responsible for all examined de-crosslinking agents.

Table 7.18 De-crosslinking parameters, sol fraction, CLD, and de-crosslinking % of the microwave irradiated compounds.

Compound	De-crosslinking agent	De-crosslinking temperature (°C)	Sol fraction (%)	CLD (mol/m^3)	De-crosslinking (%)
RC	—	180	18	108.1	62
A	MBT	180	22	83.4	70.3
B	MBTS	180	20.6	80.1	71.5
C	TMTD	180	20.8	84.1	70.1
D	TBBS	180	17.4	95.9	65.8
E	DCBS	180	17.6	110.7	60.5
F	DPTT	180	24.6	88.7	68.4
G	DPDS	180	28	61.4	78.1
H	DPDS	180	20.4	85.1	69.7
I	DPDS	180	27.6	60.4	78.5
J	DPDS	180	31.6	57.8	79.4
K	DPDS	180	28	56.4	79.9
J-1	DPDS	200	24.6	47.2	83.2
J-2	DPDS	220	25.2	49.4	82.4
J-3	DPDS	240	24.4	50	82.2
J-4	DPDS	250	24.4	58.9	79

Source: Dust and Ostad Movahed (2023) / OPAST GROUP LLC / CC BY 4.0.

The highest de-crosslinking percentage of 78.1% was recorded for compound G containing DPDS in its composition. Table 7.18 indicates that the percentage of de-crosslinking for the RC was 62%. The de-crosslinking percentages of the left compounds varied from 60.5% (compound E with DCBS) to 71.5% (compound B with MBTS). It was determined that the DPDS was the most appropriate de-crosslinking agent among the agents examined based on Horikx analysis.

Generally, the reduction of CLD is associated with a rise in rubber sol content (%). The reason pertains to the degree to which the primary rubber chains and cross-links are fractured. For example, the sol fraction and CLD for the RC were measured at 108.1 mol/m^3 and 18%, respectively. They decreased and increased to the values of 61.4 mol/m^3 and 28% for compound G (Table 7.18). The comparative reduction in CLD against the sol fraction of compounds containing varying amounts of DPDS, specifically 2, 4, 6, 8, and 10 phr (compounds H, G, I, J, and K, Table 7.18), is illustrated in Figure 7.32. The de-crosslinking occurred in the

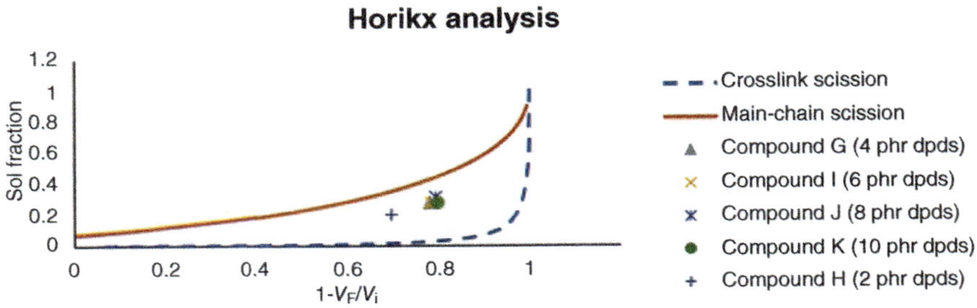

Figure 7.32 The relative decrease in CLD versus the sol fraction of the compounds with different amounts of DPDS. De-crosslinking took place inside of microwave oven with 100% output power at 180 °C. *Source*: Dust and Ostad Movahed (2023) / OPAST GROUP LLC / CC BY 4.0.

microwave oven at an output power of 100% and a temperature of 180 °C. It was noted that, with the exception of compound H, the points for the left compounds roughly overlap. Additionally, both crosslinking and main chain scission were noted for all compounds with varying amounts of DPDS in the formulation. The corresponding de-crosslinking percentages were 69.7%, 78.1%, 78.5%, 79.4%, and 79.9% for compounds H, G, I, J, and K, respectively (Table 7.18). The rise of DPDS had minimal impact on the de-crosslinking % beyond the DPDS levels of 4 phr. The sol fraction of the compounds in question also held true. Considering the potential commercial application of the research results, compound J was chosen for additional experiments. Table 7.18 presents the de-crosslinking parameters for compound J subjected to microwave irradiation at various temperatures of 180, 200, 220, 240, and 250 °C (compounds J, J-1, J-2, J-3, and J-4). It was noted that raising the irradiation temperature positively influenced the de-crosslinking %, except for the irradiated compound at 250 °C (compound J-4). It appeared that beyond this temperature, the crosslinking reactions slowly reoccur.

The de-crosslinking process reached its peak at 200 °C when compound J-1 was exposed to result in 83.2% de-crosslinking. The sol fraction decreased when temperature increased because it started at 31.6% for compound J but reached 24.6–24.4% for compounds J-1–J-4. The Horikx analysis produced similar results for all tested compounds. The study determined that temperature increases during microwave irradiation produced equivalent damage to crosslinks and main chain structures.

From this study, it was concluded that the FTIR graphs showed sufficient evidence that phenolic resin exists in the waste bladder butyl rubber powder structure. The de-crosslinking parameters showed DPDS as the suitable de-crosslinking agent for the waste bladder butyl rubber material. The compounds containing DPDS and DCBS showed the highest and lowest de-crosslinking rates, respectively.

The SEM images together with CLD measurements of reclaimed rubbers proved this. All tested de-crosslinking agents showed successful de-crosslinking results. The Horikx analysis showed that microwave irradiation broke down both crosslink bonds and rubber backbone chain structures at the same time.

References

Dust, H. A. V. and Ostad Movahed, S. Microwave-assisted reclamation of phenolic-cured butyl rubber. *Adv. Environ. Waste Manag. Recycl.* 2023, *6*(2), 416–427.

Khavarnia, M. and Ostad Movahed, S. Butyl rubber reclamation by combined microwave radiation and chemical reagents. *J. Appl. Polym. Sci.* 2016, *133*, 43363.

Miranian, Z., Ostad Movahed, S., and Ostad Movahed, N. Phenolic-cured butyl rubber reclamation using different de-crosslinking agents. *Prog. Rubber Plast. Recycl. Technol.* 2024, *40*(1), 33–47.

Mollanorouzi, M. and Ostad Movahed, S. Reclaiming waste tire rubber by an irradiation technique. *Polym. Degrad. Stab.* 2016, *128*, 115–125.

Ostad Movahed, S., Ansarifar, A., Ghanei, R., and Kermani, Y. Devulcanization of ethylene–propylene–diene waste rubber by microwaves and chemical agents. *J. Elastomers Plast.* 2016, *48*(1), 122–144.

Ostad Movahed, S., Ansarifar, A., Karbalaee, S., and Atharifar, S. Devulcanization and recycling of waste automotive EPDM rubber powder using shearing action and chemical additive. *Prog. Rubber Plast. Recycl. Technol.* 2015a, *31*, 87–116.

Ostad Movahed, S., Ansarifar, A., Karbalaei Nezhad, S., and Atharifar, S. A novel industrial technique for recycling ethylene–propylene–diene waste rubber. *Polym. Degrad. Stab.* 2015b, *111*, 114–123.

8

Advanced Techniques for Plastics Separation From a Waste Plastic Stream

8.1 Introduction

Modern society depends heavily on plastics because manufacturers produced more than 390 million tons of plastic materials during 2023, and experts predict that production will reach 500 million tons by 2030 (Plastics Europe, 2024). The combination of plastic properties including light weight, multiple uses, extended lifetimes, and affordable cost has driven fast growth in different applications including packaging, automotive parts, textile products, construction materials, medical equipment, and consumer goods (Saleh et al., 2018, 2022a, b). The environmental problems that plastics create start with their stable chemical composition and consequently, their ability to last long and their resistance to decay. The world has not recycled more than 9–12% of its plastic waste, and 60% of all produced plastics end up in landfills or the environment (Geyer et al., 2017). The increasing difference between plastic manufacturing and recycling operations has created major environmental problems, carbon pollution, resource exhaustion, and microplastic contamination.

The process of recycling high-value materials faces a major technical challenge because of the challenging task to extract plastics from diverse postconsumer waste materials. The separation of plastics from postconsumer waste becomes challenging because plastic materials share similar density values, weak spectral characteristics, identical forms, multiple additives, and complex multilayered structures. The presence of food waste, dyes, adhesives, and incompatible plastics in waste materials decreases both sorting precision and output quality. The majority of mixed plastic waste receives mechanical processing to create low-value products or gets burned for energy or sent to landfills despite having substantial potential for circular economy value creation.

The research focuses on creating efficient separation techniques which enable fast and inexpensive production of clean plastic materials. The recycling industry across the world has evolved through automated sorting systems, which Längle and Maier (2025) implemented by substituting human inspection with sensor-based separation methods that use physical, chemical properties, and solvents. The feedstock purity in high-quality mechanical and chemical recycling systems needs to reach 95–98% for food-grade polyethylene terephthalate (PET) and high-value engineering plastics. The entire recycling process depends on the exactness and speed at which plastics get separated during the separation stage.

The following sections explain current and future methods which separate plastics from different waste materials including:

- The process of manual and semiautomated sorting
- The separation methods based on density use sink–float and elutriation processes
- Triboelectrostatic separation
- The solvent-targeted recovery process
- Froth flotation and hydrophobicity-assisted separation
- Miscellaneous techniques, near-infrared (NIR), hyperspectral, and laser spectroscopy technologies. Also, artificial intelligence (AI) operates robotic sorting devices which perform optical classification tasks.

The different technologies require different levels of investment and operational difficulty while needing specific materials for processing, and they work best with particular recycling methods. The sink–float separation method works well for plastics with considerable density differences including PE and polypropylene (PP), but it fails to separate polystyrene (PS) and PET because these plastics have matching densities. The fast identification process of NIR spectroscopy faces challenges when trying to detect black plastics because carbon black material blocks its detection ability. The electrostatic separation method depends on tribocharging properties of plastics to separate them but requires exact control of moisture content and particle size. The dissolution-based (solvent) separation method shows promise for obtaining high-quality polymers but faces challenges because of expensive solvents, strict environmental standards, and high operational energy needs.

The separation process requires high purity because it establishes the quality of the output stream for different applications. The output stream requires high-purity separation to achieve suitability for different applications exampling, food-grade mechanical recycling, closed-loop recycling of PET and PE films, mono-material packaging, and chemical depolymerization of PET, polyamide, and PS.

The market value of recycled materials depends mainly on the quality of their original materials which serve as feedstock. Hahladakis and Iacovidou (2018) demonstrated that polymer performance deteriorates when contaminants enter the system and sorting errors occur because it leads to faster degradation and changes melt behavior and causes yellowing and oxidative chain scission during reprocessing. The restoration of mechanical properties to virgin resin levels becomes possible when separation achieves high selectivity rates of 98 wt.% or better.

Separation technology systems experience progress because of multiple worldwide trends.

- Policy and regulatory pressure
  The European Union together with Canada enforces strict recycling standards through their recycling quotas and extended producer responsibility (EPR) laws and mandatory recycled-content regulations. The recycling process requires exact high-purity sorting methods to meet current regulations.
- Industrial demand for high-quality recycled material
  As an illustration, the production expansion of rPET bottles and rPP automotive parts and rHDPE blow-molded containers has created a need for better plastic sorting methods.
- The industry now operates chemical recycling facilities and monomer recovery facilities.
  The chemical depolymerization process needs materials that exceed the purity requirements of mechanical recycling operations.
- Economic incentives for circularity
  The market value of pure polymers exceeds standard materials, which creates profitable recycling opportunities throughout all economic sectors.
- Business operations have undergone major transformations because of two key factors: automation technology and AI system advancements.
  The combination of AI-based optical sorting with hyperspectral imaging and robotics technology has revolutionized complex waste separation processes to deliver higher efficiency and lower personnel costs.

This chapter provides extensive details about plastic separation methods through advanced industrial examples, current research findings, and industrial applications. The chapter follows separation methods through their operational principles and mechanisms while showing their advantages and limitations. The chapter conducts a detailed examination of physical and chemical properties which affect sorting operation performance. It shows how separation methods operate with mechanical and chemical recycling systems.

Also, the chapter evaluates life cycle assessment (LCA) and total equipment assessment as evaluation methods for new industrial separation techniques.

8.2 Plastics Separation Techniques as a Crucial Part of Plastics Waste Management

The process of separating plastics stands as the essential yet most expensive operation which runs through every plastics recycling system. The entire process of recycling plastics depends on front-end separation efficiency because it determines both product quality, safety, economic sustainability, and environmental impact. Most plastic waste becomes unusable because different plastic types cannot be processed together without resulting in substantial product deterioration. The expenses of recycling operations rise because of required additional sorting procedures, filtration steps, decontamination processes, energy consumption, and personnel costs. The following section demonstrates why plastics need separation before recycling through scientific evidence, economic data, environmental impact assessments, and regulatory requirements.

The primary recycling strategy (reusing process) requires pure mono-material streams because it depends on shredding, washing, melt filtration, extrusion, and pellet forming to maintain polymer molecular structure and mechanical strength. The physical properties of different plastics differ through their melting points, viscosities, crystalline structures, and thermal resistance levels. The process of incompatible polymer melting leads to phase separation, void formation, and gelation which results in materials that exhibit poor mechanical strength.

The following situations and illustrations represent common contamination problems:

- The processing of PET bottle streams containing polyvinylchloride (PVC) results in HCl emission, which causes PET degradation and color transformation.
- The addition of PP to HDPE results in strength reduction and extended service life shortening because the materials do not mix well.
- The presence of PS in PP leads to material brittleness and hazy appearance because of its incompatible chemical structure.
- The processing of multilayer packaging materials including PE/EVOH/PET requires delamination or selective dissolution methods.
- The entire batch becomes unfit for food-grade production when PVC contamination between 0.5 and 1 wt.% appears in PET recycled materials. According to Kalauni et al. (2025), HDPE recycled material impact strength decreases by 50% when PP contamination reaches 2 wt.%.

The production of recycled materials that meet industrial and regulatory standards requires exact separation methods during the first processing stage.

The economic value of recycling depends directly on the quality of materials being processed. The market requires high prices for pure PET, HDPE, and PP materials yet refuses to accept mixed plastic streams because their processing expenses exceed their market worth. The economic success of recycling facilities depends on how pure the materials are when they receive them. The cost of sorting materials through automated NIR/optical systems becomes significantly higher when dealing with different types of materials. The process of washing and decontamination for mixed plastics requires hot water washing, caustic treatment, surfactants, and multiple rinsing steps. The process of melt filtration requires higher energy input because contaminants in the material cause the extrusion torque to rise while simultaneously increasing the total energy usage. The process of contaminant removal from waste products produces lightweight materials which cause loss of mass of recycled materials. A material recovery facility (MRF) is a centralized sorting center that processes mixed waste streams. Wang et al. (2024b) show that MRFs can increase their total plant profitability by 25% through better NIR sorting efficiency because it allows for quicker reprocessing and higher market value of clean materials. European markets pay two to three times more for high-quality rPET flakes than they do for PET flakes with mixed colors. The market value of HDPE natural bottle-grade recycled material becomes economically viable only when contamination levels reach 5% or lower. The ability to separate materials efficiently determines business profitability because it enables companies to remain competitive against virgin resin prices which follow global oil market fluctuations.

One of the main reasons for plastic separation comes from regulatory standards which control food packaging materials and medical equipment packaging. The recycling process for food-grade materials including PET bottles needs to meet absolute purity standards. Major regulatory demands include the FDA, and EFSA need documented evidence which proves postconsumer PET undergoes complete decontamination through mono-material streams. The EU Single-Use Plastics Directive (SUPD) together with REACH establishes restrictions for brominated flame retardants and pigments and legacy additives which exist in combined waste streams. Flexible PVC and electronic waste contain phthalates and heavy metals which prevent these materials from entering consumer packaging recycling systems. The presence of contaminants from other polymers, inks, adhesives, and multilayer films creates unidentifiable substances in recycled materials which violate regulations for food-grade applications. The recycling industry needs to perform complete separation of PET and HDPE materials to fulfill food-contact regulations. The recycling process demands precise separation systems to operate effectively at the beginning of the recycling system.

The chemical recycling methods (tertiary recycling strategy) of solvolysis, glycolysis, depolymerization, pyrolysis, and gasification require precise control of feedstock composition. The process of separating feedstock materials becomes necessary for chemical recycling because contaminants in the materials will change reaction paths and decrease monomer production. The chemical reactions cause HCl to be released from PVC and PVDC, which results in catalyst poisoning. The pyrolysis process produces oils that contain nitrogen and bromine compounds when it converts acrylonitrile–butadiene–styrene (ABS) and polycarbonate (PC) plastics that have flame-retardant properties. The production of high-quality BHET or EG through PET hydrolysis and glycolysis needs feedstocks which have contaminants below 1%. The broad molecular weight distribution in pyrolysis oils from various waste sources makes them unsuitable for use as cracker feedstock. The research demonstrates that PET glycolysis becomes less efficient by 40–50% when the feedstock contains 5 wt.% of polyolefins. The pyrolysis reactor needs pretreated feed materials because these materials require protection against corrosion and coking that threatens damage equipment operation. The process of separation functions as an essential requirement for mechanical recycling facilities and future chemical recycling plants.

The effectiveness of plastic recycling environmental impact decreases when separation methods improve because they boost material recovery rates and stop waste from entering landfills or incinerators. The results from LCA studies appear in Section 8.3. The production of plastics from recycled materials generates 1–3 tCO_2-eq/ton less greenhouse gas emissions than manufacturing plastics from raw materials. The process of extrusion becomes more energy-efficient by 15–30% when separation methods achieve high purity levels. The production of wastewater and washing chemicals decreases when operators maintain clean material streams. The process of poor separation produces more **microplastic** waste while creating additional sludge waste, which increases total carbon emissions. The current plastic waste problem stems from separation and recovery system inefficiencies because more than 79% of all produced plastics end up in landfills or natural environments, according to Geyer et al. (2017). The process of separation functions as a fundamental environmental solution which enables worldwide plastic recycling systems to operate effectively.

The success of design-for-recycling (DfR) strategies depends on the ability to identify various materials. Designers need to separate waste materials because this ability enables them to select materials that belong to one substance or match each other. The process of high-quality separation allows for the recycling process of PET bottles in closed loops and HDPE bottles become possible through bottle-to-bottle systems and PP packaging reuse and engineering plastic recovery of ABS, PC, PA66, PS, and PMMA upcycling and multilayer film dissolution. The quality of material separation decides which plastic materials will enter food packaging or durable products or become part of low-value products including flowerpots and benches and/or composite lumber. The ability to separate materials determines which plastic materials will find their way back into food packaging or durable products or end up in low-grade applications.

MRFs require separation to achieve their operational targets and maintain their production capacity. The lack of presorting for mixed plastics causes conveyor system overload and system variability which reduces sensing precision. The process of separation fails to produce acceptable results when problems occur including equipment clogging, lower belt speed, additional resorting cycles, higher labor demand, yield losses, and reduced plant throughput. Modern MRFs operate with modern systems which include NIR sorting, hyperspectral imaging, triboelectric separators, magnetic density separation (MDS), and AI-powered robotic optical sorters. All these systems need uniform feedstock properties to function properly, which demonstrates that separation stands as the essential starting point.

Also, several organizations need to follow international circular economy standards through proper separation methods to implement these standards including EU Circular Economy Action Plan (CEAP), EPR legislation, EU SUPD, and UNEP Global Plastics Treaty negotiations. The new international frameworks require businesses to meet main requirements, which include high recycled content levels, packaging materials should consist of single components, the system requires continuous monitoring of all materials throughout the entire system, and the design needs to function properly with automated sorting equipment. The successful implementation of these policy objectives needs suitable separation methods.

In summary, plastics need separation because of the following reasons:

- The melting process of different materials creates mechanical property degradation.
- The value of recycled plastics and economic viability depends on the purity level of materials.
- The process of accepting feedstock materials for food-grade and medical applications needs regulatory frameworks to create standards for clean materials.
- The efficiency of chemical recycling operations depends on the purity level of their feed materials.
- The environmental protection depends on the method of material separation.
- The process of achieving circularity and high-value product reuse needs streams that contain individual materials.
- MRFs operate at their best when materials get proper sorting before they start the processing stage.
- All countries must follow international waste management standards which demand waste streams that can be separated and combined properly.
- All possible plastics recycling methods need separation as their fundamental operational foundation.

8.2.1 Manual Sorting of Plastics in the Recycling Chain

MRFs use manual sorting for plastic separation because it stands as their most effective and simplest method. The recycling industry depends on human-led manual sorting for both high-income and low-income recycling facilities because new sensor-based sorting technologies and robotic pickers and AI systems have not completely replaced this traditional method yet. The research by Wang et al. (2024) shows that manual sorting serves four essential purposes in recycling operations. The process of manual sorting enables quality control and removes dangerous materials and nonrecyclable waste while improving automated sorting results and handling items that sensors fail to detect including black plastics and soiled films and multilayer packaging.

The process of manual sorting continues to dominate plastic segregation operations in developing and newly industrialized areas because it handles more than 60% of flexible films, colored plastics, mixed household waste, and packaging residues. Modern European MRFs that use multiple-stage NIR sorters require human operators at clean-up stations to achieve premium recycled material purity levels exceeding 95% for rPET and rHDPE. This section investigates the operational methods and reasons behind manual sorting as well as its restrictions, work-related factors, and evaluation criteria. The section includes two illustrative diagrams, Figures 8.1 and 8.2, which show (1) the position of manual sorters in a standard MRF production line and (2) the human sorting decision process.

The process in Figure 8.1 follows a left-to-right sequence. The waste input hopper functions as the initial processing stage. The bag opener/shredder operates as the second step. The prescreening process separates materials through either

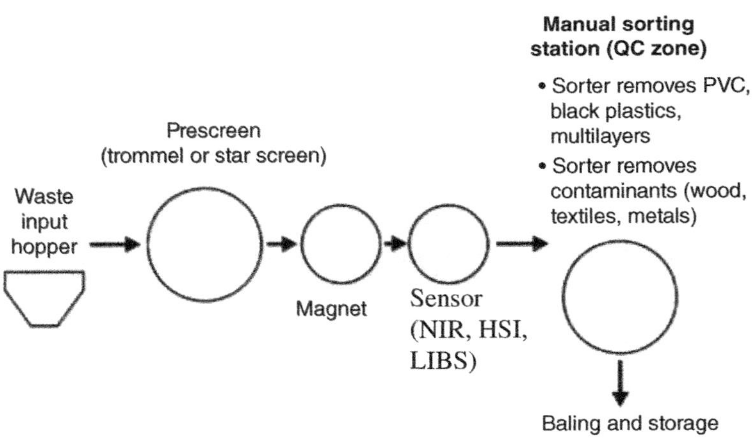

Figure 8.1 The position of manual sorters in a standard MRF production line.

Figure 8.2 The human sorting decision process.

trommel or star screen equipment. The ballistic separator function operates to distinguish between two-dimensional films and three-dimensional rigid plastic items. The system combines magnetic separation with eddy current separation. The optical sorting stage operates through the combination of NIR and high-speed imaging (HSI) technology. The manual sorting station operates as the QC zone. The sorter extracts PVC materials, black plastic items, and multilayered plastics from the stream. The sorter extracts all types of contaminants which include wood pieces and textile materials and metal objects. The system keeps baled materials available for future operational needs. The position of manual sorting depends on the specific country and technology implementation level because it occurs before or after primary sorting operations.

The process of manual sorting in plastics waste management performs three essential functions.

- Primary sorting (upstream sorting)
  The process of sorting plastics waste requires workers to identify the main categories which include polymer types, color, form, application, and contaminant levels. The initial sorting process establishes the exact composition and cleanliness level of all following processing stages which include washing, shredding, and baling operations.
- Quality control sorting (downstream sorting)
  Human operators perform two essential tasks in advanced sorting systems, which include sensor-based item detection and manual removal of off-spec materials, black plastics, multilayer, and dirty items. The process of manual quality control stations continues to play an essential role in reaching the required purity standards for contract-bale products.
- Specialist sorting tasks
  The current technology (optical sorting) fails to detect specific plastic materials which include black plastics, multilayer films, filled plastics with $CaCO_3$-modified PE, and very thin films (low reflectance). The present technology fails to achieve precise automated sorting of these materials.

The process of manual sorting depends on human ability to detect visual cues. The system uses visual elements to identify items based on their color, shape, transparency, PP cap closure in PET bottles, surface gloss, and stiffness. The system uses tactile feedback to detect material stiffness between HDPE and LDPE and to identify PET and PP materials through their crinkling sounds and to detect film flexibility between PE-based and biodegradable PLA films. Also, the human eye possesses the ability to read resin identification code (RIC) numbers which machines struggle to identify correctly.

Workers who understand products can identify them through brand recognition, packaging styles, and standard material compositions. Operators who have experience with manual sorting can achieve high accuracy in polymer identification when they work with clean and organized material streams.

The sorting rate for manual operations ranges from 150 to 400 items per minute for each operator who operates on a fast-moving belt. The system achieves more than 95% success rate when sorting PET from HDPE materials according to Wang et al. (2024). The system maintains higher than 90% accuracy when processing different types of rigid plastic materials. The human ability to sort items improves by using optical sorting systems when dealing with items that are dirty or bent or twisted.

The process of manual sorting faces multiple critical restrictions which limit its effectiveness. The manual sorting operations at MRFs are labor-intensive and require multiple staff members to operate which drives up their operational costs. Research shows that human performance in sorting tasks follows the exact patterns of decline which have been identified. The accuracy rate decreases between 15% and 25% during the first two hours of work. Also, the accuracy rate decreases by 35% when workers complete their entire shift. The workplace exposes employees to occupational health risks which can be categorized in three main hazards including biohazards from food waste, pathogens, and mechanical dangers from sharp objects, chemical substances, ergonomic problems, heat, and noise exposure. The performance of manual sorting depends on individual worker abilities which produce different levels of purity and inconsistent bale quality and unstable production rates resulting in variability in sorting results. The current sorting capacity of human operators fails to meet the needs of modern facilities which handle 10–30 ton of material per hour. The manual sorting process functions independently from automated systems because these systems exist to improve current operational procedures.

The evaluation of manual sorting performance relies on several main metrics. The percentage of correctly sorted items ending up in their designated target stream. The following values represent typical performance levels. The PET stream produces 90–98% pure output, but HDPE stream purity ranges from 85% to 95%, and film stream purity exists between 75% and 90%. The percentage of target polymer ending up in wrong processing streams. For example, the presence of PVC in PET

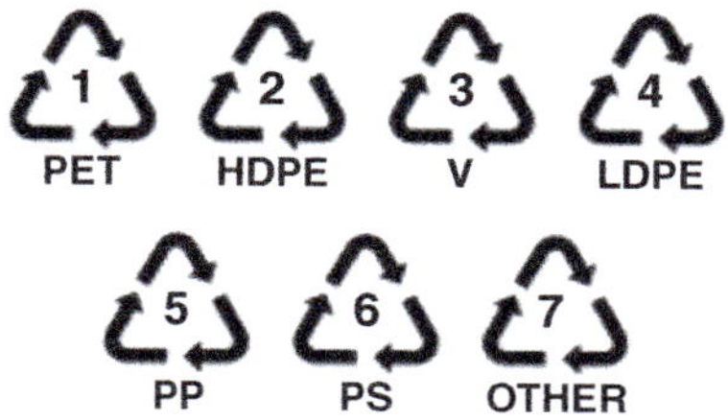

Figure 8.3 The RIC symbols. Liena / Adobe Stock Photos.

requires strict control because it should not exceed 0.05–0.1 wt.%. The evaluation of bale quality follows industry-established standards which include APR, PETCORE, and ICIS. The amount of material that manual sorting operations can process depends on worker efficiency and item dimensions. The training program enables operators to process between 200 and 350 items during each minute of operation.

A logical sequence for the human sorting decision process (Figure 8.2) will be as follows:

Pick item → inspect shape, bottle → go to PET/HDPE recognition, film → LDPE, tray → PS/PET, and tube → PP

Check color and transparency, clear → PET or PP, opaque → HDPE/PP, and black → reject stream

Check rigidity/tactile cues, hard/crinkly → PET, HDPE maintains its rigidity but becomes soft to the touch, flexible → LDPE

Look for resin symbol or label clues, the RIC symbols (Figure 8.3) which appear on the visible screen to control the sorting operation.

Final decision, the process sends unknown items to the reject stream or mixed plastics section.

Research studies show that manual sorting operations will develop into different separate operational systems. The future of manual sorting will integrate AI to provide decision support and robotic systems that duplicate human picking abilities and hybrid sorting systems which unite manual quality control with optical sorting technology and HSI with machine learning (ML) algorithms for black plastic identification.

The plastic sorting process requires manual operators to perform essential tasks because automation systems cannot replace their abilities to quality assurance, nonconforming item removal, handling irregular or dirty plastics, and visual supervision of MRF performance. Manual sorting operations will shift their primary function to monitoring and quality control activities in future operations.

8.2.2 Density Difference Separation Technique

The most common and established industrial method for plastic waste separation uses density-based separation techniques. The separation method works because different plastics have different densities which become more noticeable when using specific separation materials or when applying hydrodynamic or magnetic fields.

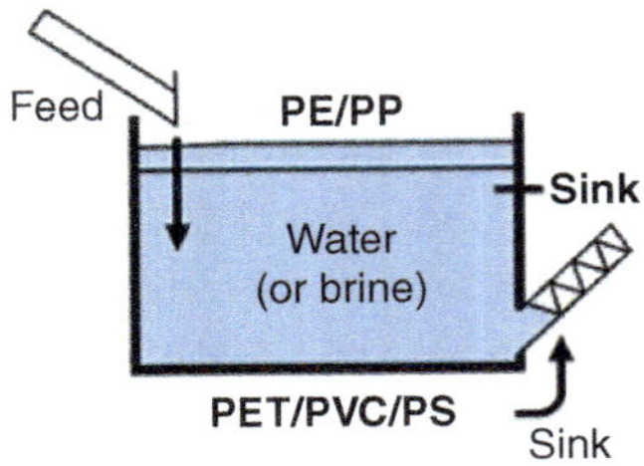

Figure 8.4 A typical plastics separation unit using density difference phenomena of the different plastics.

The process of plastic waste separation based on density difference uses several methods which include sink–float tanks, multistage density cascades, hydro cyclone, and advanced MDS systems (Taneepanichskul et al., 2022). The density-based methods work best for plastic recycling when processing streams containing polyolefin materials including PE and PP, which have densities less than 1 g/cm^3, and PET, PVC, PS, ABS, and other engineering plastics, which have densities greater than 1.03 g/cm^3. The process also functions for plastic blends through proper implementation of density gradients and multiple-stage separation systems by operators.

The system provides several benefits because it requires minimal initial investment and operates at high speeds while working well with existing washing systems. The system encounters two main problems because it struggles to separate materials with similar densities and because plastic additives and contaminants affect the measured density of plastics. Figure 8.4 shows a typical plastics separation unit using density difference phenomena of the different plastics.

This section begins by explaining density separation fundamentals before presenting industrial applications of sink–float separation, multistage density cascades, hydrodynamic devices, and MDS.

The basic principle of density separation develops from the balance between gravitational force and buoyant force which affects solid particles when they are placed in fluids:

$$F_{\text{net}} = (\rho_{\text{p}} - \rho_{\text{f}})\, V_{\text{p}}\, g$$

where

F_{net} = net implemented force on particle, ρ_{p} = particle (plastic) density, ρ_{f} = fluid density, V_{p} = particle volume, and g = gravitational acceleration. If $\rho_{\text{p}} > \rho_{\text{f}}$, the particle sinks, and if $\rho_{\text{p}} < \rho_{\text{f}}$, the particle floats.

For the plastics, the typical density ranges at 23 °C are as follows: PE (LDPE/LLDPE), 0.91–0.93 g/cm^3; PP, 0.90–0.91 g/cm^3; PS, 1.04–1.06 g/cm^3; PET, 1.34–1.38 g/cm^3; and PVC (unplasticized), 1.35–1.45 g/cm^3 (higher for plasticized grades). Water has a density of 1 g/cm^3. The water provides a natural separation boundary between polyolefins such as PE and PP and heavy polymers such as PET, PVC, PS, and ABS. This is the basis for the widespread use of simple sink–float tanks in postconsumer packaging recycling. The separation process depends on several factors which include the density difference between polymers and fluid, the dimensions and form of particles, the hydrodynamic conditions which include turbulence levels and contact duration and fluid mixing patterns, and the surface contains multiple types of contaminants which include labels, air bubbles, biological growth, and oil residue.

The main operational problem emerges because fillers and absorbed contaminants change the actual density of polymers. The behavior of heavily filled PP becomes like heavy plastic when submerged in water because it sinks, while hollow PET articles and foamed PS float on water. The process requires specific prewashing procedures, size reduction, and degassing techniques followed by using adjusted-density media or implementing multiple stages of separation.

The sink–float tank (Figure 8.4) operates as the basic density separation unit which serves as the primary choice for plastics recycling operations. A standard operation sequence follows this order:

- Feed preparation: the plastic shredder produces 5–20 mm (about 0.79 in) flakes which need washing and screening to remove small particles and organic materials.
- Feeding to tank: the flakes enter the tank surface which contains water or salt solution.
- Float fraction (light plastics): the particles which have lower density than the fluid will float to the surface where they can be removed by skimming or overflow.
- Sink fraction (heavy plastics): the denser materials (PET, PVC, and PS) sink to the bottom where they move through bottom screws, scraper chains, and sloping conveyors.
- Dewatering and drying: the two separated fractions undergo dewatering procedures before they proceed to secondary washing operations and additional sorting processes.

The current research from laboratory and pilot studies shows that water-based sink–float recovers more than 95–97% of HDPE and PP materials through floating separation while the sink fraction contains heavy plastics pending with proper washing and liberation of the feed material from contaminants. Also, 100% separation efficiency is achievable for different polymer combinations through multiple stages of water and aqueous salt solution (NaCl, NH_4NO_3) separation which produced three distinct density fractions from a six-polymer mixture PS, PMMA, two PVC grades, and two PET grades (Pita et al., 2023).

Operators achieve better discrimination between polymers with densities near 1 g/cm^3 by modifying the medium density. The density range of 1.2 g/cm^3 becomes available through NaCl or $CaCl_2$ brine solutions. The density of NH_4NO_3 solutions reaches its maximum value at 1.4 g/cm^3. Organic solvents and their blends serve as infrequent density adjustment agents, but their industrial use remains restricted because of their high cost and safety concerns. The process of increasing medium density allows users to achieve three distinct separation stages. The first separation process employs water to extract polyolefins from all plastic materials that have higher densities. The separation of PS, ABS, and PMMA from PET and PVC becomes possible through the use of denser brines which increase the medium density. The process of PET separation from PVC requires exact medium adjustments, yet the similar density levels between these materials limit this method.

The PET/PVC separation produces superior outcomes through **flotation** technique instead of density-based separation as will be described in subsequent sections.

Industrial plants operate with multiple tanks instead of using a single tank for their operations. Industrial plants implement sink–float steps which they integrate with additional processing units to create their systems. A standard integrated system consists of several main components which include:

- The first stage of the process uses water to perform a coarse sink–float separation which separates, e.g., PE/PP float materials from heavy plastic sink materials.
- The second stage operates with a brine solution to separate heavy plastics into two distinct fractions which include intermediate materials (PS and ABS and certain PVC grades) and dense materials (PET and rigid PVC).
- The third stage combines the process with hydro cyclones and flotation equipment to achieve better separation results between PS, PMMA, and PVC materials and between PET and PVC materials through surface treatment and flotation methods as describes later in this chapter.

The process requires low capital investment to operate at average costs, which enhances downstream processing efficiency according to the combination of density cascades with LCA and technoeconomic analysis (TEA).

The efficiency of density separation improves through hydro cyclones and enhanced-gravity separators which function as small-scale equipment to process increased production volumes (Taneepanichskul et al., 2022). The feed material encounters strong centrifugal forces when it enters a conical chamber which produces a swirling effect. The particles experience enhanced radial acceleration, which increases the separation effectiveness between particles with small density variations. The system achieves improved cut precision through its ability to separate materials based on their density levels. The equipment requires less space than standard large tanks. The system operates correctly when it processes both small particles and floating materials. The system experiences two main performance limitations because its operational performance decreases when viscosity levels change or when solid particle concentrations increase. Also, the system requires exact control of feed pressure and exact measurements of vortex finder dimensions and apex opening size. The system design requires multiple cyclone units to operate together to achieve industrial production levels. It delivers outstanding performance for two particular uses which involve PE/PP separation and dense contaminant extraction from polyolefin processing streams.

MDS represents a modern technology which originated in the Netherlands before its current testing phase for plastic waste from packaging and electronic devices. The separation process uses magnetic liquid (ferrofluid) which exists inside a strong vertical magnetic field. The separation medium achieves its density variation through height changes which produce a density gradient that starts at its highest point at the bottom and ends at its lowest point at the top (Wang et al., 2024). The process begins with Fe_3O_4 magnetic nanoparticles dissolved in water to form a magnetic liquid solution. The magnetic body force from the gradient magnetic field causes the fluid to experience density changes that depend on its vertical position. The plastic particles will rise through the medium until their density matches the local effective density of the fluid. The single tank containing different density particles will settle at separate heights. The system uses particular extraction points to remove particles from different density bands which exist at particular heights. The system operates without needing multiple tanks and brines because it uses a single MDS unit to perform all density-based separations of PP, LDPE/LLDPE, HDPE, PS, PET, and PVC when their densities remain distinct and the gradient reaches its optimal point.

The MDS system achieves precise separation of materials which have density differences of 0.02–0.03 g/cm^3. The system operates as a single unit which performs multiple separation cuts. It demonstrates the ability to process multiple packaging materials and plastic waste generated from WEEE and end-of-life vehicle (ELV) recycling operations. However, several challenges still exist. The management of ferrofluid requires high expenses because nanoparticles escape and the solution becomes unstable. The initial investment for magnet system equipment along with operating power expenses is high. The process requires special care to prevent plastics from becoming contaminated with ferrofluid residue. The technology exists at pilot scale or early commercial phase primarily in European markets according to Wang et al. (2024).

In summary, the separation method for plastics uses density difference techniques because the process operates with basic simplicity and durability and can handle large production volumes. The process needs average funding to operate as a sink–float system. It operates with washing and drying operations and existing processing facilities. The method effectively separates plastics into their light and heavy categories. The process works effectively for both mechanical and chemical recycling operations. The method lacks universal applicability for all plastic separation needs. The main restrictions of this method stem from overlapping density ranges. The separation of PET from PVC and different PVC grades becomes impossible when using density as the separation method. Also, the combination of plasticizers with inorganic fillers including talc and $CaCO_3$ produces density variations which increase the difficulty of separation processes. From an environmental viewpoint, brine and dense solution management requires unique treatment and recovery systems for proper management. The operation of big tanks with recirculation systems requires electricity for operation and generates wastewater effluent.

Research about MDS and sink–float systems continue to grow which indicates that future development will concentrate on:

- The combination of density separation with sensor-based sorting and flotation and solvent-based methods will become more intelligent.
- The system will achieve better separation results through dynamic media property adjustments including density, viscosity, and temperature control based on feed stream variations.
- The development of sophisticated modeling systems will improve the ability to predict separation outcomes.

The large-scale plastics recycling industry will continue to use density difference separation as its fundamental technology for processing polyolefin streams and preparing high-value engineering plastics and PET materials.

8.2.3 Triboelectric Separation Technique

The dry contact-charging method of triboelectric separation uses surface rubbing and collision to create electron transfer differences between polymers. The electric field separates particles based on their charge of intensity and sign after they receive positive or negative charges through tribocharging. The dry separation method of triboelectric separation shows promise for recycling plastic waste from waste electrical and electronic equipment (WEEE), end-of-life vehicles (ELVs), and postconsumer packaging because it operates at room temperature and does not require water while being compatible with mechanical recycling systems (Zhao et al., 2024).

Triboelectric methods operate through surface electronic properties, electron affinity, and surface chemistry and dielectric constant instead of mass and buoyancy which makes them suitable for polymer density overlap situations, likely ABS and PS but with different triboelectric behavior. The process of tribocharging electrostatic separation has proven effective for industrial and pilot-scale operations which separate PET from HDPE, extract PVC from PET streams, and convert WEEE plastic materials into pure ABS, HIPS, and PP fractions.

The process of charge transfer occurs when two different solid (polymeric) materials make contact before their separation leads at the interface. The triboelectric effect produces positive and negative charges on different materials when they touch and then separate. The charge signs and strengths depend on several factors including the position of materials in the triboelectric series, the molecule contains $-CF_2-$ groups which generate powerful negative charges and $-NH_2$ and $-OH$ groups that create positive charges, the surface texture and purity level of the materials, and the environmental conditions of temperature and humidity levels.

The triboelectric series for plastics shows PA, PMMA, PC, and PS at positive charges, while PP, PE, and PTFE exist at negative charges, but their exact order depends on the counter-surface materials, additives, and environmental conditions (Zhao et al., 2024).

The triboelectrostatic separator uses a tribocharger to charge plastic flakes through multiple collisions between the flakes and the charger walls and between each other. The separation process begins when charged particles enter a separation area which produces electric fields between 20 and 60 kV through electrode generation. The electric field drives negatively and positively charged particles to move toward different collection bins. Figure 8.5 illustrates a typical triboelectric polymeric materials separation process.

The creation of multiple configurations for tribocharging devices serves two purposes which include maximizing charge transfer and protecting particles from damage. Rotating or vibratory tube tribocharger (cylindrical tube) contains mixed plastics which become charged through particle–wall collisions that occur when the tube rotates or vibrates. The tube material selection together with rotational speed determines the charge of sign and amount. In another tribocharger type, cyclone or air-driven tribochargers, the air stream transports particles through a spiral or cyclone chamber. The system uses tangential flow and wall collisions to achieve efficient charging at high production rates for industrial applications.

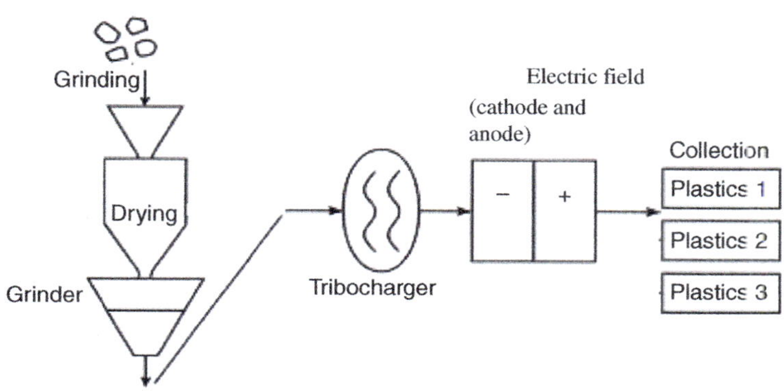

Figure 8.5 A typical triboelectric polymeric materials separation process.

In fluidized bed tribochargers, the weightless state of plastic particles occurs because air flow creates a fluidized bed which causes them to experience strong collisions. The fluidized-bed tribocharger proved best for PVC removal from PET, PP, PE, and PS streams because it recovered 96–99% PVC at > 90% purity.

In vibratory conveyors and drums for tribochargers, the system operates as a continuous process for charging granular materials. The system operates as a standalone unit which connects to commercial sorting lines for feeding materials.

The design requires a trade-off between charging performance, operational speed, equipment durability, maintenance requirements, and time spent in the system. The process becomes dangerous when charges become excessive, but the system loses its ability to separate materials when charges remain too low.

The electrostatic separators for plastics categorize as drum-type separators, plate-type or free-fall separators, and roll-type (high-tension) separators. In drum-type separators, the process involves feeding charged particles onto a grounded rotating drum which faces either one or multiple high-voltage electrodes. The drum surface attracts particles based on their charge signs, which causes them to follow different paths to the ground. The laboratory tests showed that the drum separator achieved more than 95% purity separation of PET and HDPE through adjustable electrode design. In plate-type or free-fall separators, the particles experience gravity while moving between two plate electrodes. The deflection path of particles depends on their charge-to-mass ratio, so collection bins along the bottom edge can extract different material fractions. The separation efficiencies above 90% are achievable through this method to various binary plastic mixtures. In roll-type (high-tension) separators, the process of corona charging and tribocharging operates together.

Research studies have established that operating conditions produce quantifiable relationships which affect the performance of triboelectric separation. The process depends most heavily on the current humidity levels. The adsorption of water on polymer surfaces creates a conductive layer which enables charge dissipation and results in charge loss. The study by Achouri et al. (2024) demonstrated that humidity levels between 20% and 80% reduce charge strength by more than 10 times which results in decreased WEEE plastic separation efficiency. The particles with 0.2–0.5 wt.% water content will develop opposite electric charges, which leads to agglomeration and separator misplacement. The authors suggest that predrying the material to less than 0.1 wt.% moisture content will produce stable operation.

The combination of small particle size and high charge-to-mass ratio leads to better separation, but these particles tend to merge into clusters which follow gas streams instead of electric fields. Most researchers have established that plastic flakes achieve their best separation results when their size falls between 0.5 and 4 mm (about 0.16 in). The separation process becomes challenging when working with thin foils because they tend to stick to equipment surfaces and other materials.

The amount of tribocharging between particles and charging surfaces grows until it reaches a point of saturation during multiple collision events. The research demonstrated that cyclone tribochargers achieve their best results when air velocity and residence time reach specific levels which prevent additional collisions from producing more charge but instead lead to equipment deterioration.

The material choice for tribocharger walls determines the sign of the generated charge. PET materials develop positive charges when they touch PTFE, but they become negatively charged when they contact stainless-steel or PVC surfaces. Scientists conduct experiments to study the triboelectric series by testing different polymer–wall material pairs (Zhao et al., 2024).

The composition of actual plastic waste materials includes flame retardants together with fillers, pigments, and plasticizers. The presence of brominated flame retardants together with carbon black in materials leads to modifications in their surface conductivity and tribocharging properties. Scientists discovered that triboelectric separation technology allows them to remove brominated ABS from bromine-free ABS/HIPS blends which meet RoHS compliance standards for businesses.

Materials show different charging properties after they experience UV radiation and oxidation exposure and when they develop oil or biofilm contamination. The research showed that vehicle polymer waste materials by applying corona treatment and cleaning methods enhanced both charge stability and separation performance.

The measurement of process performance depends on several factors including achievement of complete recovery of polymers into their assigned product streams, production of fractions which reach high purity levels, operation with a fully balanced mass system, and operation at a constant throughput rate of kg/h while using kWh/t of energy for its operations.

The triboelectric plastics separation technique has several advantages. The process operates without water consumption, and it does not need flotation chemicals to produce minimal waste treatment requirements. The process requires minimal energy to operate when compared to melting processing and chemical recycling methods. It operates correctly when it processes small granules which standard density tanks cannot handle. The system uses functional property differences to separate brominated ABS from nonbrominated ABS instead of density separation methods.

The system limitations show extreme vulnerability to humidity changes, surface impurities, and feed preparation requirements. The system needs precise control of dust levels, and it operates best with materials that have uniform particle sizes. The process of sorting complex polymer blends requires multiple passes because it needs multiple sorting methods. The scientific community remains uncertain about polymer charge transfer mechanisms when they contact each other or metal surfaces, which hinders tribocharger development (Zhao et al., 2024).

Modern plastics recycling facilities operate triboelectric separation as a standalone process, but most facilities use it in combination with other methods. The main ways to integrate triboelectric separation with other methods may include the aid of NIR-based sorting which divides polymers into separate families before triboelectrostatic separation generates high-purity grades.

The system employs density separation to separate materials which have similar density levels, but different triboelectric properties, namely, between ABS and HIPS. The process serves as a final stage to extract tiny amounts of PVC and brominated plastics and paper contaminants from clean material streams. The use of TEA and LCA tools at plastics recycling facilities proves that triboelectric separation works well.

8.2.4 Dissolution Separation Technique

The process of dissolution-based separation known as "selective dissolution" or "solvent-targeted recovery" or "STR" uses thermodynamic principles to separate polymers through solvent systems which dissolve one polymer while leaving others insoluble. The extraction of dissolved materials occurs through methods which include precipitation, evaporation, and antisolvent crystallization to generate pure polymer products that match virgin resin quality (Xu et al., 2025). Figure 8.6 shows typical separation of an individual plastic from a plastic mix using dissolution separation technique.

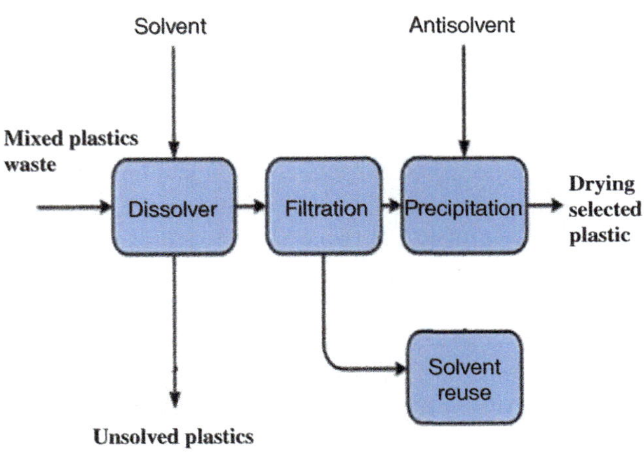

Figure 8.6 Typical separation of an individual plastic from a plastics mix using dissolution separation technique.

The process stages included shredding and size reduction. The processing of mixed waste material results in flakes which range from 5 to 15 mm (about 0.59 in) in dimensions. In the next stage, the target polymer gets dissolved through controlled solvent application at particular temperature conditions. Polymers that do not dissolve in the solution get separated through filtration or centrifugation techniques. Subsequently, the polymer resolidifies through either antisolvent addition or temperature reduction during the precipitation stage. The distillation method allows users to extract solvents at rates higher than 95% for future applications. The filtered polymer undergoes drying followed by pellet formation to produce high-purity polymer pellets. Modern industrial STR systems, including the CreaSolv® process, utilizes this multistage flow structure.

The dissolution process separates materials through natural polymer–solvent interactions by using Hildebrand solubility parameters, Flory–Huggins's interaction (χ) parameter, and crystallinity and tacticity. The development of dissolution methods becomes necessary because these methods can handle multilayer packaging materials and contaminated films and color-intensive postconsumer waste which optical and mechanical sorting systems fail to process. The physical recycling domain contains dissolution as an independent category which operates independently from chemical depolymerization. The polymer backbone stays intact because no chemical bonds break during the process, which uses physical thermodynamic selectivity to remove surrounding matrix materials and contaminants. The commercial development, e.g., APK's Newcycling®, Fraunhofer IVV's CreaSolv, and PureCycle, for PP demonstrates that dissolution produces high-purity polymer grades above 95–99% for food packaging and fiber and engineering uses.

The process of polymer dissolution depends on two factors which include the strength of solvent–polymer chain interaction and the degree of chain stiffness that develops when chains expand. The main factors that determine solubility include:

- The Hildebrand solubility parameter (δ) and Hansen solubility parameters (HSPs; δ_D, δ_P, and δ_H) serve as essential descriptors for analysis.
- The Flory–Huggins's interaction parameter (χ) shows dissolution preference through its values which stay below 0.5.
- The ability of a polymer to dissolve depends on its polymer crystallinity level because amorphous polymers dissolve more easily than highly crystalline polymers.
- The dissolution rate decreases with increasing molecular weight, but the solubility threshold remains unchanged.

The following examples show how substances behave when it comes to solubility:

- PS dissolves in limonene and ethyl acetate and toluene solutions.
  The solubility of polyolefins PE and PP depends on temperature because they exhibit different dissolution behaviors at different temperatures. The hot alkanes, xylene, and decalin solutions at 110–160 °C dissolve PE and PP polymers. PP requires higher temperature (150–160 °C) to dissolve while staying insoluble at lower temperatures. The temperature control enables researchers to achieve separate dissolution of PP and PE through temperature adjustments.
- PVC dissolves in THF and cyclohexanone solutions.
- PET requires specific conditions to dissolve in phenol and chlorophenols and ionic liquids.
- LDPE/LLDPE dissolves in hot xylene (120–140 °C).
- HDPE is partially soluble. The amount of dissolution depends on the crystalline structure of the material.

The process of selecting between two polymers becomes possible because one material dissolves, while the other remains intact during identical conditions, which allows researchers to separate PP/PE and PET/PE multilayers and PS/ABS blends.

The process of dissolution-based separation functions because polymers dissolve in solvents which have matching thermodynamic properties. Scientists determine polymer–solvent compatibility through solubility parameter measurements because these parameters enable the calculation of cohesive energy density (CED) which predicts dissolution potential (Thimmasetty et al., 2025). The two main descriptors which use in polymer recycling are Hildebrand solubility parameter (δ) and HSPs (δ_D, δ_P, and δ_H).

The Hildebrand parameter (δ) shows the square root value of CED which equals:

$$\delta = \sqrt{\frac{\Delta E_v}{V_m}}$$

where ΔE_v = energy of vaporization and V_m = molar volume.

The parameter shows the total strength of intermolecular forces which keep materials in their structure. A polymer dissolves when the polymer–solvent interaction is energetically favorable:

$$|\delta_{polymer} - \delta_{solvent}| < 1\text{–}2\,\text{MPa}^{1/2}$$

It means that a polymer solution forms when the absolute difference between polymer and solvent interaction energies remains below 1–2 $\text{MPa}^{1/2}$.

This rule results superior for nonpolar polymers including PE ($\delta \approx$ 16–17 MPa$^{1/2}$), PP ($\delta \approx$ 16–18 MPa$^{1/2}$), and PS ($\delta \approx$ 18.5–20.5 MPa$^{1/2}$). The Hildebrand method works best for nonpolar polymers, solvents that lack strong hydrogen bonding, and simple hydrocarbon systems. However, the method fails to distinguish between dispersion forces and polar and hydrogen-bonding interactions which results in poor performance for PVC, PET, PA6, PAN, and other strong polymer–solvent combinations.

The HSPs function as the main selection method for complicated dissolution systems. Hansen developed the Hildebrand concept through his introduction of three separate components which make up the total cohesive energy.

$$\delta_T^2 = \delta_D^2 + \delta_P^2 + \delta_H^2$$

where δ_D = dispersion (London) forces, δ_P = polar (dipole–dipole) interactions, and δ_H = hydrogen-bonding interactions

The interaction between polymers occurs through multiple types of forces instead of a single type. For example, the main force that allows PVC molecules to interact with each other stems from dipole forces because of its high δ_P value. Also, the ester groups in PET molecules create strong interactions between δ_P and δ_H forces. The hydrogen bonds between amide groups in polyamides generate stronger δ_H values than all other polymer systems.

Scientists apply the HSP method to predict industrial system dissolution patterns when multiple interactions take place simultaneously.

Every polymer can be represented as a solubility sphere in 3D Hansen space ($\delta_D - \delta_P - \delta_H$).

A solvent will dissolve the polymer if it falls within this sphere. The distance between polymer and solvent in this space is:

$$R_a = \sqrt{4(\delta_{D1} - \delta_{D2})^2 + (\delta_{P1} - \delta_{P2})^2 + (\delta_{H1} - \delta_{H2})^2}$$

where 4 is a weighting factor for dispersion interactions.

A solvent dissolves the polymer when:

$$R_a < R_0$$

where R_0 = polymer-specific solubility radius and R_a = polymer–solvent distance. This approach is far more accurate than Hildebrand alone. The process of dissolution recycling depends on specific solubility properties between materials. The dissolution process of PET occurs when it is combined with δ_P and δ_H values exceeding 10 MPa$^{1/2}$. PVC dissolves when it encounters solvents which have δ_P values ranging from 8 to 12 MPa$^{1/2}$. The dissolution of PP/PE happens only when they come into contact with hydrocarbon solvents that have low polarity. The dissolution of PS happens through aromatic and specific chlorinated solvents because its δ_P value is average while δ_D value is high. Engineers can use HSP to pick solvents which will break down one material without affecting the other materials. The system demonstrates two separate operational patterns when R_a reaches R_0 because it either experiences abnormal swelling, or its polymer chains undergo some degree of breakdown. The system maintains its structure because HSP matches functions correctly.

The process of precipitation happens through the addition of an antisolvent which exists outside the solubility range of the polymer. For example, the addition of low-polar IPA to PET solutions results in the development of precipitates. The reaction between ketones and alcohols results in PVC precipitation.

It is worth mentioning some industrial case studies. The APK technology uses specific solvents to extract polyamides and polyolefins from multilayer packaging materials which results in more than 98% pure PA6 and LDPE fractions. Fraunhofer IVV (CreaSolv®) process handles multiple types of packaging materials including WEEE plastics and films and sachets. The process produces food-grade polymers which meet all EFSA safety requirements. Pure Cycle Technologies (United States) employs solvent purification to clean PP materials by removing all impurities and dyes and odorants which results in material quality like virgin PP. Multiple European EPS recovery facilities now use limonene-based dissolution for their operations. Dissolution plastics separation technique has several advantages. The method produces clean polymers which fulfill all necessary standards for food packaging and film manufacturing applications. The method works well for processing complicated materials that contain different components. The process protects polymers from mechanical recycling breakdowns which occur during operational activities. The system operates through closed-loop solvent recycling, which reduces operational costs. However, it has some pitfalls. The use of solvents presents safety risks because they contain dangerous substances that can ignite and their high price points. The process requires substantial power to run its heating systems and distillation equipment. The multiple stages of processing need a large initial investment to buy necessary equipment. The dissolution process separates additives from materials, which requires additional stabilization procedures to be performed.

In summary, the dissolution recycling process requires Hildebrand and HSPs as its basic operational parameters. The process enables specific substances to dissolve while protecting all other substances from damage. Plastics streams consisting of PET/PVC/PS/PE/PP require specific solvents which dissolve individual polymers without affecting others. The method decreases the need for experimental testing. Businesses use theoretical screening methods to prevent wasting money on solvent tests for substances that will not be needed. The process enables the recovery of solvents. Engineers can select distillable or extractable solvents for their applications through knowledge of δ values. The selection process for environmentally friendly solvents becomes more effective. The selection process for environmentally friendly solvents becomes more effective through δ profile assessment. The process allows industrial dissolution systems to increase their operational capacity. Industrial dissolution systems depend on HSP maps to perform solvent choice, antisolvent selection, temperature optimization, and avoid polymer degradation.

8.2.5 Novel Flotation Separation Technique

The separation method known as flotation separation or froth flotation or surface-selective flotation or hydrophobicity-driven plastic separation uses surface free energy differences and wettability and interfacial behavior to separate plastics in water or other media. The separation process of flotation operates independently from density differences (Section 8.2.2) because it uses air bubbles to attach to plastic surfaces based on their hydrophobic or hydrophilic properties which determine their floating or sinking behavior. The method works best for plastic materials with matching densities including PP/PE, PS/ABS, PVC/PET, and various postconsumer plastic blends that resist mechanical and sensor-based sorting techniques (Negari & Ostad Movahed, 2017).

The development of flotation technology for mineral processing led to its plastic application during the 1990s and subsequent improvements through surface modifier development, pH control systems, and reagent advancements for polymer surface hydrophobicity modification. The research publications by Salari and Ostad Movahed (2023) and Shariatikia and Ostad Movahed (2023) explain the fundamental processes of interfacial behavior, surface bonding, and polymer–particle contact which directly support flotation science research. The research studies advanced our knowledge about how plastic surface chemistry interacts with fillers and network structures to control selective wetting which remains essential for plastic flotation separation.

The process of flotation depends on the varying surface properties of materials. The hydrophobic surfaces of materials bond with air bubbles which causes them to float while hydrophilic surfaces stay water-wet and sink to the bottom. Most commercial plastics including PE, PP, PS, and ABS have natural hydrophobic properties, but achieving specific separation requires additional surface modification to create different hydrophobic levels. The following methods help achieve selective separation between plastics:

- Surface hydrophilization: Several methods exist for making surfaces more hydrophilic. The process of surface hydrophilization becomes possible through using surfactants (e.g., sodium lignosulfonate [SL]), oxidizing agents, and plasma or corona treatment.

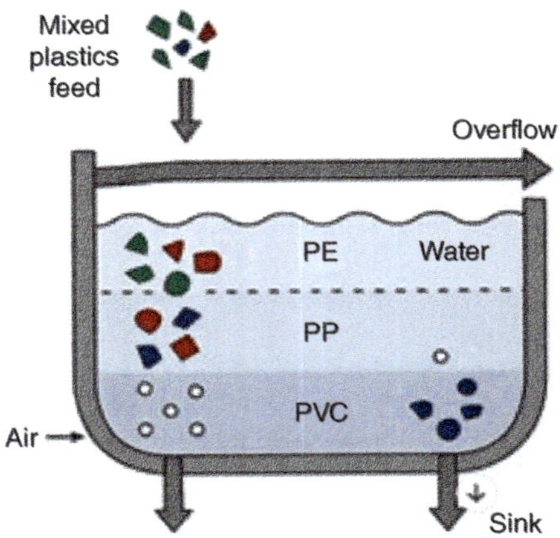

Figure 8.7 A typical flotation tank using air to create air bubbles.

- Surface hydrophobic: The process of making polar polymers such as PET, PVC, and PA more hydrophobic requires the addition of oily collectors including kerosene and MIBC, silane coupling agents (Ostad Movahed et al., 2009), and fatty acids or amine-based collectors which helps achieve this goal.

Figure 8.7 illustrates a typical flotation tank using air to create air bubbles.

A typical flotation process for mixed plastics separation included following stages:

- Grinding and washing (5–10 mm [about 0.39 in] flakes)
- Plastic surface modifications, e.g., conditioning with reagents (depressants), and UV irradiation, among others
- Media (water) operational parameters adjustment, e.g., pH, temperature, and density
- Air injection and agitation
- Bubble–particle attachment
- Froth formation
- Recovery of floated and sunk plastics fraction
- Drying and posttreatment

The success of plastics flotation depends on the ability of air bubbles to stick to plastic particles because this process allows hydrophobic polymers (PE, PP, and PS) to rise to the surface while hydrophilic materials (PET, PVC, PLA, and PA) stay submerged in water. The process of bubble attachment to surfaces depends on different main stages which are controlled by multiple thermodynamic and kinetic elements including bubble–particle collision, bubble–particle attachment (film drainage + adhesion), and bubble–particle stability. The following section details all elements which impact attachment efficiency:

- Zeta potential: The electrical surface charge of the plastic–water interface determines how bubbles and particles interact through electrostatic forces which zeta potential (ζ) measures. The surface of air bubbles in water becomes negatively charged. The high negative zeta potential of PET, PVC, and PA plastics (−50−−80 mV) creates a strong repulsive force that prevents bubble attachment. The floatability of plastics increases when their zeta potential falls between −20 and −40 mV (low negative), e.g., PE and PP plastics which have low negative or neutral charges. In other words, the floatability of PE and PP plastics is good because their zeta potential ranges from −20 to −40 mV. However, the high negative zeta potential range of −50−−80 mV in PET, PVC, and ABS plastics makes them sink. The floatability of plastics can be controlled through modifications of their zeta potential by changing media operational parameters and also, plastics surface modification.
- Surface roughness: The size of roughness at micro- and nanoscales determines how well surfaces wet and how bubbles stick to them. The surface roughness creates air pockets which enhance the hydrophobic properties of the polymer surface. The size of hydrophobic contact areas between surfaces determines how well bubbles will stick to the surfaces. The thin water film thickness leads to quick drainage which results in increased chances of bubble attachment. The surface roughness of weathered plastics increases, which makes them more buoyant. The addition of fillers to plastic materials creates surface peaks which affect how bubbles interact with the material.
- Surface polarity: The air bubble attachment on plastics surface and consequently, the floatability of plastic materials depends mainly on their molecular polarity. Hydrophobic polymers (PE, PP, and PS) have nonpolar and low surface energy. The material shows strong attraction forces with air bubbles which draw air bubbles into its structure. Polar polymers (PET, PVC, PA, and PLA) have polar groups (—Cl, —COO—, —OH, and —NH—) in their structure. The hydrophilic surface shows its ability to stop bubbles from sticking to it.
- Interfacial tension: The bubble–water–plastic system is governed by three interfacial tensions, γ_{pw}, γ_{bw}, and γ_{pb}. Attachment occurs only if $\Delta G_{adh} = \gamma_{pw} - \gamma_{pb} - \gamma_{bw} < 0$. Lower water–plastic interfacial tension → improved wettability → bubble attachment decreases.

  Higher water–plastic interfacial tension → increased hydrophobicity → enhanced bubble attachment.
- Effects of surfactants: The anionic surfactants create small bubbles which enhance the power of bubble collisions. However, cationic surfactants create surface modifications on plastic materials including PET and PVC, which results in increased hydrophobic properties.

- Mechanical energy input: The mechanical energy from agitation, impeller speed, turbulence, and aeration rate affects the air bubble attachment. The number of bubbles–particle collisions increase with higher mechanical energy which leads to better attachment probabilities. The size distribution of bubbles depends on the turbulence levels in the system. The system generates unstable large bubbles when turbulence exists at low levels. The system generates bubbles between 0.3 and 1 mm (about 0.04 in) in size when turbulence reaches its peak point, which expands the surface area. The system loses bubbles and froth when turbulence exceeds its optimal point. The hydrodynamic environment in flotation tanks needs to stay at a specific moderate level for proper operation.
- The addition of fillers including carbon black, silica, and $CaCO_3$ to plastics leads to major changes in their surface characteristics. The addition of carbon black to materials results in increased hydrophobic properties, roughness, and electrical conductivity. The flotation process benefits from these changes because the surface hydrophobic properties have been enhanced to produce superior bubble attachment. The material will sink unless flotation takes place at a high level because of the higher particle density. The addition of carbon black to PP/PE materials results in superior flotation performance than when using pure polymer materials. The addition of silica fillers to materials produces distinct polar features which become visible to the naked eye. The surface creates zones which draw in water. The surface becomes less hydrophobic in nature. The flotation process benefits from these changes because the addition of silica to plastics leads to materials that become more hydrophilic in nature. The surface shows a decreased ability to attract bubbles. The material needs collector agents to perform successful flotation operations. The addition of $CaCO_3$ to materials brings main effects including creation of sites which allow water films to form. Also, it increases the material density to the point where it starts sinking. In summary, the flotation process requires chemical treatment for successful separation of the filled plastics.

Table 8.1 summarizes the effects of mentioned factors on air bubble attachment on the plastics surface.

In summary, the factors influencing flotation efficiency included plastics particle specifications (shape, size, and density), surface chemistry, surface roughness, media (water) surface tension and pH, used mechanical devices, introduced air conditions (flow rate, air bubble size, etc.), temperature, and used additives and fillers in composition of plastic. The most suitable particle size range for flotation operations exists between 3 and 10 mm. The process of bubble attachment becomes less effective when particles exceed 10 mm (about 0.39 in) in size.

Table 8.1 The effects of different factors on air bubble attachment on the plastic surface.

Factor	Effect on bubble attachment
Zeta potential	Controls electrostatic repulsion/attraction; high negative ζ reduces attachment
Surface roughness	Increases hydrophobicity via air entrapment; boosts attachment
Polarity	Determines hydrophobic versus hydrophilic behavior directly
Interfacial tension	Governs thermodynamic adhesion energy
Mechanical energy	Controls collision frequency and detachment
Fillers	Modify polarity, roughness, and density; strongly change floatability

The left section contains two useful case studies which will be presented followed by results of discussions.

8.2.5.1 Case Study I

It appeared in *Polymers and Polymer Composites* when authors published "Study on the adsorption mechanism of the sodium lignosulphonate on the surface of the selected plastics as a potential solution for plastics waste management" in 2023.

As mentioned, in most cases, the flotation technique involves using a chemical substance which bonds physically (adsorption) to specific plastic surfaces to change their surface energy properties. The surface energy transformation of the plastic material through chemical adsorption creates a hydrophobic or hydrophilic surface depending on the chemical structure of the adsorbed agent. The treated plastic material enters a flotation vessel containing water as the liquid media. The plastic material will either float on top or sink to the bottom of the flotation vessel based on its modified surface energy properties. Multiple depressant chemicals exist for modifying the sink–float properties of chosen plastic materials.

The SL or lignosulfonic acid sodium salt emerges from needle and broad-leaved trees through the treatment of sodium sulfite. The production process also accepts pulp waste materials for SL creation. The substance exists as a brown powder which lacks any odor. The substance remains inexpensive while being nontoxic and readily available for water and alkaline solution dissolution. The substance holds its position as a vital, sustainable natural resource material. This biopolymer consists of phenylpropane and *p*-hydroxyphenyl and syringyl and guaiacyl units which form a highly branched three-dimensional structure. The substance functions as an effective dispersant in various industrial sectors including building construction, agricultural operations, dye production, and food processing.

The SL exists as a surfactant which contains both hydrophilic and hydrophobic segments in its chemical composition. The hydrophobic segments of SL create powerful surface bonding with plastic surfaces. SL produces multiple ions when it meets water because of its salt composition. The substance shows appropriate characteristics to function as a plastic flotation depressant in a water-based system. The process of chemical adsorption onto polymer surfaces represents a surface-based mechanism which allows chemicals to form a film layer on solid polymer surfaces. The adsorption–desorption process requires evaluation of multiple factors which include:

- The number of active sites on the polymer surface determines its ability to adsorb substances.
- The polymer surface supports either a single layer or multiple layers of adsorbed chemicals.
- The adsorption process needs enthalpy value measurements for its operation.
- The rate at which substances attach to and detach from polymer surfaces depends on their chemical binding and unbinding speeds.
- The Gibb's free energy values determine the adsorption process.
- The adsorbed chemical molecules interact through direct and indirect forces with their neighboring molecules.

The two models, Langmuir and Freundlich isotherms, serve as standard methods to evaluate the extent and degree of adsorption. The Freundlich model functions through data processing without needing any theoretical foundation. The Langmuir model shows that solid surface adsorption and desorption happen at equal rates between the solid surface and adsorbate (a reversible process). The solid surface contains equivalent active sites which function as single occupancy points.

The lack of any systematic research studies that investigate SL adsorption behavior on traditional plastics was the motivation of authors for this study. The study determined model parameters through SL equilibrium surface concentrations on chosen polymer materials. The research investigated six polymers which included PVC, ABS polymer, PS (general purpose PS or GPPS), PP, polyoxymethylene (POM), and PC. The research findings received validation through three conventional analytical methods, including contact angle (θ) measurement, scanning electron microscopy (SEM), and atomic force microscopy (AFM). The research presented an efficient flotation method which used SL as the flotation chemical and water as the liquid medium to extract PP from plastic mixtures. The research used commercial-grade plastics which came from local market sources. The materials exhibited different densities which ranged from 0.90 g/cm^3 for PP to 1.41 g/cm^3 for POM. The density measurements for PS, ABS, PVC, and PC showed values of 1.04, 1.07, 1.35, and 1.20 g/cm^3, respectively. The hydrodynamic behavior of these materials during flotation depends heavily on their density variations. The surface tension measurements of polymer surfaces displayed wide-ranging results with PP having the lowest value at 30.8 mN/m and POM having the highest value at 44.6 mN/m. The surface tension measurements of PS, ABS, PVC, and PC showed values of 40.7, 42.7, 41.5, and 42.9 mN/m. The flotation process depends heavily on the surface tension variations between these materials because they affect both bubble attachment, wetting properties, and separation outcomes. The polymer samples received spherical granule formation with 4 mm diameter to achieve equal hydrodynamic behavior in testing. The researchers obtained POM and PC from Korea but acquired PS, ABS, PVC, and PP from Iran. The nitrogen adsorption Brunauer–Emmett–Teller (N_2-BET method) specific surface area measurements showed unique microstructural features between different polymers. The BET surface area measurements showed ABS with the highest value at 7.29 m^2/g followed by PP at 3.66 m^2/g, PS at 3.06 m^2/g, and POM at 3 m^2/g. The BET surface area measurements for PVC and PC showed the lowest values at 1.20 and 0.80 m^2/g, respectively. The available surface area of materials determines how well collectors adsorb and how bubbles interact with particles which affects flotation results.

The used depressant consisted of lignosulfonic acid SL which TITRACHEM supplied from Iran under CAS number 8061-51-6. The chemical structure of SL consisted of $C_{20}H_{24}Na_2O_{10}S_2$. The powder showed a light tan color, and its molecular weight was measured at 535 g/mol. The plastics received deionized water washing followed by desiccator drying at room temperature before their subsequent application. The N_2-BET adsorption method from Belsorp Mini II (Microtrac Bel Corp., Japan) measured the specific surface area of plastic materials after their use. A 2 g plastic sample received 10 mL of SL solution from a glass beaker with a predetermined concentration. The UV–visible spectrophotometer (CECIL 8000 series, UK) measured the equilibrium SL concentration at 204 nm after one hour of equilibrium using the Beer–Lambert law (Figure 8.8). The measured equilibrium concentrations allowed researchers to determine the parameters of Langmuir and Freundlich isotherms. The flotation tank received 20 plastic granules for the start of the experiment. The flotation percentage calculation used the total number of floated granules. The flotation tank contained tap water at 6.5 pH during experiments which took place at room temperature.

As mentioned, the flotation experiments required 20 plastic granules to be placed in a glass flotation tank which measured 20 cm in diameter and 80 cm in height for one minute of testing. The flotation tank operated with an air-blowing system through its 13 cm circular air distributor which produced 4 L/min air bubble flow. The experiments were performed three times to obtain median results from each run.

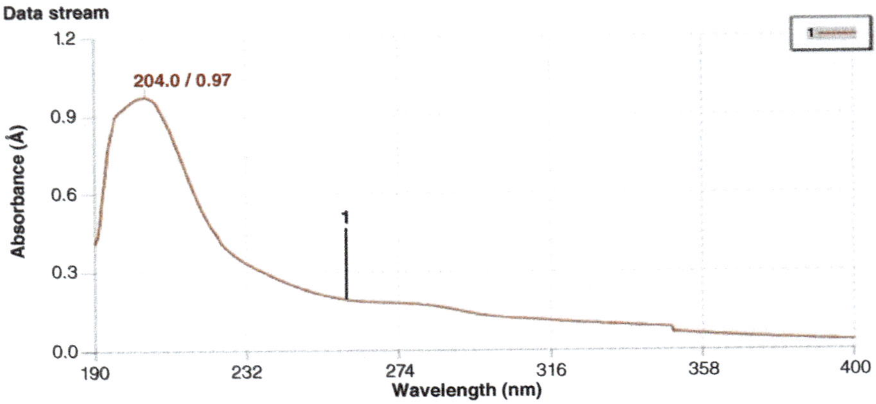

Figure 8.8 A typical equilibrium SL concentration at 204 nm.

The SZH10 stereomicroscope with Olympus DVP1 camera operated at 25× magnifications to measure contact angles (θ) between 50 μL water droplets and plastic surfaces. The MES OEL-PV 1450 field-emission gun SEM operated at 20 kV to examine plastic surface morphology with 2 nm resolution at a 12 mm working distance. The full plus series 0101 ARA Iran AFM evaluated the specifications of the plastic surface. The researchers used a cleaned glass disk as their substrate material. The researchers attached their samples to the substrate using a PLL solution before starting their experiment. The researchers needed to rinse the samples with water before drying them with nitrogen to proceed with visualization. The R_z value calculated the average height difference between peak and valley points which existed on the surface.

The equilibrium adsorption capacity (Q_e) is a useful parameter for adsorption studies through the following expression:

$$Q_e = \frac{V(C_0 - C_e)}{WS}$$

where C_0, C_e, V, W, and S were the concentrations of the initial and equilibrium SL solution (mol/L), the volume of the solution (L), initial plastic weight (2 g), and the special plastic surface area (m^2/g), respectively. The unit of Q_e is mol/m^2. By multiplying the Mw of the SL, it converts to g/m^2.

The special surface area values of ABS and PC plastics showed the highest and lowest points at 7.29 and 0.80 m^2/g, respectively. The experiments used plastics with minimal porosity which resulted in low special surface area values.

The obtained data presented the starting and final SL solution amounts and the maximum amount of SL that the plastics can absorb. The experiments took place at 25 °C. The PC material displayed the smallest special surface area (S) but achieved the highest Q_e values. The Q_e values spanned from 3.06×10^{-9}

to 31.47×10^{-9} mol/m^2 as the SL solution concentration started at 1×10^{-5} and reached 5×10^{-5} mol/L^1. The ABS material with its highest S value produced the lowest Q_e values across all tested C_0 concentrations. The Q_e values followed this sequence: PC > PVC > PS > POM > PP > ABS. The Q_e values demonstrated an expected pattern of growth when the C_0 values increased. As an illustration, the Q_e increased from 0.973×10^{-9} to 5.809×10^{-9} mol/m^2 and from 1.100×10^{-9} to 5.846×10^{-9} mol/m^2 for POM and PS, respectively, when C_0 rose from 1×10^{-5} to 5×10^{-5} mol/L. It was concluded, unlike the initial concentration of SL, that the special surface area of the plastic was not directly influential on the SL adsorption on the surface of the plastic. The adsorption process between polymers and chemicals depends on two main factors, which include polymer surface area and material chemical composition. The studied systems showed that chemical structure played a more significant role than special surface area in determining adsorption behavior.

The Langmuir isotherm functions as a basic physical adsorption model. The Langmuir theory operates based on kinetic principles which form its fundamental basis. The surface adsorption rate equals the surface desorption rate according to this theory. The model shows that adsorbed molecules stay noninteracting during adsorption when molecules adsorb in a single layer with constant adsorbent–adsorbate binding energy. The Langmuir isotherm follows this mathematical expression:

$$\frac{1}{Q_e} = \frac{1}{Q_{max}} + \frac{1}{Q_{max} \cdot K_L} . \frac{1}{C_e}$$

where Q_{max} (g/m^2) and K_L (L/g) are the maximum adsorption capacity and Langmuir isotherm constant, respectively. The Langmuir's parameters, Q_{max} and K_L, and the coefficient of determination (R^2) were obtained from the equations of the fitted lines drawn by ORIGIN 2019 software for the studied plastics developed by the authors.

The Freundlich model exists as an empirical isotherm. However, the model operates based on theoretical foundations which form its basis. The model shows solid surfaces as heterogeneous because it spreads adsorption energies throughout their entire surface area. The model demonstrates that solid surfaces consist of multiple areas (patches) which have identical adsorption sites. The model describes how material adsorption at equilibrium relates to the starting material concentration through a power law relationship as shown in the following equations:

$$Q_e = K_F C_e^{1/n}$$

or

$$\text{Log } Q_e = \text{Log } K_F + \frac{1}{n} \text{ Log}$$

Also, authors extracted Freundlich's parameters, n and K_F, coefficient of determination (R^2), and the equations of the fitted lines for the studied plastics. The R^2 values showed high numbers which confirm that both the Langmuir and Freundlich models fit the data correctly.

The experimental Q_e values showed comparisons with predicted values obtained through Langmuir and Freundlich models for PVC, PP, and PC plastics. The experimental values of PP show minimal difference from predicted values at low C_es. The SL molecules showed strong interaction with PC polar surface groups which resulted in a large deviation from experimental results. The experimental data from the other plastic material showed better alignment with the Freundlich model than the Langmuir model. The SL–PC system failed to meet the conditions of the Langmuir model because of its invalid assumptions.

The SEM images of PP and PC surfaces appear in Figure 8.9. The research used PP and PC because these materials displayed the lowest and highest Q_e values. Here, the SL concentration (C_0) was, $5 \times 10^{-5} \frac{\text{mol}}{\text{L}}$ and the plastic materials underwent treatment. The plastics maintained their smooth homogeneous appearance before starting SL treatment because they received no previous treatment. The SL treatment of plastics produced surfaces which became rough and developed coarse irregular patterns. The R_z values from AFM images showed that PP and

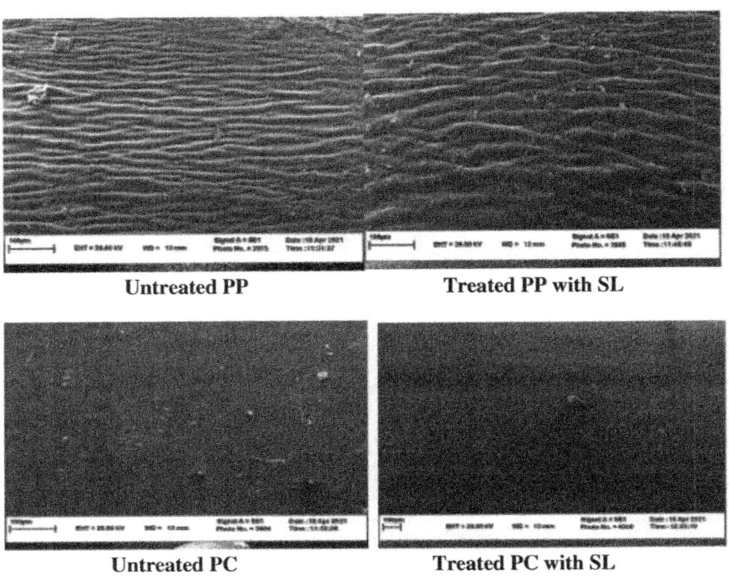

Figure 8.9 The SEM images of PP and PC surfaces before and after SL treatment, a magnitude of 100 μm.

PC surfaces had initial measurements of 334.3 and 142.3 nm, but SL treatment decreased these values to 233.9 and 8 nm. The plastic surface adsorption of SL resulted in decreased R_z values according to expectations. The SL adsorption process showed better results when it was used to treat PC surfaces.

The PP surface showed higher hydrophobicity through its contact angle measurement of 122.8° compared to the PC surface, which had a contact angle of 98.1°. The high hydrophilic nature of SL results in stronger SL adsorption to PC surfaces. The SEM and AFM images of treated plastics supported the findings.

The flotation percentage of studied plastics after they received different SL concentrations at room temperature was reported. The liquid media consisted of tap water which maintained a pH level of 6.5. The unconditioned PVC, ABS, PS, PP, POM, and PC samples showed 90%, 80%, 100%, 100%, 70%, and 75% floating behavior in tap water, respectively. The plastics PVC, POM, and PC showed density values above tap water density, yet they failed to sink in the flotation tank. The plastics failed to sink in the flotation tank because water surface tension remained too high. The plastic materials sank to the bottom of the flotation tank after receiving different SL concentrations for surface conditioning except for PP. The surface of PP did not adsorb SL effectively because its chemical structure differs from the other plastics. The Q_e values showed that the studied plastics have different adsorption properties.

From this study, it was concluded that the SL can adsorb significant amounts onto the chosen plastic surfaces. The Q_e values from measurements along with SEM and AFM images proved that SL adsorbed onto plastic surfaces. The SL adsorption sequence on plastic surfaces followed this order: PC > PVC > ABS > PS > POM > PP. The experimental results matched the data from the studied models. The Freundlich model provided the best fit for the most studied plastics. The flotation method with SL as a depressant agent successfully separates PP from plastic blends.

8.2.5.2 Case Study II

It appeared in *Polymers and Polymer Composites* when authors published "The effect of temperature on the adsorption of the sodium lignosulphonate (SL) on the surface of the selected plastics as a potential solution for the plastics waste management" in 2023.

As mentioned, plastic waste management requires the flotation technique to extract one plastic or multiple plastics from plastic mixtures. The technique depends on the surface adsorption of depressants which shows selective binding properties. The surface properties of plastics change after the adsorption process. This research examined how SL concentration levels and operational temperature settings influence lignosulfonic acid SL surface adsorption onto plastic waste materials. The research examined six plastic materials as described in an earlier case study which included PVC, ABS, PS, PP, POM, and PC. The researchers employed a floatation technique to separate plastic mixtures into individual components using SL as the floatation chemical and water as the floatation liquid. The depressant substance was lignosulfonic acid SL with CAS number 8061-51-6 which TITRACHEM supplied from Iran under the name SL. The chemical structure of SL consisted of $C_{20}H_{24}Na_2O_{10}S_2$ as expressed in earlier case study. The specifications of the studied plastics explained earlier in this text. The research used commercial plastics which local suppliers provided as their commercial products. The experiments procedure was exactly as described in an earlier case study, but the testing temperatures were 30, 50, and 70 °C.

The obtained data presented the starting and final SL solution concentrations together with the adsorption capacity at three different temperatures of 30, 50, and 70 °C. The PC showed the smallest special surface area (S) of 0.80 m^2/g, yet it achieved the highest Q_e values at all studied temperatures. The Q_e values ranged between 4.3×10^{-9} and 30×10^{-9} mol/m^2, 6.1×10^{-9} and 49.2×10^{-9} mol/m^2, and 2.2×10^{-9} and 22.5×10^{-9} mol/m^2 for studied temperatures 30, 50, and 70 °C, respectively, when the initial SL concentration in the solution increased from 1×10^{-5} to 4×10^{-5} mol/L. They found that the equilibrium adsorption capacity (Q_e) received positive effects from both higher SL concentrations and elevated operative temperatures. As an illustration for the effect of temperature, the Q_e values for SL concentration of 4×10^{-5} mol/L increased from 15.3×10^{-9} to 40.7×10^{-9} mol/m^2, from 2.9×10^{-9} to 5.8×10^{-9} mol/m^2, from 5.5×10^{-9} to 10.1×10^{-9} mol/m^2, from 4.1×10^{-9} to 6.8×10^{-9} mol/m^2, from 7.6×10^{-9} to 20.6×10^{-9} mol/m^2, and from 22.5×10^{-9} to 49.2×10^{-9} mol/m^2 for PVC, ABS, PS, PP, POM, and PC, respectively, when temperature rose from 30 to 70 °C.

As another example for the effect of SL concentration on the equilibrium adsorption capacity, the Q_e values at 50 °C increased from 2×10^{-9} to 24.7×10^{-9} mol/m^2, from 0.2×10^{-9} to 4.5×10^{-9} mol/m^2, from 0.7×10^{-9} to 8.2×10^{-9} mol/m^2, from 0.4×10^{-9} to 5.3×10^{-9} mol/m^2, from 0.7×10^{-9} to 10.1×10^{-9} mol/m^2, and from 4.3×10^{-9} to 30×10^{-9} mol/m^2 for PVC, ABS, PS, PP, POM, and PC, respectively, when SL concentration rose from 1×10^{-5} to 4×10^{-5} mol/L.

The ABS material with the highest S value of 7.3 m^2/g showed the lowest Q_e values when testing in all SL concentrations. The Q_e values followed this sequence from highest to lowest: PC > PVC > POM > PS > PP > ABS.

The research findings showed that plastic special surface area did not impact adsorption, but the initial SL concentration and operation temperature did. The adsorption of chemicals onto polymer surfaces depends on multiple factors which include both polymer surface area characteristics and chemical material properties. The studied systems showed that the chemical structure parameter produced better results than the special surface area parameter.

The data showed the experimental data from PVC measurements which were used to create the Langmuir model's fitted line at 30, 50, and 70 °C operating temperatures. Also, obtained data represented Freundlich's parameters, n and K_F, coefficient of determination (R^2), and the equations of the fitted lines for the studied plastics at studied temperatures of 30, 50, and 70 °C. The study determined that the high R^2 values indicate successful data fitting for the Freundlich and Langmuir models that were investigated.

In addition, the experimental Q_e values for PVC, PP, and PC plastics at 30, 50, and 70 °C temperatures together with predicted values from Langmuir and Freundlich models were investigated. For example, the studied models produced values that matched experimental results well for PP at 30 °C when C_e's value was below 1 g/L.

The Freundlich model produced experimental data matches when C_e values rose. The SL molecules showed increased interaction at high concentrations which violated the Langmuir model assumptions that led to this result. The experimental data for PC and PVC showed no match with predicted values from either the Freundlich or Langmuir models at different C_e values. The polar surface of the plastics caused the experimental results to differ from predicted values through the Langmuir model. The PP displayed matching results when the temperature reached 50 °C. The PVC system demonstrated excellent agreement between experimental data and model predictions when using both models at low SL concentrations. The Langmuir model failed to predict actual Q_e values when SL concentrations were increased.

The PC observation revealed that all tested models failed to estimate the measured Q_e values at different C_e concentrations. The Q_e value predictions improved when the temperature reached 70 °C for low SL concentrations. The studied plastics PP, PVC, and PC showed matching experimental and predicted C_e values. The Langmuir model generated substantial differences between predicted Q_e values and actual measured values when SL concentrations rose.

The following equations show how the Langmuir isotherm constant (K_L, L/mol) relates to the Gibbs free energy (ΔG), enthalpy (ΔH), and entropy (ΔS) changes that happen during adsorption:

$$\Delta G = -RT\ln K_L$$

$$\Delta G = \Delta H - T\Delta S$$

$$\ln K_L = \frac{\Delta S}{R} - \frac{\Delta H}{RT}$$

The equation contains R and T variables which represent the gas constant and absolute temperature (K), respectively. The vertical intercept of ln K_L versus $1/T$ graph shows ΔS values, but the slope of the line indicates ΔH for adsorbate–adsorbent surface interactions. The authors presented thermodynamic parameters which include Gibbs free energy (ΔG), enthalpy (ΔH), and entropy change (ΔS) of SL adsorption on studied plastic surfaces. The adsorption process of SL onto selected plastics occurred spontaneously because the calculated ΔG values were negative. The calculated ΔG values between −17 and −12.97 kJ/mol satisfied the physical adsorption criteria because ΔG should fall between −20 and 0 kJ/mol. The reader should evaluate these values against chemisorption ΔG values which span from −400 to 80 kJ/mol. The ΔG values decreased when the temperature rose. The adsorption process becomes easier to perform at higher temperatures because the adsorption process requires heat energy to proceed since most adsorption reactions have positive ΔH values. The ΔG values for POM decreased from −14.13 to −16.97 kJ/mol when the temperature rose from 30 to 70 °C. The ΔH values for all materials except PC showed positive small values which spanned from +0.73 kJ/mol for PP to +7.44 kJ/mol for PVC. The process showed endothermic behavior because SL molecules failed to form chemical bonds with the studied plastic surfaces. The process of chemical bond formation needs ΔH values that range from 80 to 200 kJ/mol. The PC adsorption process showed exothermic behavior because its ΔH value was determined to be −4.99 kJ/mol. The PC surface displayed superior hydrophilic characteristics than the other plastic materials which led to this result. The measured adsorption ΔS values showed small positive values which spanned from +0.030 kJ/mol for PC to +0.075 kJ/mol for PP. The adsorption process caused a minimal increase in interface irregularity (randomness).

The flotation percentages of plastics resulted from tests conducted with various SL concentrations at 30, 50, and 70 °C temperatures. The flotation tank used tap water as its liquid medium which maintained a pH level of 6.5. The plastics floated on top of the flotation media (tap water) in the flotation tank without any surface treatment using SL regardless of their density values. The high surface tension of tap water prevents plastics with densities greater than water from sinking. The flotation results showed that PP and PC floated to the surface and then sank to the bottom of the flotation tank at all studied SL concentrations and temperatures. The studied plastics showed PP and PC as their most hydrophobic and hydrophilic materials, respectively. The flotation results matched the equilibrium adsorption capacity (Q_e) values which researchers measured for the plastics at different SL concentrations and temperatures. The surface treatment of plastics with different SL concentrations led to negative effects on the flotation process of ABS, PS, and POM plastics to varying degrees. The flotation tests showed positive results when using specific SL concentrations, but the temperature measurements did not reveal any patterns. The studied plastics showed POM as the plastic with the lowest flotation percentage at all tested SL concentrations and temperatures.

The flotation behavior of PVC remained unaffected by SL concentration at 30 °C and at higher SL concentrations during 50 °C testing. The flotation behavior of PVC showed no dependence on SL concentration at 30 °C but became dependent on SL concentration when the temperature reached 70 °C.

Based on findings of this case study, the SL adsorption process took place on plastic surfaces at various levels based on the measured equilibrium adsorption capacities (Q_es). The adsorption parameter demonstrated positive changes when the temperature rose. The SL adsorption pattern on plastic surfaces followed this order: PC > PVC > POM > PS > PP > ABS. The experimental data matched the predicted values from traditional adsorption models including Freundlich and Langmuir for all tested temperatures. The Freundlich model provided better results than Langmuir for most of the studied plastics. The thermodynamic parameter measurements demonstrated that SL adsorption took place through spontaneous endothermic reactions. The physical nature of SL adsorption on selected plastics surfaces became evident through their low thermodynamic values. The flotation process achieved complete separation of PC from plastic mixtures at all SL concentrations and temperature settings.

8.2.6 Miscellaneous Techniques

The industrial plastic separation methods primarily use density-based sink–float systems, triboelectric charging (Dani et al., 2024), dissolution–reprecipitation, and flotation as stated in previous sections. However, researchers now study additional techniques to solve problems that standard methods cannot handle. The "miscellaneous" separation methods become essential for processing plastics that share identical densities and surface modifications, aged materials and plastic blends with additives and fillers, multilayer packaging, and engineering plastics with small physical differences.

The new separation method known as froth-less hydrophobicity tuning (FHT) uses surface wettability gradients to eliminate froth formation while separating plastics on inclined or oscillating plates. The separation process functions independently from surfactants to reduce surface tension of water and air bubbles. The separation process depends on main factors which include the surface adsorption modifiers which function similarly to lignosulfonate systems learned about in previous case studies. The separation process depends on how plastics behave with water because of their hydrophobic nature, which causes them to either "roll" or "stick" to each other. The research by authors (Chapter 9—case studies) conducted about microwave irradiation-assisted and chemical depressants affecting hydrophobicity serves as the foundation for FHT. The system operates with low water movement that leads to decreased energy usage. Also, the system operates successfully when processing plastic materials with substantial amounts of impurities. However, it has some limitations including the technology that exists only in experimental stages, and the process needs exact control over surface treatment methods.

The focused laser pulse in LIBS (laser-induced breakdown spectroscopy) sorting technology creates plasma emission lines which identify elements through their characteristic spectral patterns. The method generates vital data which supports PVC detection (chlorine line at 837 nm), flame-retarded plastics, engineering plastics with halogens or metals, and multilayer composites. The advantage of LIBS technology lies in its ability to detect brominated flame retardants through its identification process.

The main sorting technology of advanced MRFs operates through NIR hyperspectral sorting. The NIR technology detects plastics through C—H stretches which produce vibrational overtones. However, the detection of black plastics through IR becomes difficult because these materials absorb IR radiation. Also, plastic materials become less detectable through NIR technology when they experience weathering.

Electrostatic corona roll charging (nontribo variant) is another miscellaneous technique. The corona charging process enables direct ion deposition onto plastic materials which results in, for example, the negative charge of PP becoming extremely strong.

Authors recommend adding a note about the air classification technique at the end of this section. The air classification technique solely is not a separation technique. However, it assists and prepares a clean feed for the plastics separation techniques that have been mentioned in previous sections.

Air classification is a physical separation method. Air classification uses the differences in particle mass, aerodynamic drag, and shape-dependent terminal velocity to separate the mixed plastic flakes. The dry operation of air classification methods prevents the problems which density-based wet separation methods encounter because it does not require water consumption or effluent treatment or wetting-agent compatibility. The air classification makes the plastics mix simpler and cleaner. I have seen air classifiers separate plastics from plastics in the waste management facilities. The air classifiers remove film contaminants and premixed shredded streams before the high precision sorting steps such as triboelectric sorting or optical sorting. Recent studies show that how well air classification works has improved because of the advances in blower design, adjustable baffles and real-time sensor monitoring.

The separation process depends on the equilibrium between gravitational pull and the resistance force that air streams apply when they move either upward or horizontally. The terminal velocity of each plastic particle depends on its density, projected area, and shape factor. The upward movement of particles occurs when they have low density and large surface area and high drag coefficients such as films, foils, and thin packaging materials. The heavy fraction may consist of PET fragments and rigid ABS and PC granules because these particles keep their dense compact shape and thickness.

8.3 How Can AI Help in Manual Sorting of Plastics?

MRFs now use AI through computer vision, ML, and robotics to boost human operator performance. AI systems work alongside human operators to improve sorting accuracy and consistency and processing speed in tasks that humans struggle with due to exhaustion and incorrect identification and slow work rates.

AI vision systems analyze items on conveyors to display immediate classification results through screen or heads-up display interfaces. The system enables human sorters to perform their tasks better through two main functions. The system enables users to identify polymers between PET, PVC, and PS and detects multi-layer packaging and dangerous items including batteries, needles, and electronic devices. The system displays two types of errors to users, which include color identification mistakes and resin identification mistakes. The system protects users from making errors because of fatigue during their work activities. The system improves human accuracy rates from 85–90% to 95–98% when sorting items that look similar to each other.

The AI system tracks belt movement to determine when items out of place will arrive at the manual worker's station. For example, the system alerts an approaching PVC item which will arrive at lane 3 in 2 seconds. The system enables workers to respond quickly while reducing PVC item detection failures, which results in better quality PET and HDPE bales. The system operates at PVC contamination levels between 500 and 1,000 ppm throughout the entire process.

AI systems monitor employee work performance and simultaneously track their fatigue status.

AI cameras with AI technology enable the system to track employee body movements and identify delayed responses and wrong selection choices and determine optimal worker shift patterns and stop accidents by tracking employee posture and reaching distances. The system brings multiple advantages to the process. The system helps workers fight fatigue, which produces higher quality products and protects their safety at work. The AI system delivers training instructions to new staff members who require learning about sorting operations. The AI simulation system provides new employees with interactive training through its simulation-based training system. The system shows workers images of various plastic materials for their training process. The system delivers instant feedback to workers whenever they choose wrong bins during their interactive labeling work. Also, the system tracks user performance through feedback mechanisms. It leads users through a step-by-step process of skill development stages.

The system enables new employees to learn sorting skills which usually require 2–3 months of training within a shorter period of 2–3 weeks.

AI vision systems detect objects which robotic arms use to execute fast micro-sorting operations while humans handle difficulty identifying items. For example, the AI robot system identifies PET bottles, but human staff members must remove PVC parts and dangerous substances from the remaining waste items. The system operates through robots that execute repetitive work, but humans conduct quality control assessments. Hence, the system enables robots to execute rapid repetitive operations on extensive product batches under human supervision for critical decision-making tasks.

Standard auto-sorters fail to detect black plastics because these materials absorb NIR light. The system uses different methods to solve this problem. It uses hyperspectral imaging to analyze samples through deep-learning classifiers which evaluate their shape and texture and edge detection features. The system uses RGB (red, green, blue) + thermal fusion models. Consequently, the system achieves 80–90% accuracy in black PP/PS recovery, which was previously unattainable.

AI systems possess the ability to identify main categories of contaminants, food residues, oil, grease, soil, and dust. The system detects multiple packaging material layers and identifies both additives and shading patterns. Hence, it produces alerts for operators to remove products showing contamination signs which require disposal through the residue line.

AI systems operate to achieve both maximum belt speed and uniform load distribution through their system functions. The AI system monitors conveyor belt density to determine the best belt speed and item placement, drop points, and robotic picking duration. The system operates automatically to distribute weight evenly, which reduces operator fatigue. The system generates precise sorting results when it processes big amounts of material.

AI systems conduct quality assurance inspections of sorted materials before baling them. The AI system detects material defects which activate notification alerts for bales which have color contamination amounts that exceed their established limits. The system identifies bales that have polymer ratios which differ from established specifications. It detects the presence of metals and glass and paper fragments in the material. The system identifies bales that have moisture levels which do not match the established standards. The system checks all materials against industry standards which include APR, PETCORE Europe, and ICIS.

References

Dani, C., Nagy, Á., Kovács, T., and Fekete, I. Effect of moisture content on the triboelectric charging of polymers. *J. Phys.: Conf. Ser.* 2024, *2702*, 012001.

Geyer, R., Jambeck, J. R., and Law, K. L. Production, use, and fate of all plastics ever made. *Sci. Adv.* 2017, *3*(7), e1700782.

Hahladakis, J. N. and Iacovidou, E. Closing the loop on plastic packaging materials: What is quality and how does it affect their circularity? *Sci. Total Environ.* 2018, *630*, 1394–1400.

Kalauni, K., Vedrtnam, A., Sharma, S. P., Sharma, A., and Chaturvedi, S. A comprehensive review of recycling and reusing methods for plastic waste focusing the Indian scenario. *Waste Manag. Res.* 2025, *43*, 1378–1399.

Längle, T. and Maier, G. *Sensor-based Sorting for Waste Plastics Recycling*. In Springer Handbook of Circular Plastics Economy; Springer Nature Switzerland: Cham, 2025; pp. 375–390.

Pita, F., Castilho, A., Ferreira, A., and Rocha, F. Separation of plastic mixtures by sink–float combined with froth flotation. *Polímeros* 2023, *33*(1), e20220094.

Plastics Europe. *Plastics—the Facts: An Analysis of European Plastics Production, Demand, and Waste Data*: Plastics Europe: Brussels, Belgium, 2024.

Salari, A. and Ostad Movahed, S. Effect of temperature on the adsorption of sodium lignosulphonate on the surface of selected plastics as a potential solution for plastics waste management. *Polym. Polym. Compos.* 2023, *31*, 09673911231181850.

Saleh, N. S., Ostad Movahed, S., and Attar Bashi, F. Anti-biofouling effects of grafted polyamide 6 fibers using vinyl chemicals. *J. Appl. Polym. Sci.* 2018, *135*, 46760.

Saleh, N. S., Ostad Movahed, S., and Attar Bashi, F. Effects of operative parameters on grafting of selected agents on polyamide 6 fiber surfaces. *J. Appl. Biomater. Funct. Mater.* 2022a, *20*, 2147483647.

Saleh, N. S., Ostad Movahed, S., and Attar Bashi, F. Functionalization of selected chemical agents on polypropylene fiber and investigation of physical and antibiofouling properties of functionalized fibers. *SPE Polym.* 2022b, *3*, 128–141.

Shamsaddini Negari, M., Ostad Movahed, S., and Ahmadpour, A. Separation of polyvinyl chloride, polystyrene, and polyethylene terephthalate granules using various chemical agents by flotation technique. *Sep. Purif. Technol.* 2017, *194*, 368–376.

Shariatikia, F., Ostad Movahed, S.; and Ostad Movahed, N. Study on the adsorption mechanism of sodium lignosulphonate on the surface of selected plastics as a potential solution for plastics waste management. *Polym. Polym. Compos.* 2023, *31*, 09673911231189645.

Taneepanichskul, N., Purkiss, D., and Miodownik, M. A review of sorting and separating technologies suitable for plastic waste streams. *Front. Sustain.* 2022, *3*, 901885.

Thimmasetty, J., Shwetha, S. K. K., Sharda, S., Shashank, N. N., Srinivas, H., Bhishambar, S., and Paramita, D. Comprehensive review of solubility parameters: theoretical foundations, determination methods, pharmaceutical applications, and limitations. *Russ. J. Phys. Chem. B* 2025, *19*(3), 632–645.

Wang, M., Gao, Y., Yuan, S., Deng, J., Yang, J., Yan, J., Yu, S., Xu, B., and Ma, D. Complete hydrogenolysis of mixed plastic wastes. *Nat. Chem. Eng.* 2024b, *1*(5), 376–384.

Xu, Z., Sanchez-Rivera, K., Granger, C., Zhou, P., Munguia-Lopez, A.D.C., Ikegwu, U.M., Avraamidou, S., Zavala, V.M., Van Lehn, R.C., Bar-Ziv, E. and De Meester, S. Solvent-based plastic recycling technologies. *Nat. Chem. Eng.* 2025, *2*(7), 407–423.

Zhao, R., Zhang, Z., Bai, X., Wang, H., Zhang, H., Hao, J., and Wang, C. A review of the research on triboelectric separation: principles, devices, and applications in resource recycling. *Powder Technol.* 2024, 118889.

Zhao, R., Zhang, Z., Bai, X., Wang, H., Zhang, H., Hao, J., and Wang, C. An innovative magnetic density separation process for enhanced recycling of plastics and electronic waste. *Recycling*. 2024a, *9*(3), 48.

9

Case Studies in Waste Plastics Separation Using Novel Flotation Technique

9.1 Separation of Polyvinylchloride, Polystyrene, and Polyethylene Terephthalate

The case study appeared in *Separation and Purification Technology* when authors published "Separation of polyvinylchloride (PVC), polystyrene (PS) and polyethylene terephthalate (PET) granules using various chemical agents by flotation technique" in 2017.

In this research, the treatment process involved standard virgin plastics from municipal waste which included polyvinylchloride (PVC), polystyrene (PS), and polyethylene terephthalate (PET) plastics that received separate or combined chemical agents from polyvinyl alcohol (PVA), polyethylene glycol (PEG), methyl cellulose (MC), tannic acid (TA), and methyl isobutyl carbinol (MIBC). The impacts of different factors on the floatability of used plastics, including the type and concentration of chemical agents, conditioning temperature and duration, and the pH of the liquid medium, were examined and the findings were analyzed. Commercial granule grades of PVC, PS, and PET were obtained from the local market. PEG (chemical agent, PEG-400, Shree Chem, India), PVA (chemical agent, PVA BP 20, Liwei Chemical Co. Ltd., China), MC (chemical agent, 274429, average molecular weight of 40,000, Sigma Aldrich, United States), TA (chemical substance, 100773 tannic acid, Merck, Germany), and MIBS (chemical agent and froth, Merck, Germany) were purchased and used. The plastics used for testing had matched spherical shapes and 5 mm diameter granules, but they differed in color for easier hand sorting after the flotation test. The process began with deionized water pretreatment followed by room temperature drying of the material before it entered the flotation tank. A column flotation cell made of glass, measuring 20 cm in diameter and 80 cm in height, was utilized for flotation experiments. The thickness of the glass wall was 6 mm, and the flotation tank had a drainage valve at the bottom along with a circular air blowing system featuring a 13 cm air distributor diameter and a flow rate of 4 L/min. The research used tap water as its water source which maintained a standard pH level of 6.5. The surrounding temperature was 25 °C. The temperature of the liquid media was modified using a portable electric coil heater, and the pH level was recorded with a microprocessor pH meter (HANNA instruments). The researchers used sulfuric acid at 2 mol/L and sodium hydroxide at 2.5 mol/L to adjust the pH levels in the liquid media. Initially, a solution of the chemical agent was created with a defined concentration in a 1 L glass jar. Subsequently, 100 granules (to simplify calculations) of each polymer were immersed in the prepared solution for a designated time and temperature, and then introduced into the flotation tank to about two-thirds of the tank's height from the bottom. Subsequently, the treated polymers (plastics) were added to the flotation tank while air circulation was functioning for one minute. Every experiment was conducted a minimum of three times, and the median values were presented. Figure 9.1 represents the flotation tank.

All utilized chemicals (PVA, TA, PEG, and MC) can be dissolved in water, and except for MIBC, they remain solid at room temperature. The organic structure of these materials seems suitable for adsorption onto used plastic surfaces either individually or in combination, which results in surface energy modifications. The chosen chemical agents have minimal effects on the environment and human health, with some also serving as food additives and being utilized in medical treatments. Nevertheless, sufficient attention must be given when using and reusing them multiple times. For nonwetting, wetting, and perfect wetting, the contact angles for θ should be 180°, less than 90°, and 0°, respectively. Table 9.1 presents the surface tension of water along with the surface energies of the plastics in contact with air at room temperature.

As noted, there is a significant contrast between the surface tension of water, which is 71.97 mN/m, and that of the plastics used, ranging from 40.7 to 44.6 mN/m. As anticipated, used plastics do not wet or are at least partially wet in contact with water. To sink, the plastic surface must be entirely wet ($\theta = 0°$); otherwise, it will float on the liquid's surface. Nonetheless, it discovered that plastics flotation is influenced not just by their surface characteristics and θ angle with the liquid, but also by gravitational factors as per the subsequent equation:

$$\rho_{pb} = \frac{1.59\ \varnothing \rho_p d_{pe}}{1.59\ \varnothing d_{pe} + 4Kd_b}$$

where ρ_{pb}, $\varnothing$, ρ_p, d_{pe}, K, and d_b represent particle–bubble density, particle shape parameter, particle density, diameter of particle equivalent sphere, bubble coverage on the particle surface (a larger contact angle, θ, leading to increased K value), and bubble diameter, respectively. The densities of the employed plastics were higher than that of used water (Table 9.1), and all of them sank to the

Figure 9.1 The used flotation tank. *Source*: Negari et al. (2017) / with permission of Elsevier.

Table 9.1 Water surface tension and used plastic surface energies in contact with air at ambient temperature.

Material	Water	PET	PVC	PS
Surface energy (mN/m)	71.97	44.6	41.5	40.7
Density (g/cm^3)	1	1.40	1.44	1.05

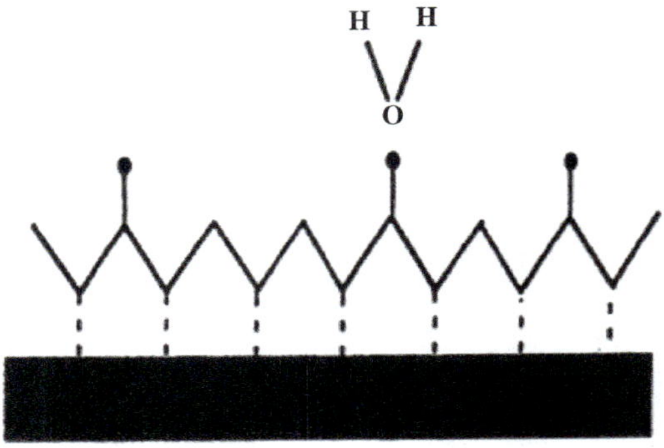

Figure 9.2 Adsorption mechanism of a typical depressant on a nonpolar polymer surface.

bottom of the flotation tank in the scenario of "without conditioning" without using chemical agents. The adsorbent (polymer surface) starts a cycle of adsorption and desorption with the adsorbate (chemical agent) when they first touch, which might lead to an equilibrium point. The adsorption process of a depressant on nonpolar polymer surfaces occur as shown in Figure 9.2 when PS interacts with water solutions. It is evident that the adsorption–desorption kinetic and equilibrium will be influenced by the nature of depressant (polar and/or nonpolar) and polymer surface.

Typically, the adsorption of utilized chemicals on the examined plastic surfaces (PET, PS, and PVC) enhanced the hydrophobicity of these surfaces, leading to increased K values in the equation stated earlier. The buoyancy of PET, PS, and PVC in relation to different concentrations of TA, MC, PEG, and PVA was examined. Also, the same for MIBC at three concentrations: 1, 2, and 3 mL/L performed. Since MIBC is a liquid at room temperature, expressing concentration in mL/L was more practical. The experiments were conducted at room temperature and a pH of 6.5 (tap water) for 15 minutes of chemical agent conditioning. The results demonstrated that without employing a chemical agent, PVC, PET, and PS were entirely submerged at the bottom of the flotation tank. This was because their densities exceeded that of the liquid media (Table 9.1). Nonetheless, raising TA to 1,200 mg/L resulted in complete PVC flotation. Outside this concentration, PVC flotation decreased significantly to 75% and 55% at TA concentrations of 1,600 and 2,000 mg/L, respectively. PET granules stayed at the bottom of the flotation tank across all TA concentrations, whereas PS floated at TA levels exceeding 1,600 mg/L, with a rate of 5%. TA is a macromolecular depressant that contains an excessive number of hydroxyl (—OH) groups and a benzene ring in its structure. It was determined that TA was successful in enhancing PVC flotation by adsorbing onto the surface of this plastic. It raised the hydrophobicity of the PVC surface to 1,200 mg/L, but surpassing the TA concentration beyond this level accelerated the desorption process as the main phenomenon.

Adsorption and desorption compete with one another, and their advancements rely on the conditions within the reaction vessel, such as depressant concentration. The solubility of TA in water is significant (2,850 g/L). This is because water molecules can form strong dipole or hydrogen bonds with TA molecules. Conversely, PVC contains chlorine atoms within its composition. Chlorine possesses the highest electron affinity and ranks third in electronegativity, following only oxygen and fluorine. At a TA concentration of up to 1,200 mg/L, TA molecules tend to attach to the PVC surface because of the chlorine present in the PVC structure and the organic characteristics shared by both TA and PVC. As TA concentration rises, TA molecules will fill all active sites on the PVC surface, leading to complete PVC flotation. Further increasing TA concentration did not improve the adsorption process since all effective sites were already filled by TA molecules. In this context, the increased presence of TA molecules in water facilitates the desorption of adsorbed TA molecules from the PVC surface, allowing them to merge. The process can be characterized by aqueous boundary layer diffusion (ABLD). Clearly, two comparable molecules (TA:TA) exhibit greater compatibility with one another than TA:PVC molecules. For PET, TA molecules showed a preference for interacting with water molecules instead of adsorbing onto the PET surface. This process is additionally facilitated by the simpler migration and interaction of light water molecules relative to heavy TA molecules on the PET surface, owing to the presence of oxygen atoms in the PET composition. Both PS and TA contain a benzene ring in their structures. This fact provides a rational explanation for the adsorption of this chemical on the PS surface when TA concentration surpassed 1,600 mg/L. However, the interactions between water and TA molecules remained stronger compared to those between TA and PS molecules, particularly for TA concentrations below 1,200 mg/L.

The results showed that MC significantly impacted PS flotation. The floatability values of this plastic at concentrations of 400, 800, 1,200, 1,600, and 2,000 mg/L MC were 85%, 85%, 90%, 95%, and 95%, respectively. In the meantime, the PET flotation showed no sensitive effect except at the concentration of 2,000 mg/L, which had a value of 5%. The peak floatability for PVC was at concentrations of 1,600 and 2,000 mg/L, both showing values of 15%. MC contains multiple oxygen atoms and methyl groups in its structure. The presence of the methyl group allows MC to easily and effectively adsorb onto the PS surface. Nonetheless, the presence of oxygen atoms in the MC structure facilitates interaction with water molecules when contrasted with the PVC surface, which contains highly negative chlorine atoms in its composition. The outcome will be reduced by MC adsorption on the PVC surface, leading to a smaller change in surface hydrophobicity (decreased flotation) compared to PS. Also, the impact of PEG on the floatability of the plastics is examined. Both PVC and PS were influenced by this chemical agent, as noted. Their floatability rose significantly to values of 80%, 85%, 85%, 90%, and 100% for PS, and 50%, 75%, 80%, 90%, and 100% for PVC by raising the concentration of the chemical agent to 400, 800, 1,200, 1,600, and 2,000 mg/L, respectively. PET was not floated except for PEG concentration, which was 2,000 mg/L and had a floatability value of 5%. PEG contains ethyl groups and oxygen atoms in its structure, allowing it to adhere to PS and PVC surfaces. The outcome will be an increase in hydrophobicity of the specified plastics. Nonetheless, the increase in hydrophobicity was more pronounced for PS due to its inherent non-polar surface. It was determined that PEG was more efficient for the flotation of PS compared to PVC. PVA showed no impact on the floatability of PET across all concentrations examined. However, it was successful on PVC, particularly when PVA concentration surpassed 800 mg/L. The maximum floatability for PVC reached 65% at a PVA concentration of 2,000 mg/L. The PS exhibited consistent floatability across all examined PVA concentrations (5%), regardless of the PVA concentration. PVA exhibited lower effectiveness than PEG for enhancing the hydrophobicity of PVC, leading to reduced floatability percentages at identical concentrations. MIBC typically combines with other chemical substances. The results showed that MIBC positively impacted the floatability of the examined plastics across various values. The lowest value was associated with PET, which had 50% floatability at a 3 mL/L MIBC concentration, while the highest value was for PVC, showing 95% floatability at the same MIBC concentration. The floatability percentage values for PS were situated between the values for PET and PVC.

Table 9.2 illustrates how various flotation temperatures 25, 35, 45, 55, and 65 °C influence the floatability percentage of PVC, PET, and PS concerning the examined chemical agents, TA, MC, PEG, PVA, and MIBC. In the first set of experiments, the

Table 9.2 Effect of temperature on floatability percentage of studied plastics.

	Floatability (%)					
Polymer	TA (400 mg/L)	MC (400 mg/L)	PEG (400 mg/L)	PVA (400 mg/L)	MIBC (1 mL/L)	Temperature (°C)
	50	5	50	20	75	25
	55	55	50	20	90	35
PVC	55	65	75	40	95	45
	65	70	90	95	100	55
	80	75	95	100	100	65

	Floatability (%)					
Polymer	TA (400 mg/L)	MC (400 mg/L)	PEG (400 mg/L)	PVA (400 mg/L)	MIBC (1 mL/L)	Temperature (°C)
PET	0	0	0	0	5	25
	0	0	0	0	5	35
	0	0	0	0	10	45
	0	0	0	0	20	55
	0	0	0	0	20	65
PS	0	85	80	5	85	25
	5	85	80	5	95	35
	10	90	95	5	95	45
	25	100	100	25	100	55
	35	100	100	30	100	65

Polymer	TA (1,600 mg/L)	MC (1,600 mg/L)	PEG (1,600 mg/L)	PVA (1,600 mg/L)	MIBC (3 mL/L)	Temperature (°C)
PVC	75	15	90	40	90	25
	80	50	95	45	95	35
	85	75	95	50	95	45
	85	90	95	95	95	55
	95	100	100	100	95	65
PET	0	0	0	0	50	25
	0	0	0	0	50	35
	0	0	0	0	50	45
	0	0	0	0	70	55
	0	0	0	0	75	65
PS	5	85	90	5	95	25
	10	90	90	5	95	35
	15	100	95	5	100	45
	25	100	100	25	100	55
	45	100	100	30	100	65

concentration of the chemical agent was 400 mg/L (for MIBC, 1 mL/L) at a pH of 6.5, with a conditioning duration of 15 minutes. In general, except for PET, higher temperatures enhanced the floatability of all the examined plastics for the utilized chemical agents, TA, MC, PEG, and PVA. The heat of adsorption sign (ΔH^0) indicates whether the adsorption process is endothermic or exothermic. Adsorption is a process that reaches equilibrium, and by understanding the sign of ΔH^0, one can anticipate how temperature influences this process. Nonetheless, the sign of ΔG^0 also relies on the change in entropy during adsorption. Regardless of the sign of ΔH^0, additional studies noted that raising temperature could enhance or hinder the adsorption process based on the examined conditions. The floatability of PET stayed the same and was not affected by operating temperature. For MIBC, the floatability of PET rose from 5% to 20% as the temperature climbed from 25 to 65 °C. As the operating temperature rose from 25 to 65 °C, for TA, the floatability of PVC and PS improved from 50% to 80% and from 0% to 35%, respectively. Similar trends were seen for MC (5–75% for PVC and 85–100% for PS), PEG (50–95% for PVC and 80–100% for PS), PVA (20–100% for PVC and 5–30% for PS), and MIBC (75–100% for PVC and 85–100% for PS). It was determined that as temperature rises, the physical interactions between the examined chemical agents and plastic surfaces enhanced, leading to greater hydrophobicity of the plastic surface and improved flotation. Table 9.2 again illustrates the impact of identical flotation temperatures on the floatability percentage of PVC, PET, and PS for the examined chemical agents at a different chemical agent concentration of 1,600 mg/L (for MIBC, 3 mL/L) at a pH of 6.5. Assessing the floatability percentage of each plastic at identical temperatures and using the same chemical agent indicates increased values for a chemical agent concentration of 1,600 mg/L in contrast to the values for 400 mg/L.

For example, the floatability percentage of PVC at 400 mg/L TA and 55 °C is 65% (Table 9.2), which can be contrasted with 85% for 1,600 mg/L TA at the same temperature (Table 9.2).

The liquid conditioning pH levels of 4.5, 6.5, 8.5, and 10.5 influence the floatability percentage of PVC, PET, and PS for the analyzed chemical agents, TA, MC, PEG, PVA, and MIBC. In these experiments, the concentration of the chemical agent was 800 mg/L (with MIBC at 2 mL/L) at room temperature (25 °C) and a conditioning duration of 15 minutes. Both economically and environmentally, minimizing the use of chemical agents is advantageous, specifically when the reader focuses on the fact that the aim of this project is to enhance the outcomes for industrial use. Therefore, after thorough evaluation and considering the outcomes for all agents, 800 mg/L was selected as the concentration for the agent. It was noted that raising the pH of the conditioning liquid, except for MIBC, did not impact the floatability of PET. For MIBC, the floatability percentage of PET achieved values of 35%, 40%, 50%, and 50% at pH levels of 4.5, 6.5, 8.5, and 10.5, respectively. MIBC is a type of alcohol that has limited solubility in water.

At decreased pH levels, MIBC molecules participate in an ester formation reaction, which decreases their adsorption on the plastic surface. Raising pH decreases the esterification reaction and facilitates the adsorption of MIBC molecules on the polymer surface. The result will be an increased MIBC concentration on the PET surface, leading to improved PET flotation. Raising the pH facilitated flotation for both conditioned PVC and PS with TA. The floatability percentage measurements were 55%, 75%, 75%, and 80% and 0%, 5%, 10%, and 10% for PVC and PS at different analyzed pH levels, respectively. In contrast to PVC, the rise in pH negatively impacted the conditioned PS with MC. The floatability percentage measurements were 75%, 85%, 80%, and 70% as pH rose from 4.5 to 6.5, 8.5, and 10.5, respectively. The decrease in floatability percentage from 80% to 70% might be linked to increased liquid surface tension at acidic pH. The use of PEG for conditioning improved PVC, leading to an increase in floatability percentage from 70% to 90% at pH levels of 4.5 and 10.5, respectively. Nonetheless, it reduced the floatability percentage of PS to 90% and 75% for the specified pH levels. Once again, the reason pertains to the two competing factors present during the flotation process: the activation/deactivation of chemical agents and the dependency of liquid surface tension on pH. The conditioned PS with PVA exhibited the smallest variation in floatability percentage, showing 0% for pH 4.5 and 5% for the left studied pH values, whereas PVC had values of 40% and 65%. The floatability percentage values of conditioned PVC and PS with MIBC were significantly higher compared to the values for TA, MC, PEG, and PVA.

The influence of conditioning durations (times) of 5, 10, 15, and 30 minutes on the floatability percentage of PVC, PET, and PS for the examined chemical agents, TA, MC, PEG, PVA, and MIBC, was examined. The concentration of the chemical agents in these experiments was 800 mg/L (for MIBC, 2 mL/L) at room temperature (25 °C) and a pH of 6.5. Once again, and as anticipated, prolonging conditioning time, except for MIBC, did not influence the floatability of PET. For MIBC, the flotation percentage of PET rose from 20% with 5 and 10 minutes of conditioning to 40% after 15 minutes of conditioning. Additional conditioning time of 30 minutes negatively impacted the floatability percentage of PET, decreasing it to 15%. The presence of the hydroxyl (—OH) group in MIBC's structure makes it proper for adsorption on the PET surface which contains multiple oxygen atoms in its chemical structure. Extending the conditioning time to 15 minutes enhances the hydrophobicity of the PET surface, leading to improved flotation. However, the MIBC desorption process from PET surfaces starts to dominate the system after this point, which results in decreased PET flotation. The floatability percentage for treated PVC with TA rose from 40% to 75% as conditioning time extended from 5 to 15 minutes. The floatability percentage stayed constant at its initial value because the length of conditioning time did not affect it. After 15 minutes, the active sites on the PVC surface were fully occupied by TA molecules, leading to a time-independent behavior of PVC following the 15 minutes of conditioning. Extending the conditioning time had no significant impact on the floatability percentage of PS for up to 15 minutes. Nonetheless, extending the duration to 30 minutes resulted in 5% flotation for this plastic. The floatability percentage showed the smallest change when researchers extended the conditioning period for conditioned PVC and PS with MC which resulted in 10% and 5% variations. The values for PS were significantly greater, ranging from 85% to 90%, in contrast to the values for PVC, which were between 5% and 15%. This finding indicated that more MC molecules were absorbed on the PS surface. The floatability percentage values of treated PVC and PS with PEG were similar, falling within the ranges of 70–85% and 80–85% as the conditioning time increased from 5 to 30 minutes for PVC and PS, respectively. For conditioned PVC and PS with PVA, it was noted that a longer conditioning time improved floatability percentage, particularly when the duration extended from 15 to 30 minutes. Extending conditioning time with MIBC negatively impacted the floatability percentage of both PVC and PS. The floatability percentage values decreased from 100% to 85% as conditioning time rose from 5 to 30 minutes, respectively. It was shown that the rate of desorption surpassed that of adsorption after 30 minutes had elapsed.

Tables 9.3 and 9.4 illustrate the impact of a binary mixture of various examined chemical agents on the floatability of PET, PS, and PVC. The conditioning time, temperature, pH, and overall chemical agent concentration (ratio) were maintained at 15 minutes, 25 °C, 6.5, and 800 mg/L (50–50) for all investigated binary experiments, respectively. It was noted that when the chemical agent ratio was maintained at 50–50 (equal ratio, Table 9.3), the flotation process did not influence PET flotation, resulting in all PET granules settling at the bottom of the flotation tank throughout all trials. Nonetheless, for TA:PVA, TA:PEG, PVA:PEG, TA:MC, PVA:MC, and PEG:MC, the floatability percentage values for PVC were 50%, 85%,

Table 9.3 Percentage of floatability for binary mixtures of utilized flotation agents. The conditioning duration, temperature, pH, and total chemical agent concentration (ratio) were maintained at 15 minutes, 25 °C, 6.5, and 800 mg/L (50–50), respectively.

	Floatability (%)					
Plastic	**TA:PVA**	**TA:PEG**	**PVA:PEG**	**TA:MC**	**PVA:MC**	**PEG:MC**
PET	0	0	0	0	0	0
PS	0	0	40	5	0	65
PVC	50	85	40	0	35	75

Table 9.4 Percentage of floatability for binary mixtures of utilized flotation agents. The conditioning duration, temperature, pH, and overall concentration of chemical agents were maintained at 15 minutes, 25 °C, 6.5, and 800 mg/L with various ratios of chemical agents, respectively.

	Floatability (%)		
Plastic	**TA:PEG + MIBC**	**TA:PVA + MIBC**	**TA:MC + MIBC**
PET	0	0	0
PS	100	20	90

40%, 0%, 35%, and 75%, respectively. The values related to PS were 0%, 0%, 40%, 5%, 0%, and 65% for the previously mentioned binary mixtures, respectively.

Unexpectedly, for TA:PEG, 85% of PVC was floated and separated from the mixture of plastics (PET, PS, and PVC). It was determined that TA:PEG possessed sufficient ability to enhance the hydrophobicity of the PVC surface when they adhered to the plastic surface.

Table 9.4 illustrates the impact of binary mixtures of TA:PEG at different ratios on the floatability of PET, PS, and PVC. It was evident that 100% of both PS and PET stayed unfloated in every examined ratio. Increasing the PEG ratio in the binary mixture from 0.2 to 0.4, 0.5, 0.6, and 0.8 resulted in an increase in PVC floatability percentage from 75% to 85%, 85%, 90%, and 100%. It was disclosed that all PVC can be extracted from a PET:PS:PVC mixture by employing combined chemical agents, TA:PEG in a 0.2:0.8 ratio under the previously examined conditions.

Table 9.5 illustrates how conditioning a blend of PET and PS with a binary combination of chemical agents, TA:PEG, TA:PVA, and TA:MC, at an equal ratio along with 2 mL/L MIBC influences the outcome. Surprisingly, by increasing 2 mL/L MIBC to match the TA:PEG ratio and then conditioning the mixture of plastics (PET and PS), all PS floated completely. In conclusion, PVC and PS can be fully separated from PET in their mixture using a TA:PEG ratio of 0.2:0.8 and a TA:PEG equal ratio combined with 2 mL/L MIBC in two stages. The left chemical agent combinations, TA:PVA and TA:MC with MIBC, floated 20% and 90% of the PS content from a binary mixture of PS and PET, respectively.

Table 9.4 additionally displays the impact of a binary mix of TA:PVA and TA:MC in different ratios on the floatability percentage of a mixture consisting of PET, PS, and PVC. It was noted that TA:PVA allowed PVC to float with percentages of 35%, 65%, 50%, 70%, and 60% for TA:PVA ratios of 0.8:0.2, 0.6:0.4, 0.5:0.5, 0.4:0.6, and 0.2:0.8, respectively. It had no impact on PS flotation in all examined ratios. From

Table 9.5 Floatability percentage for a binary mixture of PET and PS employing various flotation agents. The overall concentration of the chemical agent is 800 mg/L with an equal ratio of 50–50, in addition to 2 mL/L of MIBC. The conditioning duration, temperature, and pH were maintained at 15 minutes, 25 °C, and 6.5, respectively.

	Floatability %				
Plastic	**0.8:0.2 (TA:PEG)**	**0.6:0.4 (TA:PEG)**	**0.5:0.5 (TA:PEG)**	**0.4:0.6 (TA:PEG)**	**0.2:0.8 (TA:PEG)**
PET	0	0	0	0	0
PS	0	0	0	0	0
PVC	75	85	85	90	100
	0.8:0.2 (TA:PVA)	**0.6:0.4 (TA:PVA)**	**0.5:0.5 (TA:PVA)**	**0.4:0.6 (TA:PVA)**	**0.2:0.8 (TA:PVA)**
PET	0	0	0	0	0
PS	0	0	0	0	0
PVC	35	65	50	70	60
	0.8:0.2 (TA:MC)	**0.6:0.4 (TA:MC)**	**0.5:0.5 (TA:MC)**	**0.4:0.6 (TA:MC)**	**0.2:0.8 (TA:MC)**
PET	0	0	5	0	0
PS	5	0	0	0	0
PVC	5	5	0	0	5

Table 9.4, it's evident that TA:MC was ineffective as a floater for the PET:PS:PVC mixture and did not influence the flotation of the examined mixed plastics.

Unexpectedly, comparing the author's findings with other studies showed that the form and dimensions of the plastics utilized were crucial elements in the separation of plastics. In this research, spherical granules measuring 5 mm in diameter were utilized, whereas most researchers employed shredded flake-shaped plastics of different sizes. Some of their findings align with the results obtained by other researchers.

The research findings resulted in the followings:

- In the majority of experiments, PET maintained the lowest flotation change for the examined chemical agents and settled at the bottom of the flotation tank.

- TA effectively improved PVC flotation via surface adsorption and altered the hydrophobic properties of the PVC surface in various ways based on the TA concentration.
- Generally, except for PET, raising the temperature enhanced the floatability of all analyzed plastics for the employed chemical agents, TA, MC, PEG, and PVA.
- Raising the pH of liquid media altered the buoyancy of the examined plastics differently based on the pH level and the chemical agent used.
- The plastic flotation results showed better performance when the conditioning period was extended for up to 15 minutes.
- A key conclusion is that PVC and PS can be entirely separated from PET in their mixture using TA:PEG at a 0.2:0.8 ratio and TA:PEG at an equal ratio combined with 2 mL/L MIBC in two stages.
- The research results demonstrated that plastic separation success depended on both plastic form and dimensions according to previous scientific investigations.
- Since mixed plastic waste (MPW) has various contaminants on the surfaces of plastics, additional research and studies will be necessary for MPW. The MPW needs to undergo various pretreatment steps which begin with water cleaning followed by alkaline cleaning before it can proceed to the flotation tank.

9.2 Separation of Acrylonitrile–Butadiene–Styrene, Polycarbonate, and Polyoxymethylene

The case study appeared in *Journal of Polymers and the Environment* when authors published "The flotation by selected depressants as an efficient technique for separation of a mixed acrylonitrile butadiene styrene, polycarbonate and polyoxymethylene plastics in waste streams" in 2019.

In this work, authors took acrylonitrile–butadiene–styrene (ABS), polycarbonate (PC), and pure polyoxymethylene (POM) that were mixed in a plastic blend. They separated the ABS, PC, and pure POM from each other using the flotation technique with the aid of selected depressants. They also used Design-Expert® software to predict the floatability of the studied plastics, ABS, PC, and POM. The effective parameters studied on floatability of the plastics were the conditioning time of plastic with depressant, the flotation tank temperature, the flotation liquid pH, and the concentration of depressant during plastics surface conditioning.

The engineering polymers (plastics) exhibit superior mechanical, dynamic, and thermal properties in comparison to conventional plastics. Due to their relatively high cost, recycling them is profitable enough to conserve resources.

The production of ABS, PC, and POM engineering plastics has become more prevalent in waste streams during the last few years. This accumulation of waste requires a modern, cost-effective, safe, and simple method for separating the specified plastics to reintegrate them into the production cycle and facilitate reuse. The chosen plastics, ABS, PC, and POM were classified as engineering plastics. Due to their significantly different densities compared to commonly used plastics in waste streams, such as PE, PP, PVC, and PS, they can be isolated from these plastics using a density difference method. Nonetheless, further differentiation using the latter method is unfeasible as the chosen engineering plastics possess similar densities.

The authors conducted previous research (Section 9.1) to separate conventional plastic mixtures containing PS, PET, and PVC through flotation separation with depressant adsorption techniques. The research employed depressants which included PEG, MC, PVA, TA, and MIBC either individually or together. In this research, the identical approach with the same depressants was employed for the flotation-based separation of a mixed selection of engineering plastics, including PC, ABS, and POM. Afterward, the findings were presented and analyzed. The primary aim was the industrial and commercial use of this research. The selection process chose flotation agents which fulfill three necessary conditions of being inexpensive, nontoxic, and easily accessible in big market quantities. The local market provides three commercial granule grades which include ABS, PC, and POM. The characteristics of the utilized depressants were discussed in Section 9.1. The plastics that were used had a spherical shape with an average diameter of 4 mm. Yet, to enhance differentiation, they were in various colors. The researchers performed a thorough deionized water rinse before they allowed the samples to dry at room temperature before starting the flotation process. The flotation equipment and procedure previously described are employed (Section 9.1, Figure 9.1). A glass flotation tank with measurements of 20 cm (diameter) and 80 cm (height) was used for flotation experiments. A circular airflow system featuring a 13 cm diameter air distributor and a flow rate of 4 L/min was utilized to aid the flotation process. The flotation duration was approximately two minutes in all tests, contingent on the state of the flotation tank. The experiments were performed three times while scientists documented the median results from each test. The design of experiments (DOE) enables researchers to use modified input factors for predicting particular results. The Design-Expert® statistical software analyzed all input data which included conditioning time, flotation tank temperature, pH values, and depressant concentration levels to produce the flotation percentage of selected plastics. The data processing employed response surface methodology (RSM) through D-optimal method analysis of the previously identified factors.

The selected depressants consist of environmentally friendly materials which bond physically (adsorb) to plastic surfaces through their chemical composition to make the plastic surface more water-absorbent. The surface wettability of polymers depends on their chemical composition because this determines how well depressants can bind to the surface which affects plastic floatation or sinking behavior. The reader needs to consider multiple essential elements which affect the process including particle shape, equivalent diameter, particle density, air or gas bubble coverage on particles, bubble size, fluid velocity, and fluid (liquid) surface tension and viscosity. The surface energies of POM, PC, and ABS were 44.6, 42.9, and 42.7 mN/m, respectively. The reader should expect weak wettability and high contact angles for the mentioned plastics when they compare the values to the surface tension (energy) of water at 72 mN/m. The plastic materials settled at the bottom of the flotation tank because their densities fell between 1.05 and 1.41 g/cm^3, while water density remained at 1 g/cm^3.

The results showed that the floatability of ABS, POM, and PC plastics underwent changes when different depressants including TA, MC, PEG, PVA, and MIBC were applied. The study revealed two exceptions where TA and MIBC failed to produce any significant impact on PC and POM floatability across different depressant concentration levels. The studied plastics flotation process benefited from higher concentrations of TA, PEG, and MIBC, but the effect on MC and PVA depended on the amount of depressant used. The adsorption–desorption process requires complete understanding of its mechanisms together with all influencing factors which determine how a depressant affects plastic surfaces. The depressant adsorption–desorption process requires evaluation of these specific factors:

a. The plastic surface which functions as an adsorbent needs appropriate sites to effectively adsorb the depressant substance (adsorbate). The number of active sites remains restricted because it depends on the specific properties which adsorbent surfaces exhibit.
b. The sites of adsorption can hold one adsorbate molecule at a time, but the adsorbed molecules do not form a single layer.
c. The enthalpy of adsorption for sites may not be equivalent.
d. The enthalpy of adsorption depends on the degree of depressant molecules coverage on the plastic surface.
e. The system contains two competitive processes which involve adsorbate adsorption and desorption events. The rates of adsorption and desorption maintain equal values when the system reaches its equilibrium state.
f. The rate of adsorption remains low because the activation energy amounts to 1 kcal/mol while the process operates as a reversible system which differs from chemisorption.
g. The Gibb's free energy of the process needs to be negative because it operates as a spontaneous thermodynamic process. The sign and value of Gibb's free energy depend on the sign and values of enthalpy and entropy variation of the system during adsorption and or desorption.

h. The adsorbed molecules establish direct and indirect contact with neighboring molecules as well as with unadsorbed molecules in the system.

The main structure of TA consists of hydroxyl and carboxyl groups together with a benzene ring system. The numerous hydroxyl groups in TA structure enable fast hydrolysis reactions with water molecules because TA dissolves at 2,850 g/L in water. The presence of carbonate ester groups in PC indicates that this plastic material will absorb water. The water molecules showed greater bonding strength with TA molecules than TA molecules did with PC molecules because PC failed to float by TA at any of the tested concentrations. The POM test revealed that TA concentrations above 1,600 mg/L would produce 5% flotation results. The ABS floated at 10% when TA levels hit 1,600 mg/L before the desorption process accelerated, and the flotation falls to 5%. The process occurs through ABLD mechanism, while TA molecules in liquid media tend to form clusters with adsorbed TA molecules. The MC and PVA tests showed that the studied plastics did not follow any predictable pattern of floatability behavior.

The floatability of the samples showed an increase for specific MC and PVA concentrations, but it decreased for all other tested MC and PVA concentrations. The floatability of ABS reached 65% and 70% when MC concentrations reached 400 and 1,600 mg/L, but it decreased to 55% and 60% at 800 and 1,200 mg/L and reached 35% at 2,000 mg/L. The flotation percentages of POM showed the following results when PVA concentrations reached 400, 800, 1,200, 1,600, and 2,000 mg/L: 80%, 85%, 70%, 65%, and 75%, respectively. The adsorption–desorption process includes at least three different mechanisms which operate in this system. The plastic surface shows depressant adsorption because its chemical structure matches the MC compound which contains oxygen and methyl groups. The solid surface and aqueous boundary layer experience competition between water molecules and depressant molecules for adsorption sites while boundary layer diffusion of aqueous substances occurs. The study produced distinct flotation patterns for used plastics when researchers applied different concentrations of depressant. The PEG demonstrated outstanding ability to float all plastic materials which researchers studied. The flotation of POM, ABS, and PC reached 80%, 60%, and 55%, respectively, when PEG concentration reached 400 mg/L. The PEG concentration needed to be increased for plastics flotation to achieve any additional benefits because the flotation percentage only rose by 5–10%. The main factor which caused PEG molecules to adsorb onto plastic surfaces was their structural similarity to chemicals (PEG contains ethyl groups and oxygen atoms in its composition). MIBC functions primarily as a froth agent instead of a depressant and it requires use with additional agents for its application. The results showed that POM flotation did not work but the ABS and PC materials floated at 65–70% and 20–25%, respectively. MIBC exists as a small molecule which contains a hydroxyl functional group. The high hydrolysis ability of MIBC and its fast water molecule absorption rate which competes with MIBC molecules for POM surface sites resulted in no significant impact from this depressant on POM flotation.

The study revealed that all plastics except PC and POM achieved high flotation percentages using depressants except for TA and MIBC.

Table 9.6 shows how different flotation temperatures from 15 to 55 °C affect the floatability of POM, ABS, and PC when using TA, MC, PEG, PVA, and MIBC as depressants. The experiments used 400 and 1,600 mg/L depressant concentrations (MIBC at 1 and 3 mL/L) with 15 minutes of conditioning at pH 6.5.

The results indicated that TA concentrations at 400 mg/L did not affect PC and POM flotation when temperatures ranged from 15 to 25 °C. The research found that POM and PC showed 20% and 30–65% flotation rates at different temperature points. The ABS flotation process using this depressant did not show any regular response to rising temperatures. The flotation percentage reached 85% when the temperature reached 35 °C before it decreased to 45% at 55 °C. The 1,600 mg/L TA concentration showed a distinct response to temperature changes. The flotation process for all plastic samples achieved its highest point at 35 °C before the temperature rose to 55 °C which caused the process to decrease. The increase in temperature produced negative results for all plastics flotation tests which used MC at 400 mg/L concentration and PVA at 400 and 1,600 mg/L concentrations. The MC flotation values spanned between 65% and 20%, while ABS and PC and POM values reached 50–0% and 35–0%, respectively, when the temperature rose from 15 to 55 °C. The PVA solution at 400 mg/L showed three different value ranges which included 80–10%, 60–0%, and 60–0%, while the 1,600 mg/L solution showed three different value ranges from 65–10% to 60–0% and 55–0% for the same plastics. The data showed no consistent pattern when the MC concentration reached 1,600 mg/L. The left depressants PEG and MIBC showed no consistent patterns in plastics flotation during experiments with different depressant concentrations. The recovery rate reached 90% for ABS and PC when PEG concentration reached 1,600 mg/L during 15 °C testing.

The other research studies demonstrated that temperature elevation produced better results in specific situations although it led to negative effects on adsorption in different cases. The Gibbs free energy change in the following equation needs to have a negative sign for spontaneous adsorption to occur:

$$\Delta G_{ad} = \Delta H_{ad} - T\Delta S_{ad}$$

where ΔG_{ad}, ΔH_{ad}, ΔS_{ad}, and T represent Gibbs free energy, enthalpy, entropy changes of the system during adsorption, and absolute temperature, respectively. The enthalpy change exists as a temperature-dependent property which falls under thermodynamic properties. The sign of this property is negative for exothermic processes while it is positive for endothermic ones. The property values

Table 9.6 The effect of temperature on floatability percentage of studied plastics. The conditioning time and pH were 15 minutes and 6.5, respectively.

	Floatability (%)					
Polymer	**TA (400 mg/L)**	**MC (400 mg/L)**	**PEG (400 mg/L)**	**PVA (400 mg/L)**	**MIBC (1 mL/L)**	**Temperature (°C)**
ABS	5	65	80	80	70	15
	25	50	20	20	85	25
	85	25	25	15	80	35
	55	25	35	15	75	45
	45	20	40	10	50	55
PC	0	50	80	60	20	15
	0	20	10	5	75	25
	65	15	15	0	70	35
	45	10	25	0	25	45
	30	0	45	0	20	55
POM	0	35	60	60	0	15
	0	0	0	0	20	25
	20	0	0	0	35	35
	20	0	25	0	0	45
	20	0	30	0	0	55

Polymer	**TA (1,600 mg/L)**	**MC (1,600 mg/L)**	**PEG (1,600 mg/L)**	**PVA (1,600 mg/L)**	**MIBC (3 mL/L)**	**Temperature (°C)**
ABS	10	70	90	65	65	15
	25	55	25	45	85	25
	30	50	20	25	80	35
	25	65	55	15	80	45
	10	60	60	10	55	55

Polymer	TA (1,600 mg/L)	MC (1,600 mg/L)	PEG (1,600 mg/L)	PVA (1,600 mg/L)	MIBC (3 mL/L)	Temperature (°C)
	0	50	90	60	20	15
	20	15	0	5	50	25
PC	60	25	10	0	75	35
	50	40	20	0	75	45
	20	30	30	0	60	55
	5	20	70	55	0	15
	5	0	0	25	30	25
POM	30	10	0	0	45	35
	20	20	5	0	40	45
	15	25	10	0	20	55

change when temperature increases while its sign might also transform during this process. The sign of the above equation depends on both the absolute temperature, and the entropy change which occurs during the adsorption process. The entropy change which occurs during adsorption processes results in negative values. The enthalpy change sign stands as a vital element which determines the process. The adsorption process will stop when the sign of ΔG_{ad} becomes positive because the system will start desorbing substances. The observed results show that temperature produced distinct effects on the flotation behavior of the investigated plastic materials. The adsorption process became more effective when temperature levels rose but this change worked only for specific temperature ranges and substance types including depressants and plastics. The desorption process became more active when temperature levels increased for particular systems. Authors' research results matched the findings which other studies had already reported.

The ABS, POM, and PC flotation results against liquid conditioning pH values from 4.5 to 10.5 were also examined for different studied depressants at 800 mg/L (2 mL/L for MIBC) depressant concentration with a 15-minute conditioning time at 25 °C. The chosen operating conditions proved beneficial for economic operations within the flotation industry sector. The experiments for TA and MIBC demonstrated plastic flotation reduction at pH levels between 6.5 and the starting point. The studied plastics showed improved flotation when the pH level exceeded this point, which resulted in 55%, 50%, and 45% flotation rates for TA and 55%, 75%, and 60% flotation rates for MIBC when testing POM, ABS, and

PC materials. The H^+ ion from acidic conditions interacts with OH groups found in TA and MIBC which results in their inability to adsorb onto plastic surfaces. The flotation process of studied plastics became more effective when the pH level reached 8.5 according to MC. The flotation values spanned from 5% to 30%, from 25% to 60%, and from 40% to 75% for POM, PC, and POM, respectively. The PEG experiment showed no pattern of plastic flotation occurred when researchers raised the pH levels. The highest flotation values for POM, ABS, and PC were 50%, 85%, and 55%. The adsorption process of depressant onto plastic surfaces involved multiple mechanisms which operated at the same time. The flotation process of PVA-conditioned plastics showed increased performance until pH reached 6.5, but then it started to decrease significantly after this point. The maximum flotation values reached 85% for ABS, 55% for PC, and 50% for POM at the different pH levels which were tested.

The floatability results of POM, PC, and ABS particles under different depressant conditioning times from 5 to 30 minutes were reported. The solution contained 800 mg/L of depressant (2 mL/L of MIBC) at 6.5 pH while maintaining at 25 °C. The floatability of all plastics conditioned with TA became worse when the conditioning time reached 15 minutes. The parameter showed improvement when the conditioning time exceeded this value. The process at its beginning stage depended on water molecules which caused TA to hydrolyze quickly while preventing TA molecules from binding to the plastic surface. The MC conditioning time extension did not produce any significant changes in POM and PC floatability because the depressant completely covered all active sites on these plastics surfaces during the first five minutes for PC and 10 minutes for POM. The same trend was also observed for PEG. The plastics required five minutes to achieve their highest flotation point but extending the conditioning period to 30 minutes resulted in only a 10% change in this measurement. The data showed no consistent patterns when PVA and MIBC were used to condition plastics during various conditioning periods. The adsorption–desorption process underwent changes because of various factors which were previously explained in the text.

The results from Table 9.7 demonstrate how the combination of two different depressants affects the ABS, PC, and POM flotation process. The experiment maintained 25 °C temperature and 15 minutes conditioning time and 800 mg/L total depressant concentration and 6.5 pH while adjusting the depressant ratios. The depressant ratios in dual systems ranged from 0.8 to 0.2, from 0.6 to 0.4, and from 0.5 to 0.5. The ABS flotation of the TA:PEG dual system reached 25% when the PEG ratio was raised to 0.4 according to Table 9.7. The flotation process experienced a 15% reduction when the system received additional input. The PEG ratio of 0.8 produced the highest flotation point which reached 30% during the test. The floatability of this plastic became worse when PEG levels were raised during PC processing. The flotation process achieved a maximum of 10% flotation. The same

Table 9.7 The flotation percentage for dual systems with various depressant ratios. The conditioning time, temperature, pH level, and total depressant concentration were 15 minutes, 25 °C, 6.5, and 800 mg/L, respectively.

	Floatability (%)				
Plastic	**0.8:0.2 (TA:PEG)**	**0.6:0.4 (TA:PEG)**	**0.5:0.5 (TA:PEG)**	**0.4:0.6 (TA:PEG)**	**0.2:0.8 (TA:PEG)**
ABS	15	25	15	15	30
PC	10	5	0	0	5
POM	10	10	5	0	5
	0.8:0.2 (TA:PVA)	**0.6:0.4 (TA:PVA)**	**0.5:0.5 (TA:PVA)**	**0.4:0.6 (TA:PVA)**	**0.2:0.8 (TA:PVA)**
ABS	5	5	5	20	5
PC	0	0	0	0	0
POM	0	0	25	30	10
	0.8:0.2 (TA:MC)	**0.6:0.4 (TA:MC)**	**0.5:0.5 (TA:MC)**	**0.4:0.6 (TA:MC)**	**0.2:0.8 (TA:MC)**
ABS	15	10	15	60	25
PC	0	0	0	15	20
POM	0	5	5	10	5

trend was also observed for POM with increasing PEG in TA:PEG dual system. The floatability of ABS did not show any significant change when PVA concentrations were raised except for the 0.4–0.6 (TA:PVA) range. The method proved to be completely useless for PC floatation. The floatability of POM increased to 30% when PVA concentrations reached 0.6 but exceeding this value it started to decrease flotation to 10%. The dual system of TA:MC achieved effective floatability results for ABS which reached 60% floatability when the MC ratio in dual TA:MC system reached 0.6. The system failed to produce effective flotation results when processing PC and POM materials. The flotation process of studied plastics through these dual systems appears to involve complex mechanisms which dominate their separation behavior.

The last few years have brought multiple outstanding software tools which help scientists perform design work and analyze their research data. The statistical software package Design-Expert provided general-purpose statistical analysis capabilities to users through its excellent data analysis features which supported both fixed and random factor experiment analysis. Design-Expert operates as a dedicated experimental design software package which provides complete design construction tools and evaluation capabilities together with powerful analysis functions. The used RSM serves as a valuable method for analyzing the collected data. For example, the software generated the following second-order polynomials to predict individual plastic flotation based on input data (conditioning time, flotation tank temperature and pH, and MIBC concentration) for MIBC conditioning of plastics:

$$\begin{aligned}\text{Flotation \% (ABS)} = {} & 82.68 + 0.03T + 2.09t + 47.23C - 21.57\text{pH} + 0.03Tt \\ & + 0.09TC + 0.05T\text{pH} + 0.35tC + 0.55t\text{pH} + 1.88C\text{pH} \\ & - 6.69 \times 10^{-3}T^2 - 0.23t^2 - 17.68C^2 + 0.54\text{pH}^2\end{aligned}$$

$$\begin{aligned}\text{Flotation \% (POM)} = {} & -41.91 + 2.31T - 1.02t + 41.80C + 2.50\text{pH} + 0.02Tt \\ & + 0.02T\text{pH} + 0.20tC + 0.23t\text{pH} + 0.83C\text{pH} \\ & - 0.04T^2 - 0.05t^2 - 12.39C^2 + 0.54\text{pH}^2\end{aligned}$$

$$\begin{aligned}\text{Flotation \% (PC)} = {} & -20.32 + 1.45T + 1.46t + 23.24C + 9.04\text{pH} + 0.015Tt \\ & + 0.06TC + 0.20tC + 0.30t\text{pH} + 1.25C\text{pH} \\ & - 0.02T^2 - 0.14t^2 - 9.08C^2 - 1.01\text{pH}^2\end{aligned}$$

The previous equations contain T and C variables which represent temperature in °C and MIBC concentration in mL/L, and t and pH variables which represent conditioning time in minutes and flotation vessel pH. The process of finding the best floatation percentage (optimum) requires differentiating above equations while varying each independent parameter including T, C, t, and pH. The derivation of 12 equations (four from each equation) resulted in solving the equations to find the best values for the four independent variables which are shown in Table 9.8. The calculated floatation percentages for ABS, POM, and PC was determined by substituting the obtained optimum parameters into equations, which resulted in 96%, 78%, and 91%, respectively.

Table 9.8 Optimum studied parameters, vessel pH, conditioning time, vessel temperature, and MIBC concentration along with optimum floatation percentage for studied plastics, ABS, PC, and POM.

Plastic	pH	*t* (min)	Temperature (°C)	Concentration (mL/L)	Floatation (%)
ABS	6.8	16.3	79	1.94	96
PC	7.9	17.9	46	2.17	91
POM	8.5	23.2	21	1.20	78

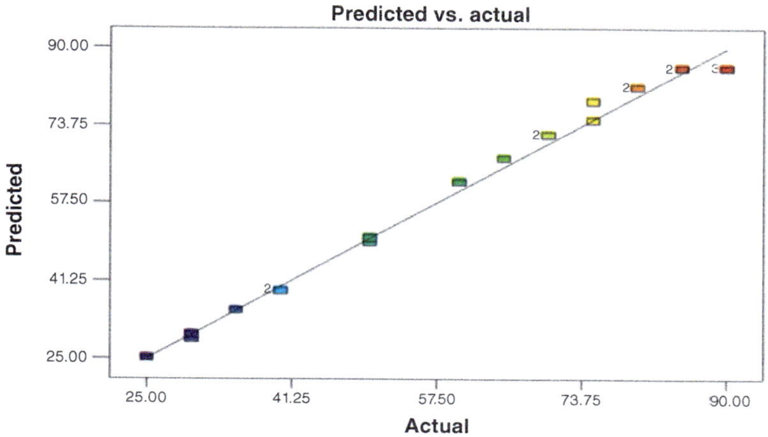

Figure 9.3 The predicted values by software versus actual values for PC.

The graph in Figure 9.3 shows the predicted values by the above-mentioned software against the actual values for PC. The measured values showed a small difference between them. The research results demonstrated that the developed equations could accurately predict how the studied plastics would float under various experimental conditions.

Concluding remarks:

- The flotation technique successfully separated ABS from PC and POM through the use of specific depressants.
- The adsorption–desorption process requires complete understanding of its mechanisms together with all influencing factors which determine how a depressant interacts with plastic surfaces.
- The floatability of PC and POM remained unaffected by TA and MIBC throughout all tested depressant concentration levels.
- The MC and PVA tests showed that plastic floatability did not follow any predictable pattern when the depressants concentration changed.
- The increase in temperature produced negative results for MC at 400 mg/L concentration and PVA at both 400 and 1,600 mg/L concentrations during all plastic flotation tests.
- The flotation process of PVA-conditioned plastics with TA and MIBC reached its highest point at pH 6.5, but it started to decrease significantly after this value.
- The floatability of all plastics studied under TA conditioning became worse when the conditioning time reached 15 minutes.
- The complex phenomenon which controls flotation of studied plastics occurs through the selected dual systems.

- The Design-Expert software generated equations which produced plastic flotation values that matched the experimental results from the corresponding tests.
- The research team conducted all their experiments using plastic materials which had not undergone any previous use. The research needs to continue studying actual plastic waste materials which exist in waste management systems.

9.3 Microwave-assisted Flotation Technique for Separation of PVC, PS, and PET

The **first case study** appeared in *Progress in Rubber, Plastics and Recycling Technology* when authors published "Effect of pre-microwave irradiation on the flotation of polystyrene, polyethylene terephthalate, and polyvinyl chloride using traditional depressants" in 2023.

They stated waste management faces a major challenge because there exists no reliable method to extract single plastic items from plastic mixtures. The floatation technique which scientists recently developed enables engineers to extract traditional plastics such as PS, PET, and PVC from each other. The floatation process received assistance from traditional depressants which functioned as chemical agents. The research investigated how different microwave output powers from 20% to 100% affect the floatation behavior of plastics after surface premicrowave irradiation. The research investigated how different concentrations of depressant from 400 to 2,000 mg/L would affect plastic floatation. Also, the results received confirmation through the combination of contact angle (θ) measurement with scanning electron microscope (SEM) images and attenuated total reflectance Fourier transform infrared spectroscopy (ATR-FTIR) spectra analysis. The driven equations by a DOE software (Design-Expert) presented to show suitable conformity between the predicted and actual plastics flotation values.

The surface engineering process uses microwave irradiation as an effective method to transform solid materials through surface modification. The frequency range of this band extends from 300 MHz to 300 GHz, while its wavelength spans from 0.001 to 1 m. The microwave treatment process modifies both the surface texture and chemical composition and bonding patterns of functional groups which exist on plastic surfaces. The process also works for rubber recycling because it breaks the crosslink bonds which exist between rubber molecules in rubber products. Authors' current understanding indicated that there are no published research studies which employ sink–float plastic separation through microwave irradiation in their experiments. The industrial and household waste sector uses PVC, PS, and PET as their most common available plastic materials. All of them are thermoplastic. The materials work best for recycling programs if they can be separated efficiently from each other within waste streams.

The research used preirradiated plastics whose surfaces received microwave exposure at different power levels. The following popular depressants received surface treatment before being added to the flotation tank at different concentrations ranging from 400 to 2,000 mg/L: PVA, MC, PEG, and TA. The sink–float results were recorded and discussed for their application in commercial plastics waste management operations. All used depressants in this study are non-toxic substances which do not harm the human body. Most of them use as food additive. For example, the results of subchronic toxicity and genotoxicity studies demonstrate that PVA presents no risk to human health when people encounter it through different exposure routes during their normal activities.

The used plastics in Table 9.9 display their characteristics through PVC, PS, and PET. The materials used were commercial grades which the team obtained from the nearby market. The object had a spherical shape with a diameter of 5 mm. The granules underwent washing with deionized water followed by oven drying at 40 °C before microwave pre-irradiation. PEG has a molecular weight range of 380–420 g/mol and a density of 1.13 g/mL (PEG, depressant, PEG-400, Shree Chem, India). PVA has a molecular weight of 22,000 g/mol and a density of 1.19 g/mL (PVA, depressant, BP 20, Liwei Chemical Co., China). TA has a molecular weight of 1,701 g/mol and a density of 2.12 g/mL (TA, depressant, 100773, Merck, Germany). MC has NAMW of 40,000 g/mol and a density of 1.33 g/mL (MC, depressant, 274429 Sigma, Sigma Aldrich, USA). They were prepared and used.

The laboratory microwave oven (Gosonic, GMO-530, Turkey) operated at 30 L volume to perform premicrowave surface irradiation of plastics. The device operated at 2,000 MHz while producing 900 W of maximum output power at 100% frequency. The used floatation tank and experiments procedure already explained (Section 9.1 and Figure 9.1). The experiment involved 20 granules from each plastic material which made a total of 60 granules that received

Table 9.9 The characterization of the used plastics, PVC, PS, and PET.

Plastic	PVC	PET	PS
Density (g/cm^3)	1.35	1.40	0.98
Surface energy (mN/m) at ambient temperature	41.5	44.6	40.7
Granule shape	Sphere	Sphere	Sphere
Diameter (mm)	5	5	5
Supplier	Iran	Iran	Iran

20 seconds of microwave oven irradiation at specific power levels. The process of extra irradiation will cause the granule temperature to increase while simultaneously raising the probability of plastic granule deformation. Hence, it should be avoided. The researchers then added selected depressants at different concentrations to the irradiated granules before placing them into the flotation tank. The flotation media consisted of tap water which maintained a pH level of 6.5. The experiment required the collection of sunk and floated granules from the flotation tank after one minute of operation followed by result documentation. The flotation tank consisted of a glass-made structure. The object had a $20 \times 20\ cm^2$ square base which supported an 80 cm tall structure. The system includes an air blowing system which operates at the tank bottom to produce 4 L/min air bubble flow (Figure 9.1). The adsorption status of the depressant on plastics surfaces was evaluated through ATR-FTIR analysis which used the AVATAR 370 FTIR instrument from Thermo Scientific Nicolet in the United States. The system included a ZnSe ATR crystal which operated as a high throughput ATR crystal. The instrument resolution was 4 1/cm. The base lines received correction for all spectra to enable their evaluation against each other. The MES OEL-PV 1450 field-emission gun SEM employed to study the surface morphology of plastics. The instrument operated at 20 kV while providing 2 nm resolution through its 12 mm working distance. The research on plastic surface hydrophilic properties used contact angle (θ) measurements to determine how water droplets (50 μL) interact with plastic surfaces. The stereomicroscope (SZH10) with Olympus DVP1 camera operated at 25× magnification to complete the task. The θ angle received measurements at five different points which produced a median value for the results.

As stated in previous sections, the sink–float behavior of plastic particles in liquid media depends on multiple parameters which include the depressant chemical structure, plastic particle shape and size, air bubble flow rate and size distribution, and plastic surface properties. The selection of an appropriate depressant requires evaluation of multiple factors which include depressant-plastic structural compatibility, commercial availability and price affordability, and safety considerations. The selection of depressants for this study depended on all the parameters which were evaluated. The sink–float behavior of plastics depends on their surface characteristics which include their roughness and the specific functional groups that exist on their surface. The microwave irradiation method provides an affordable way to modify plastic surface properties. The process operates as an environmentally friendly method.

The microwave treatment process produces changes which affect how the plastic surface appears and what properties it has. The process of plastic surface adsorption becomes challenging for depressants based on both the extent of their changes and their overall quality. The hydrophobic–hydrophilic properties of a solid surface depend on its surface morphology. The contact angles (θ) between water droplets and PS surface were measured to assess how premicrowave irradiation affects these properties for four different PS samples. The PS samples consisted of three groups which included samples without microwave preirradiation and samples that received 100% output power preirradiation and samples that underwent irradiation with PVA at two different concentrations of 400 and 2,000 mg/L (Figure 9.4). The unirradiated PS sample showed a θ value of 118° according to Figure 9.4a. The θ value indicates that plastics (PS) surfaces exhibit hydrophobic properties because it exceeds 90°. The contact angle reached 122° after microwave irradiation according to Figure 9.4b. The surface of PS became more hydrophobic after irradiation according to this finding. The surface of irradiated PS underwent PVA treatment which resulted in θ angle reductions to 99° and 90° when using 400 and 2,000 mg/L PVA concentrations, respectively. The hydrophilic properties of PS surfaces became more favorable when researchers used higher concentrations of PVA during the process. The O—H groups of PVA adsorbed on PS surface made the plastic material more hydrophilic according to the findings. These findings were in conformity with the sink–float test results for treated PS with PVA. The PS samples without microwave preirradiation, with 100% output power preirradiation, and irradiated samples conditioned with PVA at two different PVA concentrations, 400 and 2,000 mg/L, had the floatation percentages of 95%, 100%, 65%, and 45%, respectively. The decrease in floatation percentage requires an increase in sink percentage which results in higher hydrophilic properties of the plastic material.

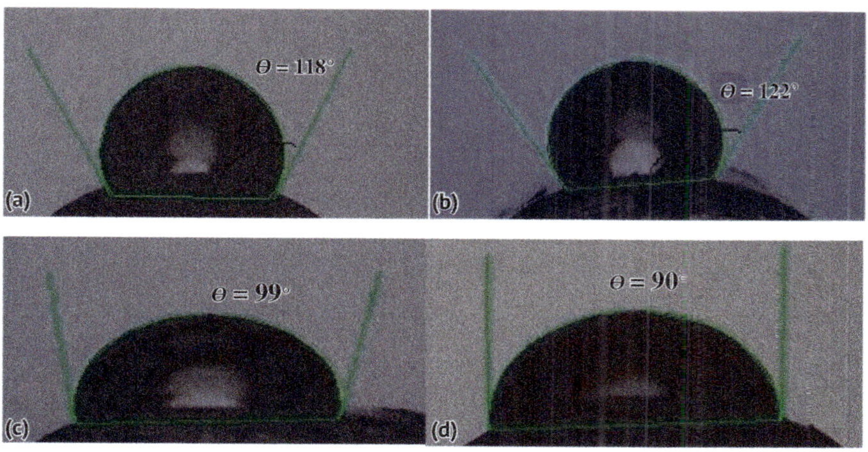

Figure 9.4 The contact angles (θ) for four PS samples: (a) no microwave irradiation, (b) subjected to 100% output power microwave irradiation, (c) treated with PVA (400 mg/L) at 100% output power microwave irradiation, and (d) treated with PVA (2,000 mg/L) at 100% output power microwave irradiation. *Source*: Pakbar et al. (2023) / with permission of Sage Publications.

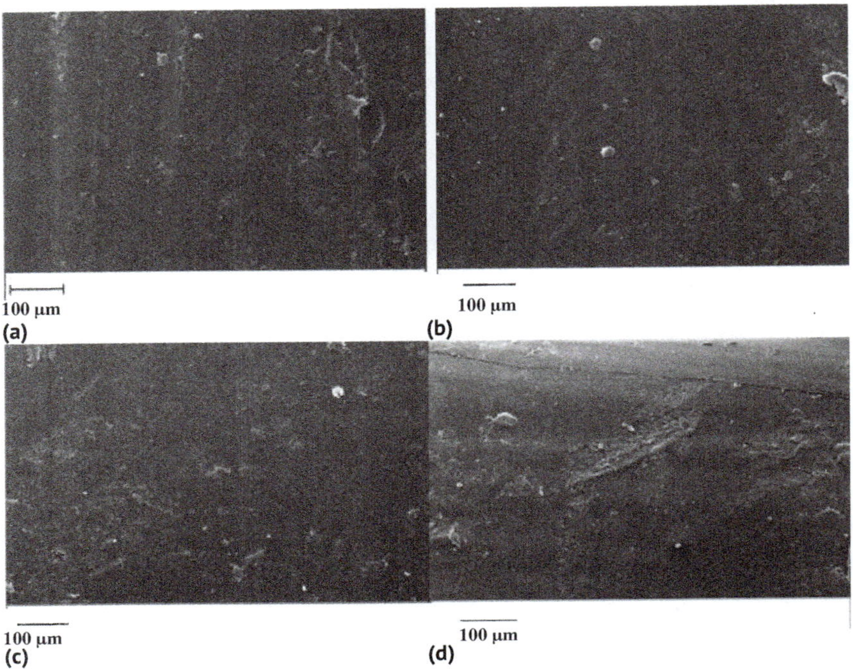

Figure 9.5 The SEM images of four PS samples: (a) without microwave exposure, (b) subjected to 100% output power microwave exposure, (c) treated with PVA (400 mg/L) at 100% output power microwave exposure, and (d) treated with PVA (2,000 mg/L) at 100% output power microwave exposure. *Source*: Pakbar et al. (2023) / with permission of Sage Publications.

The SEM images from Figure 9.5a–d show the appearance of the PS samples. The analysis of Figures 9.5a and 9.5b demonstrates that microwave treatment successfully modified the surface structure of the plastic material (PS). The surface roughness underwent natural changes following microwave irradiation exposure. The plastics surface shows a smooth appearance after microwave irradiation according to Figure 9.5b. The surface developed different levels of roughness after PVA conditioning at two separate concentrations which depended on the amount of PVA that adsorbed to the PS surface (Figure 9.5c,d). The results from the ATR-FTIR spectra of the mentioned samples supported the findings. The ATR-FTIR spectra in Figure 9.6 presents the PS samples which received no microwave treatment (brown) and complete output power microwave treatment (blue).

The spectra comparison reveals that microwave treatment produced changes in both functional group amounts and their chemical identities. The irradiation process caused changes in the absorption intensity of multiple chemical groups.

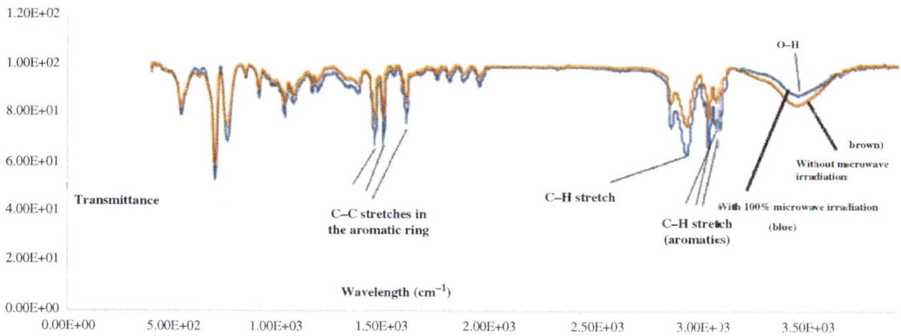

Figure 9.6 Comparison of ATR-FTIR spectra for PS samples subjected to 100% output power microwave irradiation and those without irradiation.

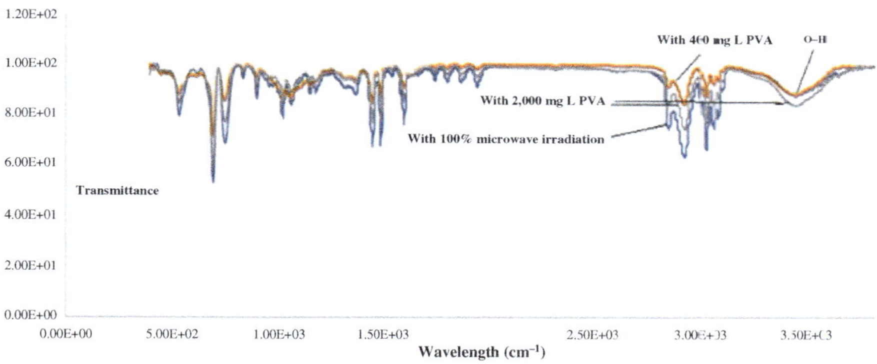

Figure 9.7 The comparison of ATR-FTIR spectra for PS samples, subjected to 100% output power microwave irradiation without PVA conditioning and conditioned with PVA at concentrations of 400 and 2,000 mg/L under 100% output power microwave irradiation.

The three dedicated peaks showing C—C stretches in the aromatic ring between 1,465 and 1,614 1/cm, C—H stretch at 1,614 1/cm, and three additional peaks for C—H stretches in aromatic ring between 3,032 and 3,099 1/cm experienced modifications in their intensity following irradiation. The ATR-FTIR spectra from Figure 9.7 show the microwave preirradiated PS samples before and after PVA conditioning at 400 (brown) and 2,000 mg/L (gray). The peak intensity at 3,400 1/cm, which represents the O—H group band, became more intense when PVA concentration in solution rose from 400 to 2,000 mg/L. The increase in PVA adsorption onto PS surfaces became evident because the majority of O—H groups existed as alcoholic functional groups in PVA.

The microwave irradiation speed which hits the plastics surface depends on the microwave output power level. The PVC, PS, and PET flotation percentages measured at different concentrations of PVA and MC and PEG and TA while using different microwave output power levels (%) were evaluated by the authors. The pH level of flotation tank media kept 6.5 when tests were conducted under ambient temperature conditions. The plastics underwent 20 seconds of microwave preirradiation before being placed into the flotation tank for study. The microwave output power operates at 20%, 40%, 60%, 80%, and 100% power levels. The depressant concentrations followed a sequence of 400–2,000 mg/L at five different levels. Table 9.9 shows that PET, PVC, and PS have densities which measure 1.40, 1.35, and 0.98 g/cm^3, respectively. The surface energy measurements for PET, PVC, and PS showed values of 44.6, 41.5, and 40.7 mN/m, respectively. The plastic density serves as the single factor which determines whether plastic objects will sink or float in water. The PET and PVC materials would sink to the bottom while PS would float on top of the flotation tank based on density alone. Surprisingly, the sink–float test revealed that 90% of the PVC remained afloat, while only 10% descended to the bottom of the flotation tank. Conversely, if surface energy were the primary factor influencing the sink–float behavior of a plastic, and taking into account the significant difference between the surface tension of water (72 mN/m) and that of the analyzed plastics, it is anticipated that all plastics would remain atop the flotation tank. Surprisingly, once more the sink–float observation revealed that 100% of the PET settled at the bottom of the flotation tank. It was concluded that while the density and surface energy of plastics were significant factors in predicting sink–float behavior, other elements also contribute to this phenomenon. The results demonstrated that PET flotation remained unaffected when the premicrowave treatment was applied without any depressant at various microwave power levels. The method proved successful for left plastics which included PVC and PS materials. The flotation process achieved 85–95% separation for PVC and 85–100% separation for PS based on the applied microwave output power levels. The surface treatment of PET with PVA resulted in higher floatation percentages for the specified plastic materials. The PVA concentration range from 400 to 2,000 mg/L did not affect floatation percentage when no premicrowave irradiation was applied because the floatation percentage remained constant at 5%. The floatability of PET underwent a change when irradiation treatment was combined with PVA depressant conditioning. The maximum floatation percentage should be allocated to the conditioned PET which contains 2,000 mg/L PVA when the microwave output power reaches 60% at a value of 30%. The findings demonstrated that PVA conditioning of PVC and PS materials resulted in decreased floatability of these plastics particularly for nonirradiated plastics. The first floatation process achieved 90% separation for PVC and 95% separation for PS. The floatation percentages of mentioned plastics decreased to 10–40% (PVC) and 25–30% (PS) when no irradiation occurred and depressant conditioning was applied. The floatability of PCV and PS showed the smallest decrease when irradiation was combined with depressant conditioning. The study demonstrated that microwave preirradiation surface modification of these plastics created a barrier which prevented PVA from adsorbing onto their surfaces. The O—H group of PVA which adsorbs to plastic surfaces makes the studied plastics more hydrophilic. The floatation process of PET underwent a significant change when irradiation was combined with depressant adsorption (MC). The floatation percentage of PET remained at zero during both experiments which used no microwave power and all tested microwave power levels without depressant conditioning. However, it increased to the value of 65% at 60% irradiation power with MC concentration of 1,200 mg/L. The floatability of the PVC and PS changed after irradiation and using MC as a depressant. The data shows no consistent pattern. PS showed the smallest variation in floatation percentage among all the plastic materials studied. The results from the PEG-treated plastics surface showed similar patterns when compared to the irradiated and conditioned plastics. The highest PET floatation percentage occurred in irradiated samples which received 60% microwave power during two PEG concentration tests at 1,200 and 1,600 mg/L while achieving a floatation value of 50%. The floatation percentages of unirradiated and irradiated studied plastics with TA also were examined. The floatability of all plastic materials showed positive results when TA was used. The results showed that PET delivered the highest level of effectiveness. The floatability of PVC and PS became less sensitive to irradiation and TA treatment but at a reduced extent. The maximum PET floatation percentage occurred in irradiated samples which used 40% microwave power at 400 mg/L TA concentration to achieve a 65% floatation rate.

In general, the experimental results produced various floatation percentages which depended on the plastic material, microwave power level, and depressant selection and dosage amount. No regular trend observed. The study demonstrated that plastic sink–float behavior depends on multiple factors which include the surface adsorption of depressants onto plastic materials. The descrption–adsorption process of chemicals on solid surfaces requires detailed examination to understand this phenomenon. A chemical substance will bond to a solid polymer surface through surface adsorption which creates a chemical film that accumulates on the solid surface. The following aspects need to be considered when performing desorption–adsorption operations:

- The accessibility and the quantity of appropriate active sites on the solid surface
- The direct and indirect interactions among neighboring molecules of the adsorbed substance, as well as with the molecules present on the solid surface.

- The heat of adsorption and the variation in the heat of adsorption across various effective sites.
- The Gibb's free energy associated with the adsorption process, along with its negative value indicating spontaneous adsorption.
- The single or multiple layers of the attached substance on the solid surface
- The speed at which the chemical is adsorbed onto and desorbed from the solid surface
- The constant of equilibrium for the adsorption and desorption processes

The research established that microwave preirradiation produced different effects on mono or multilayer adsorbed depressant capacity and active site numbers and types which existed on plastic surfaces under various operating conditions. The system experienced a change in its parameters which resulted in a shift of its entropy sign. The Gibb's free energy sign transition occurred which led to spontaneous adsorption and desorption processes. Future research must validate the previously mentioned findings through additional studies.

The DOE serves as a practical method which enables users to forecast desired output properties through modifications of their chosen input parameters. The Design-Expert statistical software served as the tool for this process. The system received depressant concentration (C, mg/L) and microwave output power (P, %) as inputs to produce plastics floatation (%) as output. The research used RSM with D-optimal method as its analytical approach. RSM identifies the connections between independent variables C and P and the dependent variables which include predicted floatation percentages. The software produced Table 9.10 which contains the derived equations that predict plastic floatation percentages when using specific depressants at various C and P concentrations.

From this study, it was concluded that the microwave preirradiation process successfully affected the sink–float behavior which the researchers observed in their plastic samples. The PET results showed no evidence that stimulant use alone would produce this effect. The floatability of plastic did not show any consistent pattern when researchers raised microwave power levels. The value of the mentioned parameter depended on both the amount of irradiation and the strength of the depressant. The used depressants adsorbed successfully onto the plastic surfaces which were studied. The plastic surface underwent changes which affected its hydrophilic–hydrophobic properties. The research used three standard characterization methods which included SEM, ATR-FTIR, and contact angle measurement to obtain funding. The Design-Expert software used for DOE analysis produced equations which demonstrated proper alignment between predicted and measured plastics flotation results.

The **second case study** appeared in *Polymers and Polymer Composites* when authors published "Application of a dual depressant system and microwave

Table 9.10 The recommended formulas by the experimental design software (Design-Expert) for the flotation percentage of the analyzed plastics in relation to microwave output power percentage (*P*) and concentration of depressant (*C*, mg/L).

Depressant	Plastic	Equation
PVA	PET	Floatation (%) $= -1.35913 + 0.284226P + 0.013554C + 7.14286\text{E}{-}06PC - 0.002604P^2 - 5.11533\text{E}{-}06C^2$
	PS	Floatation (%) $= 75.16865 + 0.254464P - 0.072820C - 0.000054PC - 0.000521P^2 + 0.000029C^2$
	PVC	Floatation (%) $= 50.802151 + 1.51114P - 0.009577C - 0.000165PC - 0.011458P^2 + 4.27827\text{E}{-}06C^2$
MC	PET	Floatation (%) $= 8.60119 + 0.141220P + 0.080089C + 7.14286\text{E}{-}06PC - 0.001674P^2 - 0.000032C^2$
	PS	Floatation (%) $= 86.99688 + 0.147959P - 0.027684C - 0.000030PC - 0.001190P^2 + 0.000011C^2$
	PVC	Floatation (%) $= 92.43339 - 0.042368P - 0.011173C + 0.000173PC - 0.000930P^2 - 1.48810\text{E}{-}06C^2$
PEG	PET	Floatation (%) $= 0.684524 + 0.107292P + 0.054985C + 0.000018PC - 0.000335P^2 - 0.000021C^2$
	PS	Floatation (%) $= 90.06094 + 0.109099P + 0.007293C - 0.000074PC - 0.000149P^2 - 1.76711\text{E}{-}06C^2$
	PVC	Floatation (%) $= 92.42489 - 0.058270P + 0.010784C - 7.65306\text{E}{-}06PC + 0.000409P^2 - 4.27827\text{E}{-}06C^2$
TA	PET	Floatation (%) $= 5.93963 + 0.3350553P + 0.72877C + 0.003795PC - 0.003795P^2 - 0.000031C^2$
	PS	Floatation (%) $= 76.40731 + 0.154443P - 0.024369C + 0.000138PC + 0.001228P^2 - 2.79018\text{E}{-}06C^2$
	PVC	Floatation (%) $= 80.27069 + 0.078168P - 0.005459C - 0.000023PC + 0.000484P^2 + 1.48810\text{E}{-}06C^2$

irradiation for flotation-based separation of polyethylene terephthalate, polyvinyl chloride, and polystyrene plastics" in 2024.

The research involved applying microwave preirradiation to plastic substrates (PVC, PS, and PET) through different microwave power settings which spanned from 20% to 100%. The surfaces received preirradiation treatment before researchers applied specific dual depressants which included PVA, MC, PEG, and TA at different concentrations between 200 and 1,600 mg/L before they entered the flotation tank. The sink–float analysis results received detailed documentation which led to an extensive discussion about their effects on commercial plastic waste management systems. All employed depressants in this study have nontoxic properties which make them safe for human consumption, as stated earlier. Many of these compounds exist as food additives, which include PVA as an example. Multiple studies about subchronic toxicity and genotoxicity have proven that PVA remains safe for human contact through different exposure routes which people experience during their normal daily activities. The used materials and experiments procedure described early in this section for first case study.

As stated, the sink–float behavior of plastic particles in liquids depends on multiple factors which include the chemical composition of the depressant, the size and shape of plastic particles, air bubbles and their flow rates and size variations, and plastic surface characteristics. The selection process for a suitable depressant requires evaluation of four essential elements which include plastic material compatibility with the chemical agent structure (depressant), market accessibility, price affordability, and protection of human health. The research examined all parameters which affect depressant selection according to the previous list. The sink–float behavior of plastic particles depends on two main factors which include the surface characteristics of plastic materials and the specific functional groups which are attached to their surfaces. The process of using microwave irradiation proved to be an affordable method which allowed researchers to modify plastic surfaces. The method stands out because it operates as an environmentally friendly process.

Research shows that microwave radiation causes plastic materials to experience surface property changes which affect their morphological structure. The extent of surface changes determines how well they can bond with other materials, including depressants. The surface properties of hydrophobicity and hydrophilicity on solid materials depend on their specific surface structure.

The research team measured contact angles (θ) between water droplets and PS surface areas of four samples to assess premicrowave irradiation effects on these properties. The PS samples included three groups which were nonirradiated samples, samples that received 100% microwave power preirradiation, and samples treated with TA:PVA at 800 and 1,600 mg/L concentrations. The microwave irradiation process ran for 20 seconds. The plastic surfaces received simultaneous exposure to a depressant solution which they stayed in at room temperature for 30 minutes. The contact angle (θ) measurement of the untreated PS sample showed 91° which indicated its hydrophobic surface because the angle exceeded 90°. The surface hydrophilic properties of PS became more pronounced when researchers applied microwave radiation because the angle between the surface and water decreased to 88.4°. The PS surface, which received irradiation treatment, became more hydrophilic after TA:PVA solution application at 800 and 1,600 mg/L concentrations, which decreased contact angles to 72.8° and 70.8°, respectively. The plastic surface received improved hydrophilic properties through the application of microwave irradiation. Unexpectedly, the results demonstrate that PS flotation percentages increased from 55% to 100% for unconditioned PVA:TA samples which received no treatment and those that underwent microwave irradiation.

The plastic surface showed slightly better hydrophilic properties after microwave treatment, but this change made the plastic floatability worse. The observation shows that hydrophilicity elements alone do not determine plastic flotation in water because surface roughness and other elements play a role. The hydrophilic properties of PS surface became more favorable when researchers increased the concentration of TA:PVA in their solution. The existence of O—H groups from adsorbed PVA and TA on PS surfaces leads to increased hydrophilicity of the material. The contact angle results match the sink–float test results which were obtained from conditioned PS samples that used PVA:TA solutions. The flotation rates for the specified samples at PVA:TA concentrations of 800 and 1,600 mL/L were 35% and 0%, respectively. The decrease in flotation percentage shows that sink percentage has increased, while the hydrophilic properties of plastic have become more pronounced. The SEM images in Figure 9.8a–d show the surface morphologies of the PS samples which were previously discussed.

The comparison between Figure 9.8a and b demonstrates how microwave treatment affects the surface appearance of PS samples which were not treated with PVA:TA solution. The surface roughness of the material underwent natural changes when it underwent irradiation, as Figure 9.8b demonstrates the plastic surface became rougher after microwave treatment. The surface became rougher after PVA-TA conditioning at two different concentrations which produced different levels of surface roughness depending on the amount of

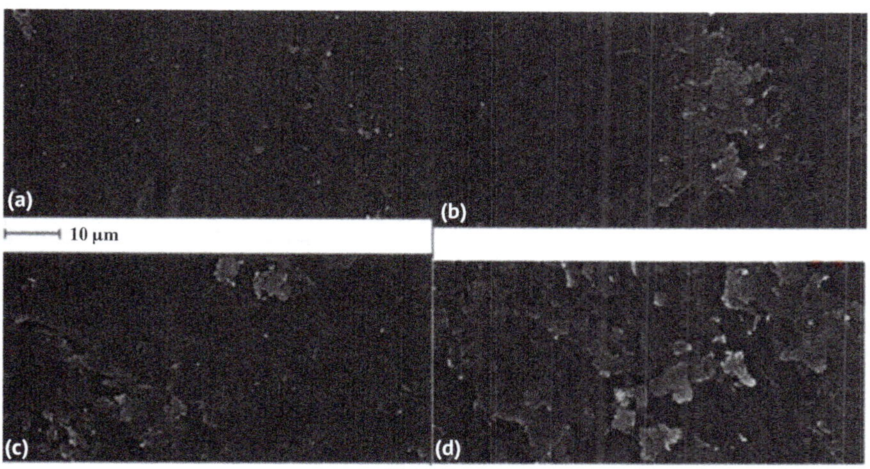

Figure 9.8 The SEM images depict four PS samples: (a) no microwave irradiation, (b) subjected to 100% power microwave irradiation, (c) treated with PVA:TA (800 mg/L) at 100% power microwave irradiation, and (d) treated with PVA:TA (1,600 mg/L) at 100% power microwave irradiation. Microwave irradiation was performed for 20 seconds. *Source*: Pakbar et al. (2023) / with permission of Sage Publications.

PVA:TA that bound to the PS surface (Figure 9.8c and d). The surface roughness increasement stems from the adsorption of depressants, which include PVA and TA. Aligned with the results from contact angle assessments, the increased hydrophilicity of the discussed plastic surfaces can be explained by the data derived from ATR-FTIR analyses of the plastic specimens. Figure 9.9 shows the ATR-FTIR spectra of PS without microwave exposure (a) and with irradiation at 100% output power (b). A comparative analysis of the spectra shows that microwave irradiation caused changes in the concentration and types of functional groups.

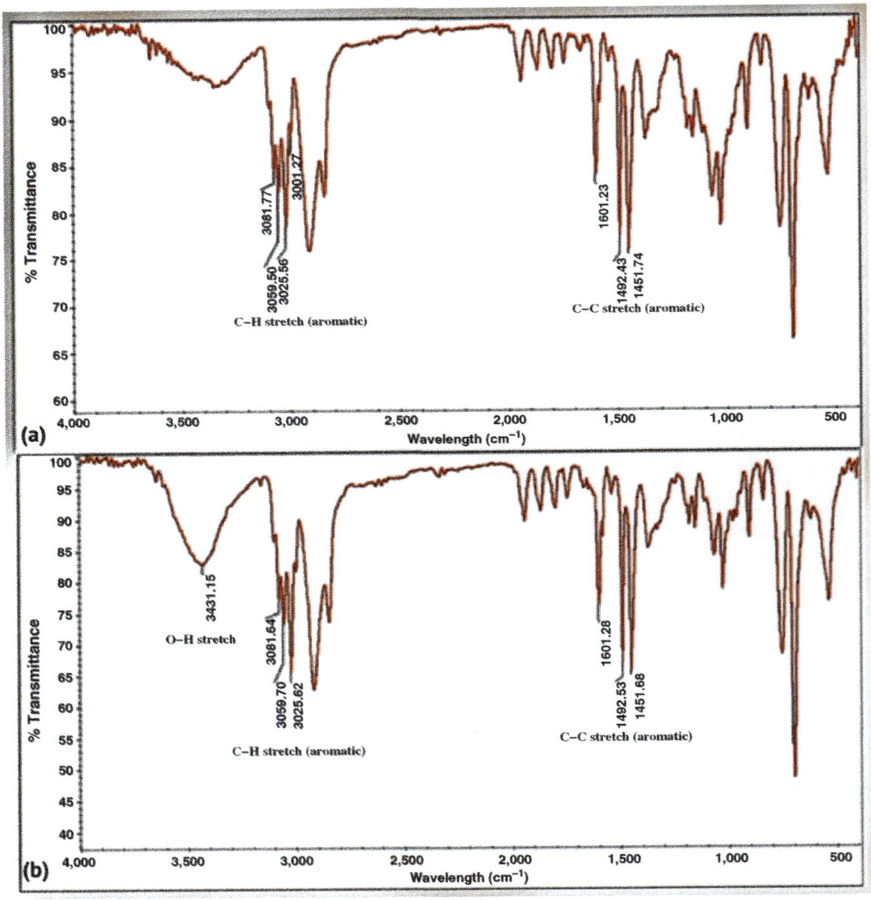

Figure 9.9 Comparison of ATR-FTIR spectra for PS samples, displayed (a) without and (b) with 100% output power microwave treatment. The irradiation in the microwave lasted for 20 seconds. *Source*: Pakbar et al. (2023) / with permission of Sage Publications.

The different groups showed changes in their absorbance levels following the irradiation process. The three distinct peaks which showed C—C stretch in the aromatic ring between 1,452 and 1,602 1/cm, O—H stretch near 3,440 1/cm, and three additional peaks for C—H stretches in the aromatic ring between 3,025 and 3,082 1/cm experienced intensity variations following irradiation. Also, another ATR-FTIR spectra demonstrate the comparison between PS samples which received microwave preirradiation and PVA:TA solutions at 800 and 1,600 mg/L. The peak intensities between 3,433 and 3,440 1/cm which correspond to the O—H group showed an upward trend when PVA:TA concentration increased from 800 to 1,600 mg/L. The results demonstrate that PVA:TA adsorption onto PS surfaces increases because the majority of O—H groups exist as alcoholic functional groups in PVA and TA.

The effects of microwave irradiation on plastic surfaces become more defined through the process of adjusting the microwave output power levels. The flotation rates of PVC, PS, and PET using different concentrations of dual depressant systems including PVA:TA, MC:TA, PEG:TA, PEG:PVA, MC:PEG, and MC:PVA at various microwave power levels (%) were examined by authors. The flotation tank media (liquid, water) operated at a pH of 6.5, while plastic surfaces received microwave preirradiation for 20 seconds before they entered the flotation tank. The microwave power settings operated at 20, 40, 60, 80, and 100% power, while the depressant concentrations were set at 200, 400, 800, 1,200, and 1,600 mg/L. Table 9.8 shows that PET, PVC, and PS have densities of 1.40, 1.35, and 0.98 g/cm^3, respectively, while their surface energies measure 44.6, 41.5, and 40.7 mN/m, respectively. The sink–float observations showed that density alone would cause PET and PVC to sink while PS would float. The experiment showed that 30% of PVC material stayed on top of the water, but 70% of it sank to the tank floor. The PS flotation rate reached 55% which resulted in 45% of the material settling at the bottom. Conversely, if surface energy were the main factor determining sink–float behavior, given the significant disparity between the surface tension of water (72 mN/m) and that of the plastics examined, it would be expected that all plastics would float. Surprisingly, the sink–float tests revealed that all PET sank to the bottom of the flotation tank. This indicates that, although plastic density and surface energy are important in forecasting sink–float behavior, other elements are clearly at play in this occurrence as represented in the first case study in this section. The float–sink results clearly show that premicrowave irradiation, in the absence of a depressant, at different microwave output powers has no noticeable effect on the flotation characteristics of PET. On the other hand, it works well for other plastics, specifically PVC and PS. The flotation rates for PVC and PS vary between 30% and 80% at microwave irradiation power levels of 0% and 60%, and between 55% and 100% at microwave output powers of 0%, 80%, and 100%, respectively. Treating PVC and PS with a dual system of PVA:TA, along with prior microwave irradiation, has a significant impact on their flotation. Significantly, PVC obtains the minimum flotation values of 5% and 10%, and at a PVA:TA concentration of 400 mg/L, it attains 0%, leading to all PVC granules settling at the base of the flotation tank. For PS, the related values vary from 100% flotation at a PVA:TA concentration of 200 mg/L with a microwave power of 40%, to diminished values reaching 5% for different PVA:TA concentrations and microwave irradiation powers. The best separation outcomes for PVC and PS occur at a PVA:TA concentration of 800 mg/L without prior microwave irradiation, resulting in 0% flotation for PVC and 90% for PS. Unlike PS, PVC has natural polarity, which enables PVA and TA to adsorb onto the plastic surface more readily. The presence of hydroxyl (—OH) groups in these depressants creates a stronger attraction to PVC, causing it to sink easily in the water. Once again, the integrated system of MC and TA showed greater effectiveness on PVC compared to PS, while exhibiting minimal effect on PET. Flotation percentages for PVC varied from 0% to 30% after premicrowave irradiation was utilized on the plastic surface. For PS, flotation values ranged from 30% to 85%, depending on the concentration of the depressant and the intensity of microwave irradiation. In a similar manner, the situation for the PEG:TA dual system resembled that of earlier studied dual systems. A depressant concentration of 200 mg/L at all microwave irradiation powers showed increased effectiveness for PVC, leading to flotation values between 0% and 5%. For PS, the maximum and minimum flotation values were 100% and 5%, respectively. Notably, a 15% flotation was recorded for PET at a PEG:TA concentration of 400 mg/L when using 80% and 100% microwave irradiation powers. As shown, the combined system of microwave-assisted MC:PEG demonstrates reduced effectiveness in affecting the flotation behavior of the analyzed plastic materials, especially when compared to earlier studied dual systems in this text. This is seen more distinctly for PS. Flotation rates for PVC and PS range from 5% to 80% and from 15% to 100%, respectively, at different microwave preirradiation power levels. Importantly, PET shows no flotation percentage with the given MC:PEG concentrations and the premicrowave irradiation powers applied. This result matches expectations, as the hydroxyl (—OH) group in any of the employed depressants cannot be compared to TA. The dual system that includes microwave-assisted MC:PVA shows comparable results to MC:PEG but is more efficient with PVC than MC:PEG. Flotation percentages for PVC and PS ranged from 5% to 80% and from 50% to 100%, respectively, across different microwave preirradiation power levels. Once more, PET shows no flotation percentage for the used MC:PVA concentration and premicrowave irradiation power levels.

The research results demonstrate that plastic type and microwave power level and depressant type and concentration affect flotation percentages. The data shows no recurring patterns which can be identified. The research shows that plastic sink–float behavior depends on multiple elements which include how a depressant substance interacts with plastic surface molecules. The complete analysis of surface chemical desorption–adsorption processes require scientists to study the solid surface chemical interaction mechanism. A substance will adsorb onto solid polymer surfaces through surface adsorption because it forms a chemical layer which accumulates on the solid material. The process of desorption and adsorption depends on multiple elements including, the speed of adsorption and desorption of the chemical onto and off of the solid surface, the presence and quantity of appropriate active sites on the solid surface, and the adsorbed substance molecules interact with each other through direct and indirect contact while also interacting with solid surface molecules. Also, the heat of adsorption and the disparity between the heat of adsorption of various effective sites, the constant of equilibrium for the processes of desorption and adsorption, the Gibbs free energy associated with the adsorption process and its negative value indicating spontaneous adsorption, and the single or multiple layers of the adsorbed substance on the solid surface may be the other parameters.

The research results led the authors to conclude that microwave preirradiation transforms the mono- or multilayer properties of adsorbed depressants while changing both the number of active sites and their types on plastic surfaces according to different operational parameters. The modification of the system parameters leads to a change in the entropy sign of the system. The Gibbs free energy sign change results in spontaneous adsorption and/or desorption events. The research requires additional studies to validate these results which will be conducted in future studies.

As mentioned, the application of the DOE approach is key in predicting a targeted characteristic (output) by adjusting chosen variables (inputs). Statistical software Design-Expert was utilized for this objective, using depressant concentration (C, mg/L) and microwave output power (P, %) as inputs, while plastics flotation (%) was the output. The utilized strategy encompassed RSM employing the D-optimal technique. RSM aids in examining the connections between explanatory variables (C and P) and response variables (predicted flotation percentage). Table 9.11 displays equations generated by the previously mentioned software to forecast the flotation percentage of the examined plastics at different concentrations of depressants (C) and microwave output powers (P).

For example, Figure 9.10 shows a three-dimensional (3D) depiction of response surfaces for conditioned PS alongside PEG:PVA. This figure evaluates the combined effect of depressant concentration (PEG:PVA, mg/L) and microwave output power (P) on PS flotation percentage in accordance with the derived equations (Table 9.11) for PS and PEG:PVA.

Table 9.11 The proposed equations from a DOE software (Design-Expert) for flotation percentage of the analyzed plastics, PVC, and PS in relation to premicrowave output power percentage (*P*) and dual depressant concentration (*C*, mg/L).

Depressant	Polymer	Equation
MC:PEG	PVC	Floatability (%) = $32.64936 + 0.166903P - 0.016273C - 0.000119PC - 0.000037P^2 + 9.10238\text{E}{-}06C^2$
	PS	Floatability (%) = $98.86487 - 0.801585P - 0.049455C + 0.000340PC + 0.004613P^2 + 0.000024C^2$
MC:PVA	PVC	Floatability (%) = $60.60229 - 0.763167P - 0.027329C - 0.00153PC + 0.008110P^2 + 5.62799\text{E}{-}06C^2$
	PS	Floatability (%) = $74.470181 + 0.151100P + 0.010635C + 2.25564\text{E}{-}06PC - 0.000074P^2 - 5.98910\text{E}{-}06C^2$
PEG:PVA	PVC	Floatability (%) = $52.11906 - 0.199846P - 0.025628C - 0.000307PC + 0.003646P^2 + 0.000019C^2$
	PS	Floatability (%) = $54.33696 + 0.648421P + 0.023425C - 0.000248PC - 0.001786P^2 - 4.15756\text{E}{-}06C^2$
PEG:TA	PVC	Floatability (%) = $36.64349 + 0.017387P - 0.72183C - 1.87970\text{E}{-}06PC + 0.001042P^2 + 0.000035C^2$
	PS	Floatability (%) = $74.24140 - 0.517541P - 0.022107C - 0.000084PC + 0.005432P^2 + 0.000014C^2$
MC:TA	PVC	Floatability (%) = $32.14691 + 0.227580P - 0.083503C - 0.000094PC - 0.000632P^2 + 0.000044C^2$
	PS	Floatability (%) = $72.60880 - 0.256382P - 0.020409C - 0.000089PC + 0.002604P^2 + 0.000014C^2$
PVA:TA	PVC	Floatability (%) = $37.09452 + 0.349156P - 0.102896C - 0.000027PC - 0.003088P^2 + 0.000052C^2$
	PS	Floatability (%) = $65.19272 + 0.081748P - 0.072784C - 0.000303PC - 0.000744P^2 + 0.000037C^2$

Source: Pakbar et al. (2023) / with permission of Sage Publications.

As a conclusion, the studied plastics show different sink–float behavior after microwave preirradiation treatment except for PET. The plastic floatability shows no consistent pattern when the microwave output power increases. The parameter value depends on both the amount of radiation exposure and the amount of depressant substance present. The **dual depressants** used in this study successfully bonded to plastic surfaces which resulted in changes to their hydrophilic–hydrophobic properties. The results match the data which researchers obtained through SEM, ATR-FTIR, and contact angle measurement techniques. The Design-Expert software generated equations which demonstrate an excellent relationship between predicted and actual plastic flotation values.

Depressant	Polymer	Equation
MC:PEG	PVC	$\text{Floatability}\ (\%) = 32.64936 + 0.166903P - 0.016273C - 0.000119PC - 0.000037P^2 + 9.10238\text{E}{-06}C^2$
	PS	$\text{Floatability}\ (\%) = 98.86487 - 0.801585P - 0.049455C + 0.000340PC + 0.004613P^2 + 0.000024C^2$
MC:PVA	PVC	$\text{Floatability}\ (\%) = 60.60229 - 0.763167P - 0.027329C - 0.00153PC + 0.008110P^2 + 5.62799\text{E}{-05}C^2$
	PS	$\text{Floatability}\ (\%) = 74.470181 + 0.151100P + 0.010635C + 2.25564\text{E}{-06}PC - 0.000074P^2 - 5.98910\text{E}{-06}C^2$
PEG:PVA	PVC	$\text{Floatability}\ (\%) = 52.11906 - 0.199846P - 0.025628C - 0.000307PC + 0.003646P^2 + 0.000019C^2$
	PS	$\text{Floatability}\ (\%) = 54.33696 + 0.648421P + 0.023425C - 0.000248PC - 0.001786P^2 - 4.15756\text{E}{-06}C^2$
PEG:TA	PVC	$\text{Floatability}\ (\%) = 36.64349 + 0.017387P - 0.72183C - 1.87970\text{E}{-06}PC + 0.001042P^2 + 0.000035C^2$
	PS	$\text{Floatability}\ (\%) = 74.24140 - 0.517541P - 0.022107C - 0.000084PC + 0.005432P^2 + 0.000014C^2$
MC:TA	PVC	$\text{Floatability}\ (\%) = 32.14691 + 0.227580P - 0.083503C - 0.000094PC - 0.000632P^2 + 0.000044C^2$
	PS	$\text{Floatability}\ (\%) = 72.60880 - 0.256382P - 0.020409C - 0.000089PC + 0.002604P^2 + 0.000014C^2$
PVA:TA	PVC	$\text{Floatability}\ (\%) = 37.09452 + 0.349156P - 0.102896C - 0.000027PC - 0.003088P^2 + 0.000052C^2$
	PS	$\text{Floatability}\ (\%) = 65.19272 + 0.081748P - 0.072784C - 0.000303PC - 0.000744P^2 + 0.000037C^2$

9.4 Microwave-assisted Flotation Technique for Separation of ABS, PC, and POM

The **first case study** appeared in *Journal of Polymers and the Environment* when the author published "Effect of microwave irradiation on the flotation of selected polymers as a potential solution for plastic recycling" in 2021.

In this study, the chosen plastics from municipal and industrial waste, such as PC, POM, and ABS, were preirradiated using microwaves at varying microwave power percentages for different irradiation durations. The irradiated plastics were then treated with chosen depressants, MC, TA, PVA, and PEG, individually at varying concentrations before being placed in the flotation tank. The impacts of the parameters mentioned above were assessed regarding the float–sink behavior of the analyzed plastics. The main research goal involved studying how commercial applications of discovered methods could achieve efficient plastic waste stream separation for specific engineering plastics.

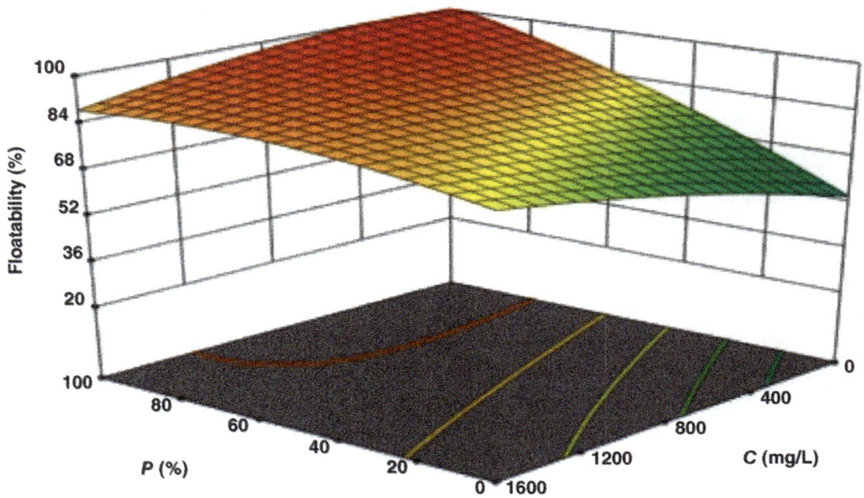

Figure 9.10 The 3D response surfaces assess the combined impact of microwave power (*P*, %) and depressant concentration (PEG:PVA mg/L) on the flotation % of the PS. *Source*: Pakbar et al. (2023) / with permission of Sage Publications.

The used plastics and depressants characteristics were previously detailed in Section 9.2. Additionally, the microwave irradiation instruments and techniques along with experimental procedures were explained in Section 9.3.

As mentioned in previous sections, the depressants show different surface absorption patterns on plastic materials through their chemical properties and the chemical makeup of plastic surfaces. The treatment method changes the surface properties of plastic materials between hydrophobic and hydrophilic, which enables the separation of specific plastics or multiple plastics from a plastic mixture through floatation in tap water. The float–sink behavior of plastic in a liquid media with a selected depressant depends on the surface characteristics of the plastic including its surface roughness and attached functional groups. The left factors include plastic particle shape, density, and size (diameter) and air bubble coverage on the particle surface, air bubble rate, and size. The microwave treatment process creates changes to plastic surface roughness and shape which affects how the depressant substance binds to the surface based on the specific processing parameters. Additionally, the process changes both the amount and nature of functional groups which bond to polymer surfaces. The SEM images in Figure 9.11 show the surface of POM and PC before and after microwave power application at 100% power level. The plastic surface roughness showed strong sensitivity to microwave irradiation according to the observations. The microwave

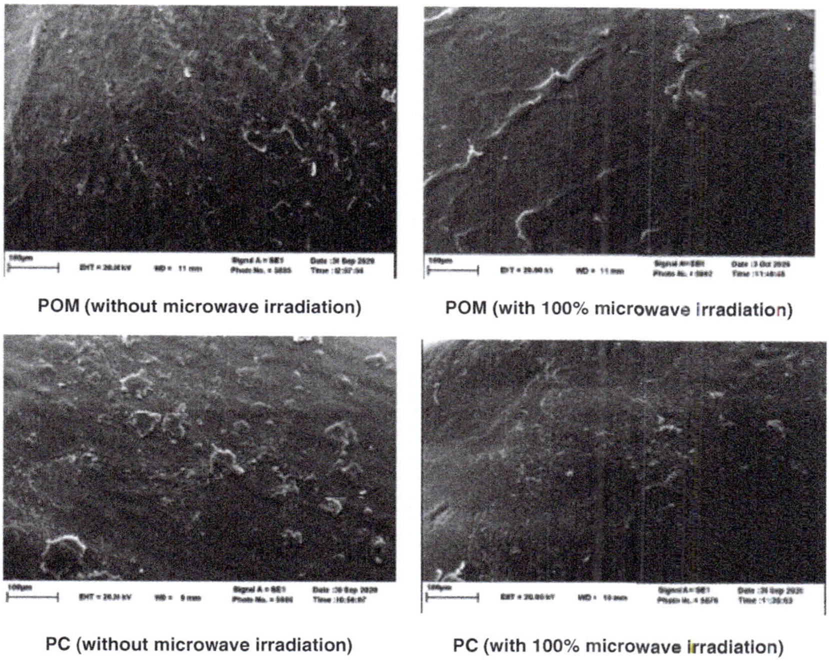

Figure 9.11 The SEM images of POM and PC at 100% microwave power and without it.

treatment created a more even plastic surface by making its surface area smoother. The surface modification treatment creates different effects on depressant adsorption–desorption behavior on plastic surfaces based on operating conditions.

The contact angle (θ) of water droplets on plastic surfaces without depressant treatment underwent measurement before and after microwave irradiation at 60% power for 20 seconds. The θ values for POM, PC, and ABS corresponded to 113.1° and 108.8°, 106.8° and 102.4°, and 113.2° and 108.3°. The microwave treatment produced lower θ values on plastic surfaces which occurred for all plastic types before the depressant treatment. The microwave treatment successfully enhanced the surface hydrophilicity of plastic materials. The reader verified this fact by examining how the correspondent flotation percentage values decreased from 70% to 35%, from 75% to 40%, and from 85% to 30% for POM, PC, and ABS, respectively.

The microwave power percentage served as the evaluation standard to determine how microwave radiation affects plastic surface degradation. The findings showed that the POM, PC, and ABS flotation percentages depended on depressant amounts at different microwave power levels. The flotation tank media operated at a pH of 6.5 while maintaining ambient temperature conditions. The plastics underwent microwave irradiation for 20 seconds before being added to the flotation tank. The research showed that microwave irradiation successfully separated all plastic types from their float–sink tests when using various depressants at different concentration levels. The plastic flotation results showed improvement for some powers and depressants but these substances did not enhance plastic flotation in other instances and most plastic materials settled at the tank bottom. For example, the highest flotation of POM depicted for various TA concentrations, specifically 400, 800, 1,200, 1,600, and 2,000 mg/L, was recorded at 60% microwave power percentage, yielding values of 90%, 90%, 75%, 70%, and 85%, respectively. The relevant value without the use of depressant (TA) was 70%. In contrast to POM, the results showed a significant decline in the flotation of PC and ABS as TA concentration increases. At the same power percentage (60%), the flotation percentages of PC and ABS decreased from 55% and 90% without TA conditioning to values of 5% and 20% at a 1,200 mg/L TA concentration, respectively. The minimum flotation values for various power percentages at the same TA concentration (1,200 mg/L) varied from 5% to 25% for PC and from 15% to 20% for ABS. The minimum PC flotation percentage was 5% at 20%, 40%, and 60% microwave power levels for 1,200 mg/L TA concentration and at 100% power for 2,000 mg/L TA concentration. The identical minimum value was noted for ABS at 40% power and 2,000 mg/L TA concentration. The highest flotation value for POM noted at 20% power for 1,200 and 1,600 mg/L TA concentrations reached 90%. As previously mentioned, the reduced floatation values for PC relative to POM indicate stronger bond strengths of the former in relation to the latter with TA molecules. Due to the multiple hydroxyl groups present in its structure, TA quickly hydrolyzes in water. When conditioned PC and POM with TA are immersed in liquid media (water), the desorption rate of TA molecules adsorbed on the POM surface was faster than that on the PC. As a result, the hydrophilic property of treated POM decreases, causing more POM granules to remain on the surface of the flotation tank. The authors think that microwave irradiation had a greater impact on the quantity, capacity (individual or multilayer adsorbed depressant molecules), and concentration of the active sites. Nevertheless, more in-depth research should be conducted on this subject in the future.

The findings demonstrated noteworthy flotation for all plastics examined when subjected to microwave irradiation and subsequent treatment with PEG. The POM values varied from 65% for the sample lacking PEG and microwave irradiation to 95% for a PEG concentration of 1,600 mg/L at 60% microwave irradiation power. The corresponding values for PC were 55%, while without PEG, they ranged from 60% power to 90% for 800 and 1,200 mg/L PEG with no microwave irradiation. The associated ABS values were 70% for 2,000 mg/L PEG at 40% microwave irradiation, compared to 100% for 1,200 and 1,600 mg/L PEG without microwave irradiation. Microwave irradiation appeared to be ineffective for PC flotation, particularly at PEG concentrations above 800 mg/L. The same observation applied to ABS flotation with PEG levels greater than 400 mg/L. Nonetheless, microwave irradiation proved advantageous for the flotation of all plastics (POM, PC, and ABS) in samples not conditioned with PEG. The results distinctly demonstrated that the combination of microwave irradiation and conditioning PC and ABS with PVA negatively impacted their flotation and led to them sinking to the bottom of the flotation tank. Microwave irradiation appeared to enhance active sites on the plastic surface, boosting the adsorption capacity of PVA on that surface. The adsorption of PVA diminished the hydrophobic characteristics of the PC and ABS surfaces. The highest and lowest PC flotation values were recorded at 100% power without PVA, reaching 85%, while for 400 and 800 mg/L PVA, the flotation percentage was 0% at varying power levels. The related values for ABS were 85% with no PVA and microwave irradiation and 5% at 40% power for 2,000 mg/L PVA, respectively. The POM flotation showed less sensitivity to PVA conditioning and microwave irradiation compared to the values for PC and ABS. The flotation of POM decreased from 65% to 30% when plastic granules were irradiated at 40% power and treated with a 2,000 mg/L PVA solution. The flotation of POM, PC, and ABS was also studied at various microwave irradiation powers across different MC concentrations. It was noted that there wasn't a consistent pattern, and the flotation showed varying values based on the MC concentration and microwave power percentage. The POM, PC, and ABS flotation varied between 55% (60% power, 800 mg/L MC) and 95% (60% power, 1,600 mg/L MC), from 40% (20% power, 2,000 mg/L MC) to 85% (no irradiation for 800 mg/L MC), and from 45% (20% power, 2,000 mg/L MC) to 95% (100% power without MC), respectively. The study found that using microwave irradiation alongside varying depressant concentrations was effective for the flotation of the analyzed plastics with distinct adsorption mechanisms.

The flotation of the investigated plastics, POM, PC, and ABS, was examined at varying concentrations of depressants, TA, PEG, PVA, and MC, for different durations (times) of microwave irradiation. The pH of the flotation tank media was maintained at 6.5 at room temperature. The chosen microwave power is set at 60%. It was noted that the flotation of POM improved when it was exposed to microwave irradiation prior to conditioning with all the examined depressants. Nonetheless, prolonging the irradiation duration enhanced POM flotation for TA and MC concentrations exceeding 1,200 and 2,000 mg/L, respectively. At lower TA and MC concentrations, no clear trend was observed in the flotation enhancement of this plastic. In contrast to TA and MC, extending the irradiation duration worsened the POM flotation for treated plastics with PEG and PVA when depressant concentrations exceeded 1,600 mg/L. For lower concentrations of PEG and PVA, the scenario was quite similar to TA and MC. The highest and lowest POM flotation percentages were 95% and 25%, achieved with 20 seconds (1,600 mg/L PEG) and 80 seconds (2,000 mg/L MC) of irradiation time and a 1,200 mg/L TA concentration without microwave assistance, respectively. The results distinctly indicate the minimal flotation variation for conditioned PC with PEG and MC at the examined irradiation durations of 20, 40, 60, and 80 seconds. Notably, a significant decrease in flotation was noted for conditioned PC with TA and PVA as the irradiation duration increased from 20 to 80 seconds. The top and bottom PC flotation percentages were associated with a 1,200 mg/L MC concentration for a 20-second irradiation time and 1,600 and 2,000 mg/L TA concentrations for an 80-second irradiation time, yielding values of 95% and 0%, respectively. The ABS flotation percentages were not significantly affected by extending irradiation time for unconditioned samples. The values varied from 75% to 90%. The previously stated fact was generally applicable to conditioned ABS samples with PEG and MC under various irradiation durations. The ABS flotation percentages varied from 75% to 100%. Nonetheless, for conditioned ABS samples containing TA and PVA, the flotation interval was broadened to 10–85%. The highest and lowest flotation percentages were associated with 2,000 mg/L PVA without microwave irradiation and 2,000 mg/L TA for an irradiation duration of 80 seconds.

The Gibbs free energy change in the below equation needs to be negative for spontaneous adsorption to occur:

$$\Delta G_{\text{ad}} = \Delta H_{\text{ad}} - T\Delta S_{\text{ad}}$$

The equation includes four variables which represent the Gibbs free energy change (ΔG_{ad}) and entropy change (ΔS_{ad}) and enthalpy change (ΔH_{ad}) and temperature (T) of the system during adsorption. The microwave treatment seemed to modify the system entropy through its impact on the plastic surface active sites which it altered. The system develops positive entropy when microwave radiation is applied because the entropy increase results in a positive ΔS_{ad} value. The negative sign of ΔG_{ad} emerges when T and ΔH_{ad} maintain their current values. The system's entropy showed different responses to microwave irradiation depending on the specific characteristics of the plastic surface. The authors established that essential elements which include microwave power strength and irradiation velocity and a complicated process determine the flotation behavior of selected plastic materials.

Table 9.12 displays the proposed equations from an experimental design software (Design-Expert statistical software) relating the floatation percentage of examined plastics to microwave power (P, %), the concentration of depressant used (C, mg/L), and microwave irradiation duration (T, seconds). The DOE is an effective method for forecasting a specific property, such as flotation percentage, as an outcome while varying the previously mentioned chosen input parameters.

Table 9.12 The equations proposed by Design-Expert statistical software for the flotation percentage of the analyzed plastics in relation to microwave power (P, %), employed depressant concentration (C, mg/L), and microwave irradiation duration (T, seconds).

Depressant	Polymer	Equation
TA	POM	$\text{Floatation}\ (\%) = 40.81056 + 0.632210P + 0.002168C + 0.792439T - 0.000035PC + 0.000000PT + 0.000261CT - 0.005191P^2 - 2.84091\text{E}{-}06C^2 - 0.010955T^2$
	PC	$\text{Floatation}\ (\%) = 86.21642 - 0.141963P - 0.054963C - 0.657488T - 3.21782\text{E}{-}06PC + 0.000000PT + 0.000080CT + 0.000135P^2 + 0.000013C^2 + 0.004218T^2$
	ABS	$\text{Floatation}\ (\%) = 95.91642 + 0.086277P - 0.041801C - 0.781427T - 0.000172PC + 0.000000PT - 0.000078CT - 0.001067P^2 + 0.000014C^2 + 0.006759T^2$
PVA	POM	$\text{Floatation}\ (\%) = 70.79661 - 0.280136P + 0.010507C - 0.783006T - 0.000121PC + 0.000000PT - 0.000160CT + 0.003051P^2 - 1.72484\text{E}{-}06C^2 + 0.011086T^2$
	PC	$\text{Floatation}\ (\%) = 71.77381 - 0.394048P - 0.060455C - 0.685417T - 0.000021PC + 0.000000PT - 0.000064CT + 0.003125P^2 + 0.000022C^2 + 0.009449T^2$
	ABS	$\text{Floatation}\ (\%) = 112.46667 - 0.662113P - 0.063100C - 1.41492T - 0.000102PC + 0.000000PT - 0.000036CT + 0.005492P^2 + 0.000026C^2 + 0.018281T^2$
PEG	POM	$\text{Floatation}\ (\%) = 71.74922 - 0.088863P + 0.034020C - 0.636943T - 0.000131PC + 0.000000PT - 0.000034CT + 0.002018P^2 - 0.000011C^2 - 0.005981T^2$
	PC	$\text{Floatation}\ (\%) = 81.19930 - 0.174209P + 0.020837C - 0.567562T - 0.000079PC + 0.000000PT + 0.000113CT + 0.001992P^2 - 8.16761\text{E}{-}06C^2 + 0.004258T^2$
	ABS	$\text{Floatation}\ (\%) = 93.86398 - 0.208646P + 0.012047C - 0.402113T - 0.000044PC + 0.000000PT - 0.000086CT + 0.002181P^2 - 5.12378\text{E}{-}06C^2 + 0.005706T^2$
MC	POM	$\text{Floatation}\ (\%) = 69.83481 - 0.021291P + 0.005175C - 0.197834T + 0.000048PC + 0.000000PT + 0.000220CT + 0.000120P^2 - 3.50041\text{E}{-}06C^2 - 0.000269T^2$
	PC	$\text{Floatation}\ (\%) = 72.66612 - 0.031494P + 0.005603C - 0.199393T + 0.000052PC + 0.000000PT + 0.000220CT + 0.000306P^2 - 4.76867\text{E}{-}06C^2 + 0.000401T^2$
	ABS	$\text{Floatation}\ (\%) = 87.11233 + 0.028370P - 0.023114C - 0.175527T + 0.000013PC + 0.000000PT + 0.000166CT + 0.000140P^2 + 6.29058\text{E}{-}06C^2 - 0.000624T^2$

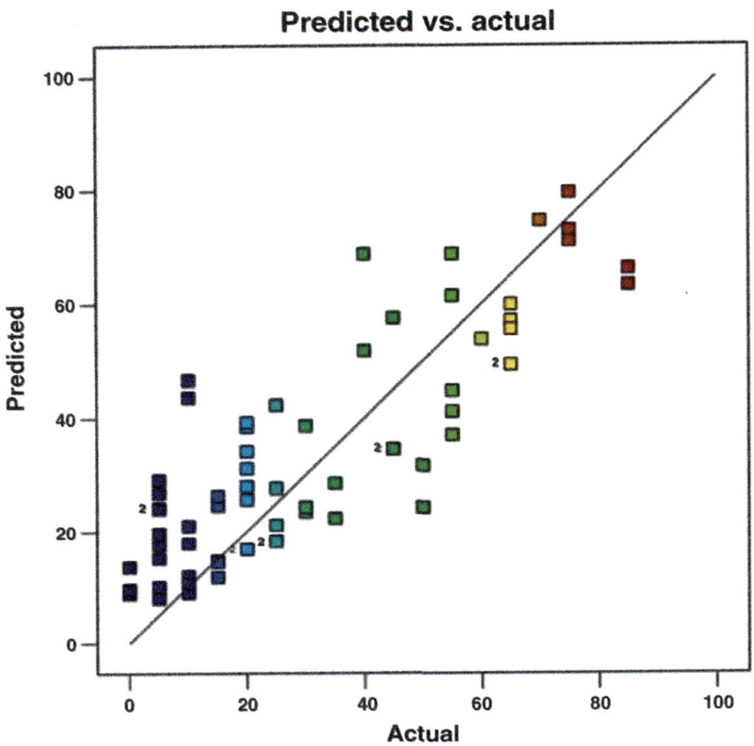

Figure 9.12 The expected values compared to the real experimental values for conditioned PC with TA.

The D-optimal method within RSM was utilized to analyze the data. The anticipated values aligned well with the experimental values.

For example, Figure 9.12 contrasts the expected values with real experimental values for conditioned PC with TA. The anticipated floatation percentage figures were like real values.

9.5 Final Observations

From this study, it was concluded that the plastic surface's premicrowave irradiation significantly impacted the float–sink behavior of the plastics examined. The microwave irradiation influenced the quantity, capacity (mono- or multilayer adsorbed depressant molecules), and concentration of active sites. Microwave irradiation decreased the θ (contact angle) values of the untreated plastic surfaces with the examined depressants for all plastics, leading to an enhancement in their hydrophilic characteristics. Nonetheless, to understand depressant adsorption on a plastic surface, the reader must possess a comprehensive knowledge of the adsorption–desorption mechanism. Preirradiation of certain plastics and depressants in the microwave was advantageous for plastic flotation, while for other specimens, it caused the plastic to settle at the bottom of the flotation tank. The equations proposed by Design-Expert software for forecasting the plastic flotation percentage in relation to the examined parameters aligned well with the experimental outcomes.

The **second case study** appeared in *Journal of Polymers and the Environment* when authors published "Float–sink behavior of selected pre-microwave-irradiated plastics by surface adsorption of dual depressants" in 2022.

In this research, the chosen engineering plastics present in municipal and industrial waste include ABS, PC, and POM, which were microwaves irradiated at various microwave power levels. The irradiated plastics treated with the chosen **dual depressants** at varying concentrations were subsequently placed into the flotation tank filled with tap water as the liquid medium. The influence of microwave power and depressant concentration on the hydrophilic–hydrophobic (sink–float) characteristics of the plastic surface was assessed. The used plastics, depressants, and experimental procedures were the same as described in the first case study in the current section.

The contact angle measurements of water droplets on POM surfaces showed and assessed how microwave radiation affects the surface water affinity of the plastic material. The θ values of the plastic material measured 122° before microwave irradiation and 118° after 100% power microwave treatment. The microwave treatment produced a decrease in θ value, which indicated that it made the POM surface more hydrophilic. The POM surface became accessible to hydrophilic groups because microwave irradiation facilitated their migration to the plastic surface. The POM achieved its highest hydrophilic properties when researchers used TA:PVA as a dual depressant at 100% power to condition the microwave-irradiated material which resulted in a θ value of 106°. Raising the TA:PVA concentration to 2,000 mg/cm^3 decreased the θ value to 105. The results stated above aligned with the flotation percentage reduction of nonirradiated POM from 85% to the values of 75%, 15%, and 0% for irradiated samples without depressant and with 800 and 2,000 mg/cm^3 TA:PVA, respectively. Figure 9.13 displays the SEM images for identical samples. It was noted that microwave irradiation was entirely effective on the texture of the plastic (POM) surface. The microwave irradiation caressed the plastic surface and decreased the roughness. Nonetheless, the altered surface could promote the adsorption of the depressant (TA:PVA) onto the POM surface. Identical observations noted from AFM (atomic force microscope) images (Figure 9.14). As noted, the R_z value for the unirradiated

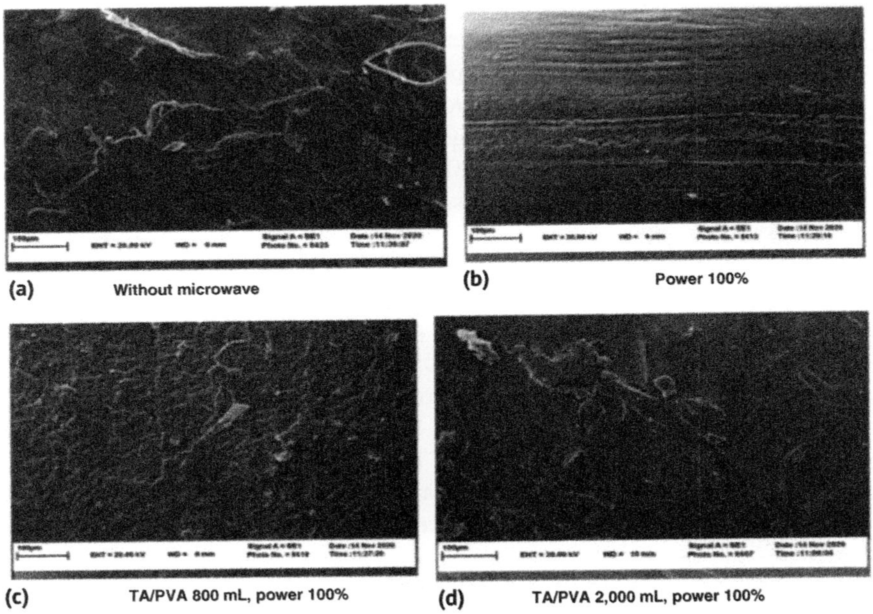

Figure 9.13 The SEM images of four POM samples: (a) without microwave exposure, (b) exposed to 100% power microwave irradiation, (c) treated with TA:PVA (800 mg/L) under 100% power microwave irradiation, and (d) treated with TA:PVA (2,000 mg/L) under 100% power microwave irradiation.

POM surface was 365.8 nm, which decreased to 97 nm following microwave irradiation. After treating the POM surface with TA:PVA, the R_z values rose again to 178 and 165 nm for TA:PVA concentrations of 800 and 2,000 mg/L, respectively. The ATR-FTIR spectra of the specified samples showed that microwave irradiation changed both the types and amounts of specific functional groups. The O—H group peak appears at 3,400 1/cm in the detected peaks which stem from water and TA and PVA materials. The stated peaks showed different peak intensities throughout the experiment which included irradiation and TA:PVA adsorption onto POM surfaces at 800 and 2,000 mg/L. As another example, the C—O infrared absorption bands appear across a wavelength range which spans from 900 to 1,400 1/cm. The observed peaks in the specified wavelengths showed different strength levels across all samples that were studied.

The research studied how different concentrations of MC:PVA, MC:PEG, PVA:PEG, TA:PEG, MC:TA, and TA:PVA affect the flotation of POM, PC, and ABS when exposed to different microwave power levels (%). The flotation tank media operated at a pH of 6.5 while maintaining room temperature conditions. The plastic materials underwent 20 seconds of microwave exposure before they entered

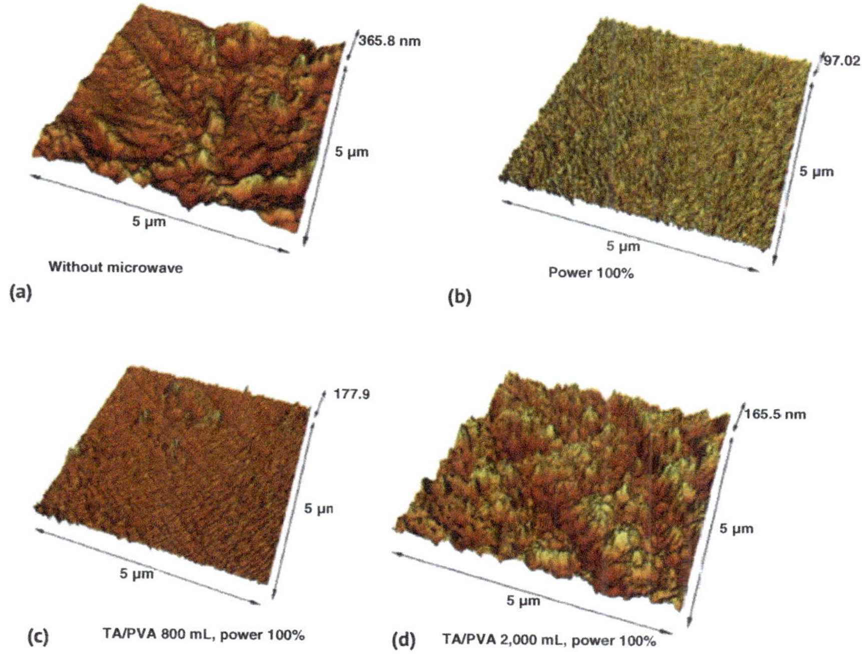

Figure 9.14 The AFM images of four POM samples: (a) without microwave irradiation, (b) with 100% power microwave irradiation, (c) treated with TA:PVA (800 mg/L) at 100% power microwave irradiation, and (d) treated with TA:PVA (2,000 mg/L) at 100% power microwave irradiation.

the flotation tank. The research results showed that plastic surface treatment with MC:PVA, PVA:PEG, TA:PEG, MC:TA, and TA:PVA depressants resulted in reduced plastic flotation for all tested materials. The buoyant plastics from the specified samples settled at the bottom of the flotation tank following microwave treatment and depressant application. The flotation of POM, PC, and ABS without a depressant demonstrated that all microwave power percentages varied from 55% to 85%, from 65% to 90%, and from 75% to 95%, respectively. Employing the MC:PVA as a dual depressant at the tested concentrations of 400, 800, 1,200, 1,600, and 2,000 mg/L decreased the flotation percentages of the specified plastics POM, PC, and ABS to 5–45%, 0–10%, and 5–45%, correspondingly. The flotation percentage values for MC:TA and TA:PVA were 0–75% and 0–40% for POM, 0–55% and 0–35% for PC, and 0–55% and 0–25% for ABS, respectively. The impact of the rest depressants, MC:PEG, PVA:PEG (for POM only), and TA:PEG (for POM only) on the flotation of the plastics analyzed was less significant in comparison to MC:PVA, MC:TA, and TA:PVA. For example, the flotation of ABS both without and with 400, 800, 1,200, 1,600, and 2,000 mg/L of depressant MC:PEG ranged from 75% to 95%, from 60% to 90%, from 55% to 95%, from 50% to 95%, from 45% to 90%, and from 75% to 95%, respectively. The initial flotation percentages of POM, PC, and ABS samples without microwave treatment and depressant addition were 85% for POM, 80% for PC, and 90% for ABS. The flotation percentages of their samples transformed into 55–85% (POM), 65–90% (PC), and 75–95% (ABS) after microwave irradiation based on the applied microwave power. The minimum flotation percentages for POM, PC, and ABS required 55% at 60% microwave power, 65% at 80% microwave power, and 75% at 60% microwave power, respectively. The research showed that plastic surface treatment with depressants and microwave irradiation produced major effects on the flotation behavior of all tested plastic materials. The system delivered different results when it processed both values and trends. The plastic flotation received enhancement from specific microwave power levels and depressants, but these substances failed to improve plastic flotation in different situations. Instead, the system helped plastics move from the top to the bottom section of the flotation tank.

References

Davari, M. and Ostad Movahed, S. Flotation using selected depressants as an efficient technique for separation of mixed acrylonitrile–butadiene–styrene, polycarbonate, and polyoxymethylene plastics in waste streams. *J. Polym. Environ.* 2019, *27*(5), 1709–1720.

Heidarpoor, M., Ostad Movahed, S., and Jourabchi, S. Effect of microwave irradiation on the flotation of selected polymers as a potential solution for plastic recycling. *J. Polym. Environ.* 2021, *29*, 1–12.

Hossine, S. H., Ostad Movahed, S., and Jourabchi, S. Float–sink behavior of selected pre-microwave-irradiated plastics by surface adsorption of dual depressants. *J. Polym. Environ.* 2022, *30*, 1–11.

Jabbari, S., Ostad Movahed, S., and Jourabchi, S. Application of a dual depressant system and microwave irradiation for flotation-based separation of polyethylene terephthalate, polyvinyl chloride, and polystyrene plastics. *Polym. Polym. Compos.* 2024, *32*, 09673911241248418.

Pakbar, H., Ostad Movahed, S., and Jourabchi, S. Effect of pre-microwave irradiation on the flotation of polystyrene, polyethylene terephthalate, and polyvinyl chloride using traditional depressants. *Prog. Rubber Plast. Recycl. Technol.* 2023, *39*, 1–12.

Shamsaddini Negari, M., Ostad Movahed, S., and Ahmadpour, A. Separation of polyvinyl chloride (PVC), polystyrene (PS), and polyethylene terephthalate (PET) granules using various chemical agents by flotation technique. *Sep. Purif. Technol.* 2017, *194*, 368–376.

10

Economic, Policy, and Industrial Perspectives of Polymer Recycling

10.1 Global Market Overview for Recycled Polymers and Rubbers

The global polymer industry has experienced a fundamental change during the last 10 years because people started paying more attention to environmental issues and new regulations appeared, while resources became scarce and businesses found economic value in circular economy systems. The transition has brought recycled polymers and rubbers into strategic material status which now appears in various mainstream manufacturing sectors that produce packaging, automotive products, construction materials, electronic devices, and consumer items. The market for recycled polymers has evolved from its niche origins into a growing industry, which transforms how companies operate their supply chains and establish prices and develop new products in the plastics and elastomers sectors (Geyer et al., 2017).

The annual production of plastics reached more than 400 million tons during the early 2020s, and the global market for recycled plastics achieved more than USD 50–60 billion during the middle of 2020s, and experts predict it will grow at 7–10% per year through specific polymer types and geographical areas (Liu et al., 2024). The market leads in volume because polyolefins including recycled polyethylene (rPE) and recycled polypropylene (rPP) continue to be popular due to their wide usage and simple recycling processes and their established market presence in packaging and consumer-product industries. The market for recycled polyethylene terephthalate (rPET) operates as the most advanced recycled polymer market because it benefits from existing beverage bottle recycling systems and food-grade processing facilities and strict governmental oversight. The circular economy framework now uses recycled polymers as essential secondary raw materials which serve as strategic materials (Plastics Europe, 2024).

Most recycled polymer output stems from mechanical recycling and reusing because this method produces more than 70% of all recycled materials which include PE, PP, and PET. The fast-growing investment in chemical recycling technologies including pyrolysis, solvolysis, and depolymerization enables their use for processing various types of waste materials including those that contain different substances and contaminants (Liu et al., 2024).

The rubber industry depends on tire recycling as its main material stream. The world faces two problems because waste tires amount to more than one billion units which become waste annually. The current applications of recovered rubber materials include energy recovery, civil engineering, reclaimed rubber manufacturing, and devulcanization-based closed-loop recycling systems. The development of new technology has not led to recycled rubber penetration in high-performance applications because the materials lose their properties and the recycling materials show unpredictable quality (Castillo et al., 2025).

The European Union (EU) has established itself as a worldwide market leader for recycled polymers because of its strict policy framework which includes extended producer responsibility (EPR) and recycling targets and landfill restrictions. Plastics Europe (2024) reports that European plastic demand reached 13% through recycled polymer usage, while packaging usage remained the leading market segment. The countries located in Northern and Western Europe achieve the highest recycling performance, but Southern and Eastern European nations depend mainly on energy recovery systems.

The Asian market shows the highest rate of growth for recycled polymers because of its extensive manufacturing sector and increasing local market demand and new recycling facilities built after the import ban took effect. China, Japan, and South Korea have started to include chemical recycling as part of their national development plans, but India leads the way with its fast-growing mechanical recycling operations in both formal and informal settings.

The region of North America shows high levels of technological progress in recycling, but the adoption of recycled polymers in markets remains slow because of affordable landfill disposal and multiple sets of regulatory rules. The current market shows rising corporate sustainability commitments and state-level mandates which drive up demand for recycled materials in packaging and automotive industries.

The main market for recycled polymers exists in five sectors which include packaging, construction, automotive, textiles, and consumer goods. The packaging industry leads all other sectors because it must follow government rules which require recycling materials in PET bottles and flexible packaging products. The main applications of recycled rubber exist in four sectors which include tire derived fuel, asphalt modification, sports surface construction, and molded rubber product manufacturing.

The economic value of recycled materials depends on three main factors which include feedstock purity, sorting efficiency, and processing technology. The market demands consistent quality and competitive prices which advanced separation methods together with digital sorting technology provide to producers (Wang et al., 2024). The recycled polymer market now shifts from its previous waste management focus to develop into a materials engineering-based value chain.

10.2 Economic Impact and Industrial Cost–Benefit Analysis

The economic feasibility of polymer and rubber recycling systems depends on multiple factors which include equipment costs, running expenses, material supply, output standards, and customer market requirements. The evaluation of recycling routes requires cost–benefit analysis (CBA) as its core assessment method which functions at both industrial and policy-based levels.

The capital expenses (CAPEX) and operating expenses (OPEX) of mechanical recycling and reusing methods remain lower than those of chemical recycling methods. The processing costs for recycling materials include both collection expenses and subsequent steps of sorting (separation) and washing and grinding and reprocessing. The economic results of mechanical recycling depend heavily on two factors which are contamination levels and polymer heterogeneity because these elements decrease both product quality and market worth (Ostad Movahed et al., 2016).

The deployment of chemical recycling systems requires major financial investments because these systems need complex reactors and catalysts and large amounts of energy to operate. Nevertheless, they offer higher flexibility in feedstock composition and the potential to produce virgin like monomers or fuels. The techno-economic assessments show that chemical recycling operations will reach financial stability because of three essential elements which consist of increasing landfill taxes, rising demand for recycled materials, and the ability to operate inside existing petrochemical facilities (Liu et al., 2024).

The economic difficulties of rubber recycling have become more complicated because elastomers exist in a crosslinked state. The devulcanization and reclamation methods increase production costs but produce valuable materials which exceed the value of standard size reduction operations (mechanical recycling). Research shows that microwave-assisted and chemical devulcanization methods will reduce both energy consumption and manufacturing expenses when they enter commercial operation (Ostad Movahed et al., 2016).

Recycling operations generate economic advantages which include job creation and reduced expenses for environmental cleanup and decreased requirements for petroleum-based materials. The economic assessments which focus on life cycle phases show that recycling operations generate lower greenhouse gas (GHG) emissions than producing new materials from raw materials for PET, HDPE, and EPDM rubber systems (Geyer et al., 2017).

The industrial use of recycled materials helps companies maintain stable supply chains because it decreases their dependence on crude oil market fluctuations. Manufacturers now see recycled feedstocks as their legal requirement for compliance and their opportunity to establish market value through brand differentiation. Organizations that use recycled polymers to create premium products will obtain sustainable cost protection and better environmental sustainability performance indicators.

The program provides multiple advantages, yet multiple economic obstacles continue to exist. The business faces multiple challenges which include unreliable waste delivery systems, unstable recyclate market values, expensive power expenses, and customers who do not want to use recycled materials for important performance needs. The process of rubber recycling faces major technical barriers because it cannot maintain its original properties, which blocks its use in industrial production at present.

The solution to these obstacles requires policy instruments which include subsidies and carbon pricing, and recycled content standards for implementation. The economic models show that small policy rewards will generate major improvements in internal rate of return (IRR) and payback period for recycling projects which fund both advanced recycling technology and other recycling initiatives.

10.3 Comparative TEA and LCA of Major Recycling Technologies

Scientists now investigate various recycling approaches because polymer and rubber recycling technology have experienced rapid development. These range from traditional mechanical reprocessing to advanced chemical and physical processes. The assessment of their potential requires two fundamental evaluation methods which consist of techno-economic analysis (TEA) and life cycle assessment (LCA) to determine financial performance and environmental effects. The following section evaluates the three main recycling methods through TEA and LCA frameworks to determine their relative advantages.

The field of recycling technology includes two main categories which are:

- The process of mechanical recycling and reusing include main steps which start with sorting followed by cleaning and shredding before the final step of remelting polymer streams.

- The process of chemical recycling uses pyrolysis, gasification, solvolysis, and depolymerization to transform polymers into their original monomers and other valuable chemicals. The process of rubber recycling/devulcanization handles crosslinked elastomers through devulcanization and reclaiming methods which include mechanical, chemical, microwave irradiation, and other approaches.

The economic and environmental results of each pathway differ because they need different amounts of energy and have varying CAPEX, OPEX, product worth, and feedstock needs.

TEA determines the economic feasibility of technologies through the calculation of CAPEX, OPEX, product revenue, and profitability metrics. The TEA standard metrics include net present value (NPV), IRR, and payback period which analysts use to assess projects during their designated period.

Mechanical recycling is the economic recycling method which has achieved the highest development level uses mechanical recycling because it operates with basic production steps and requires minimal initial investment. The main expenses in this process stem from collection activities and subsequent steps which include sorting, washing, and palletization. The economic viability of mechanical recycling for high purity streams becomes possible through PET bottle recycling because the process benefits from both plentiful feedstock availability and current distribution networks (Xu et al., 2025). The process of sorting and preprocessing becomes more expensive when dealing with contaminated and mixed polymer waste streams, which decreases the overall profit margin.

The CAPEX for chemical recycling through pyrolysis and solvolysis operations increase because these processes need sophisticated reactors and catalysts and substantial amounts of energy. The process requires additional thermal energy for operation, which together with purification procedures increases operational expenses. The process of chemical recycling enables the treatment of contaminated waste streams which mechanical recycling methods cannot handle because it produces valuable products such as monomers and fuels at higher price points per ton. The research by Liu et al. (2024) shows that chemical recycling operations will become financially viable when several conditions are met which include elevated oil market prices and organizations must use recycled materials and operate their petrochemical facilities at maximum capacity. The hybrid system, which starts with mechanical presorting before pyrolysis of leftover materials, allows better management of capital costs and material choices which leads to improved financial performance.

The recycling process of crosslinked rubbers which is a subsection of chemical recycling needs special methods to break down their network structure when working with end-of-life tires (ELTs) and rubber made articles. The process of simple mechanical size reduction which produces crumb rubber requires minimal capital investment yet generates products of limited value. The three devulcanization methods chemical, microwave irradiation, and thermo-mechanical raise operational expenses, but they enable polymer chain movement which produces materials that have better potential for reuse (Ostad Movahed et al., 2016). The TEA comparison shows that devulcanized rubber costs more in the market, but business success depends on company size and electricity costs and how customers like the finished products.

The environmental performance of LCA evaluates all stages from product origin to its complete life cycle by analyzing energy consumption, GHG emissions, water usage, and multiple other environmental effects. The standard LCA methodology follows ISO 14040/44 guidelines (Finkbeiner et al., 2011) which use functional units such as tons of recycled polymer to evaluate different pathways.

Mechanical recycling produces the smallest environmental effects because it requires minimal processing and uses less energy than producing new materials from unprocessed materials. The production of rPET from bottle streams results in a 70–80% decrease of GHG emissions because it removes all fresh manufacturing emissions which occur when using new virgin PET (Chen and Hu, 2024). The amount of benefits which can be achieved depends on three main factors which include sorting efficiency, transport distances, and energy mix.

The chemical recycling method enables researchers to extract polymers from contaminated waste streams which contain multiple impurities that mechanical recycling methods cannot handle. The process needs more thermal energy, which results in higher direct energy usage and increased carbon emissions. The process of pyrolysis produces major air pollution when operators use fossil fuels as their power source. The environmental advantages of chemical recycling emerge when facilities run on renewable power and when they substitute fossil-based materials which petrochemical production currently depends on (Xu et al., 2025). The results from LCA studies demonstrate that products such as syngas and pyrolysis oil can reduce environmental impacts when they receive proper marketing strategies.

Rubber recycling LCA varies with process type. The energy requirements for basic mechanical grinding operations stay below devulcanization levels, but the resulting materials fail to achieve required functional characteristics. The devulcanization process needs electrical power together with chemical substances to function but it enables businesses to swap their rubber inventory with fresh materials while they reduce their dependence on landfills which produces major GHG emission. Quantitative LCAs demonstrate that devulcanized rubber applications in asphalt or molded products result in better life cycle energy performance than primary rubber when feedstock processing reaches optimum.

The combined analysis of TEA and LCA shows that economic sustainability creates conflicts with environmental sustainability because of their opposing requirements. The environmental benefits of mechanical recycling are high for pure waste streams, yet the economic benefits are average, but the process fails to handle different types of waste. The process requires minimal initial investment and OPEX because it produces quick returns when businesses use stable feed materials, and they have ready markets for their recycled products. The implementation of chemical recycling technology enables businesses to use various waste materials for production while creating new market opportunities, but they need to handle its expensive construction costs and power consumption requirements. The system achieves better environmental performance through two factors which include using renewable energy and when organizations discover worth in their secondary products. The devulcanization process for rubber creates products with greater value than crumb rubber but requires additional processing expenses. The results of LCA become more favorable when devulcanized products replace virgin materials in markets which produce high value. The implementation of policy instruments which include recycled content mandates, landfill taxes, EPR, and carbon pricing across all pathways will improve both TEA and LCA results through market stabilization and external cost internalization.

10.4 Life Cycle and Techno-economic Evaluations of Separation Methods

Modern mechanical recycling systems depend on efficient sustainable plastics separation technologies which serve as their fundamental structure. Research studies identify separation quality as the primary factor which determines both environmental and economic outcomes in plastic recycling systems because it determines the quality of melt purity, polymer properties, and virgin-material displacement potential. Research studies about LCA-based mixed-plastic recycling show that separation systems reach 70% of their total GHG savings through their ability to produce streams of pure recyclable materials (Davidson et al., 2021).

The ISO 14040/44 frameworks serve as the basis for plastic separation of LCAs which use the functional unit as their basis. The process involves sorting one ton of postconsumer plastic waste into separate materials which come from different sources. The system boundaries in typical cases consist of collection and pretreatment, washing and shredding, separation process, and recovered plastic. The assessment of impact categories includes main categories such as global warming potential (GWP100), cumulative energy demand (CED), freshwater consumption, photochemical oxidant formation, and human toxicity and ecotoxicity.

Research studies using LCA methods demonstrate that different separation methods produce distinct results regarding energy consumption and emission output particularly when dissolution methods are compared to dry separation methods.

The research shows that water-based density separation systems need minimal energy to operate while producing small amounts of GHGs according to various LCA studies. The reported values for this process show energy consumption between 2 and 5 kWh/t, GWP between 0.15 and 0.25 tCO_2-eq/t, and water consumption between 3 and 6 m^3/t. The method of density separation maintains its appeal because it needs no additional substances and it introduces only small amounts of chemicals, yet it fails to distinguish between polymers that have similar density values.

The flotation studies show that flotation provides environmental benefits because microwave pretreatment decreases the need for chemicals as proven through experimental results (Pakbar et al., 2023). The energy consumption and GHG emissions of this technique fall within specific ranges which multiple LCA sources have documented. The energy consumption ranges between 5 and 12 kWh/t, and the GWP emissions range from 0.20 to 0.35 tCO_2-eq/t. The process of flotation provides an excellent solution because it enables the separation of polymers which have equal density. The dry triboelectric separation process operates without water consumption while generating moderate environmental impacts. It needs high energy, 10–25 kWh/t, for plastics separation process. The system produces between 0.15 and 0.30 tCO_2-eq/t of product each year. The operation of triboelectric systems depends on humidity levels and subsequently, it does not work well with weathered plastic materials. The dissolution–reprecipitation method for producing high-purity (>99.9%) recyclate materials requires significant environmental expenses. The GWP ranges from 0.8 to 1.6 tCO_2-eq/t, and the required energy ranges to 300–800 kWh/t. The method exists for use with expensive engineering plastic materials. Hyperspectral near-infrared (NIR) sorting needs 8–20 kWh/t energy, and its GWP is 0.10–0.20 tCO_2-eq/t. It has strong environmental performance, though NIR cannot detect black plastics.

Projected CAPEX figures for various technologies (annual capacity perspective) for density separation, flotation, triboelectric, NIR HIS, and dissolution are estimated at 200–400, 300–600, 700–1,200, 1,500–3,000, and 5,000–8,000 USD per ton annually, respectively. Furthermore, the OPEX for flotation, triboelectric separation, density separation, and dissolution are anticipated to be low–moderate, moderate, low, and very high, respectively.

In summary, the extent of separation of purity determines the alignment between environmental targets and economic performance. The combination of flotation with microwave irradiation represents the best solution to achieve environmental sustainability and economic efficiency. The triboelectric system demonstrates excellent capabilities for functioning in regions which experience dry weather conditions and have restricted access to water supply and access to cheap electricity. The method of dissolution produces the cleanest results, but its operation needs major financial resources and produces major GHG emissions. The high throughput of NIR systems requires users to implement these systems with supplementary methods for complete separation.

10.5 Policy, Incentives, and Circular Economy Models

The circular economy serves as a fundamental policy shift which creates an alternative production framework to replace the traditional linear "take–make–dispose" system because it solves environmental, economic, and social problems which occur during polymer manufacturing and waste management (Kirk et al., 2024). The worldwide plastic and rubber material consumption needs policy frameworks together with economic incentives to establish recycling performance and technology implementation and secondary polymer market stability. This section investigates how polymer recycling systems worldwide have developed through their policy instruments, incentive systems, and circular economy models.

The core design of circular economy systems for polymers exists because of official government regulations. The EU operates at the supranational level to establish a detailed regulatory system through its Circular Economy Action Plan (CEAP) and EU Plastics Strategy which contains mandatory recycling targets, recycled material requirements, and waste minimization standards. The policies require plastic packaging to become more recyclable while banning single-use plastics and establishing base levels of recycled materials for particular polymer products which drive market demand for recycled polymers. The EPR scheme functions as the primary framework which directs businesses to handle their polymer waste. The EPR system requires producers to manage waste from consumer use through financial or physical methods, which motivates them to develop sustainable products that use fewer materials and have improved recyclability. Research data shows that EPR systems which implement correct design principles will obtain better plastic and rubber waste collection results through recycling operations when their payment systems base their rates on material recyclability and actual recycled material content. The EU maintains circularity-oriented governance through its national regulations which function independently from other EU member states. The United States has created recycled content standards while the United Kingdom has introduced plastic taxes and Japan has developed packaging regulations to support circular economy principles. The current system faces main obstacles which prevent standardized polymer recycling methods from becoming worldwide accepted. The structural cost difference between recycled polymers and virgin materials requires economic incentives to function as an alternative to regulatory systems. The financial appeal of recycling investments receives support through several main tools which include subsidies, tax credits, reduced landfill fees, and green public procurement policies. The European countries reached their goals through landfill and incineration taxes which redirected waste streams toward recycling and recovery facilities that focus on treating polymer-containing waste materials. The market-based instrument of tradable recycling credits and recycled content certificates serves as a proposed flexible system to boost secondary polymer market demand. These instruments enable organizations to meet their regulatory requirements through cost-effective system implementations which stimulate developers to create new sorting, separation, and upgrading solutions. The economic assessments based on life cycle analysis show that proper incentive systems together with strong monitoring and verification systems will effectively decrease both GHG emissions and resource consumption. The market competition for recycled polymers remains at risk because oil price volatility persists in the market. The decrease in virgin polymer prices leads to decreased recycling profit margins which proves that permanent policy indicators must exist to maintain stability. The use of carbon pricing mechanisms through emissions trading systems and carbon taxes has become more popular because these systems create effective indirect motivations for polymer recycling. The polymers industry implements circular economy frameworks (models) which go past recycling to include product service systems and practices that involve redesigning, reuse, and remanufacturing. The most efficient plastic recycling technique uses closed-loop recycling to transform recycled polymers into matching products, but it requires both clean materials and advanced separation systems. The process of open-loop recycling happens often but it leads to downcycling which reduces the total material lifespan. The development of circular models for rubber products including tires and technical elastomers now uses devulcanization, reclaiming, material recovery techniques, and energy recovery systems. The policy framework supports waste rubber conversion into construction materials, asphalt modifications, and molded goods products which demonstrates industrial symbiosis achieves its two objectives of maximizing material usage and minimizing environmental damage. The present circular economy systems need complete system integration which unites mechanical and chemical recycling with bio-based alternatives through local material management networks. The hybrid models recognize that various polymer waste streams require distinct processing methods since no universal solution can manage their complete waste range. The policy demands organizational operational changes because it drives technological advancement through its requirement for solution development based on waste stream characteristics.

The process of polymer circularity faces various barriers which stop it from reaching full implementation. The lack of standardized regulations together with insufficient quality standard alignment for recycled polymers and inadequate material tracking systems creates obstacles for international trade and industrial adoption. Kalauni et al. (2025) support particular policy solutions which need to handle social acceptance and consumer behavior and unregulated waste management systems that exist in developing nations because these systems differ from those found in industrialized countries. The present policy strategy depends on data-based governance systems which connect digital product passports to life cycle performance metrics to create both transparency and accountability. The implementation of artificial intelligence (AI) together with blockchain technology and smart sorting systems within policy structures will create better systems for tracking and following rules and improving operational performance in polymer production networks. The path to sustainable polymer circularity needs all regulatory frameworks and economic rewards and circular economy systems to work together at three different levels of governance which include local, national, and global.

10.6 Economic and Environmental Assessment of Rubber Recycling

The recycling of rubber materials, which includes ELTs and industrial rubber waste, serves as a vital environmental protection measure that generates economic benefits through circular material systems. The disposal of rubber waste in landfills together with uncontrolled stockpiling of waste leads to longer decomposition times and possible chemical pollution and dangerous fires which generate GHG. The environmental impact of primary rubber production becomes significantly lower through recycling methods which include mechanical grinding, devulcanization, and thermochemical conversion (Ostad Movahed et al., 2015, 2016).

LCA studies show that mechanical recycling methods, which include ambient or cryogenic grinding to create crumb rubber, produce the lowest GWP for each ton of processed rubber. The process achieves this benefit because it needs less energy for operation while skipping the production of new virgin elastomer materials. The devulcanization methods which require more energy allow partial restoration of elastomeric properties, thus producing greater substitution credits when reclaimed rubber replaces virgin rubber in expensive applications including tire compounds, seals, and technical rubber products. The economic viability of rubber recycling depends on four main factors which include the amount of available feedstock, the size of the recycling operation, the cost of energy, and the market worth of recycled products. The process of mechanical grinding operates as a cost-effective method because it requires small starting costs and operates at budget-friendly rates which maintain their affordability. The process of chemical and microwave devulcanization needs substantial initial costs and precise operational management to produce recyclates which show better long-term value potential. The revenue potential of pyrolysis-based systems extends to multiple products including carbon black and oils and steel but their financial success depends on market oil prices and government support programs and industry-wide product quality standards.

The economic performance of rubber recycling becomes more competitive through three policy mechanisms which include EPR, landfill bans, and recycled content mandates that make environmental costs part of business operations. The economic evaluation of environmental benefits from CO_2 emission reduction and resource conservation and toxicity reduction shows that recycling methods generate better results than waste disposal methods.

10.7 Future Research Directions and Emerging Technologies

The future of both polymer and rubber recycling will be shaped by policy innovation, combined with digitalization and process intensification principles from materials science. Current research has concentrated on selective devulcanization where crosslinks formed by sulfur are broken while the polymer material's structure is preserved. Studies at the laboratory and pilot stage have found that new chemical compounds and supercritical fluids as well as the combination of microwave and chemical processing achieve results in a selective manner using efficient energy levels (Ostad Movahed et al., 2015).

Currently, there is growing interest in the market for the recycling of rubber waste through upcycling. Scientists are developing new materials from recycled rubbers, and these materials are evaluated on their performance. The modification of the surface of crumb rubber, and the treatment of reclaimed rubber and the production of composite filler systems enables recycled rubber to meet rigorous mechanical standards and long-lasting requirements which are needed by the automotive and construction industries. Further investigation into reactive compatibilization and interphase engineering is required by the scientific community to increase their applications.

At present, a large amount of research is being carried out into methods of improving the technology behind thermochemical processing. Effort is being concentrated on the development of low-emission gasifiers which can continually produce carbon black of high quality. The production of tires containing rCB demands that the material has standardized properties, along with an approved certification system. Currently, the recycling of blends which contain elastomers is difficult with mechanical recycling methods. Consequently, solvent-based and catalytic depolymerization processes are being researched.

At the systems level, researchers use TEA alongside LCA and circularity metrics to decide which new technologies should be developed. Organizations can speed up market entry by combining digital twins and real-time monitoring systems with these tools. This aids the organization in overcoming scale-up difficulties.

10.8 Microplastics Challenges

The scientific community defines microplastics as plastic fragments which measure less than 5 mm in size while they come from two sources: human-made small plastic materials (primary microplastics) and the breakdown of bigger plastic objects throughout their life cycle (secondary microplastics (Du et al., 2021). The two main categories of primary microplastics consist of cosmetic microbeads and industrial abrasives, and preproduction polymer pellets known as nurdles. The breakdown of large plastic items through weathering, mechanical wear, UV exposure, and chemical deterioration produces secondary microplastics The classification of microplastics extends into three categories which include their morphological forms (fibers, fragments, films, foams, and pellets), their polymer composition (PE, PP, polystyrene, polyvinyl chloride, PET, and elastomers), and their source origins (terrestrial, freshwater, and marine). The urban environment contains tire and road wear particles (TRWPs) which are elastomer-rich microplastics that result from tire wear and represent a significant yet frequently ignored pollution source. The recycling of polymers faces challenges because TRWPs contain multiple substances which make up their complex structure of rubber, fillers, metals, and adsorbed contaminants. The environment releases microplastics which start their complex journey through different environmental systems that include air, soil, freshwater, and marine ecosystems. The environmental behavior of these materials depends on their polymer density, their particle size, surface chemical properties, and their tendency to develop biofilms. The initial floating pattern of low-density polymers including PET and PP emerges before higher-density polymers and rubber-based microplastics begin their journey to settle in sediment layers. The primary route which leads to microplastic pollution now involves atmospheric transportation that enables these particles to move across extensive distances before they arrive at remote locations including polar territories and mountainous territories. The main soil sinks in terrestrial systems exist in agricultural lands because farmers use sewage sludge and compost and plastic mulching films which create problems with soil fertility and water retention and crop nutrient absorption. The continuous presence of microplastics in soils violates circular economy targets because it produces extensive plastic storage that cannot be recovered.

The physical properties and chemical composition of microplastics create two distinct threats which damage organisms in the environment. The physical process of ingestion results in tissue damage, feeding efficiency reduction, reproductive problems, inflammation, and tissue damage in both water-dwelling and land-based organisms. The chemical composition of microplastics enables them to transport plastic additives including plasticizers, stabilizers, flame retardants, environmental pollutants like PAHs, heavy metals, and persistent organic pollutants.

Scientists currently study microplastic effects on biological systems through cellular and molecular level research of their biological interactions. The breakdown of microplastics into nanoplastics through fragmentation allows these particles to pass through biological membranes, which leads to oxidative stress, inflammation, and potential genetic harm. Scientists continue to study the direct relationship between microplastics and human diseases, but research shows that these plastics exist in drinking water and food items and human body tissues which has led to increased government oversight and public concern.

The main challenge of present polymer recycling operations stems from microplastics which represent their primary problem. The mechanical recycling methods which handle large materials fail in extracting microplastic particles that exist as dispersed particles. The chemical recycling methods demonstrate ability to process contaminated plastic mixtures, yet their operations need substantial energy for pretreatment and their current economic model fails to support the treatment of weak microplastic solutions. The environmental impact of rubber recycling becomes more severe because of the small rubber particles which form during grinding operations, devulcanization processes, and tire wear. The production of fine rubber powders during ELT processing requires sophisticated systems which handle containment, filtration, and monitoring to prevent environmental contamination from secondary sources. The recycling system needs to include microplastic mitigation as an essential design element because it should not be handled as a separate environmental problem.

The technical challenges of microplastic detection and measurement stem from their tiny size and their various materials and their presence in complex environmental conditions. The research employed FTIR and Raman spectroscopy together with GC–MS and thermal extraction methods for analysis. The methods require users to make decisions about which parameters to prioritize between their ability to detect substances and their processing speed and their capacity to identify specific chemicals.

Standardization stands as the primary obstacle which needs to be addressed. The process of studying different populations through sampling methods along with size measurement criteria and reporting standards creates obstacles for researchers who want to analyze data between studies and policymakers who need to create policies. The absence of standardized microplastic emission databases for recycling-based LCAs creates environmental impact assessment challenges because it produces unpredictable results in toxicity assessments for ecosystems and human health evaluations.

The policy framework now deals with microplastics as a new type of pollution which spreads throughout the environment instead of coming from specific sources. The current regulatory framework focuses on preventing waste at its source through material transformation and plastic elimination and product longevity enhancement and biodegradable polymer implementation for specific applications. Multiple countries have established laws which prohibit the use of primary microplastics while their authorities work to create regulations which will control tire wear and textile fiber release.

The polymer recycling industry needs to develop three essential strategies which include design-for-recyclability, closed-loop systems, and advanced separation technologies that will reduce microplastic formation. The development of new solutions involves three approaches which include encapsulation technologies, functional coatings that minimize abrasion, and hybrid mechanical–chemical recycling methods which decrease the amounts of fine particles in the process. The solution to microplastics requires a complete system of integration which combines material science with recycling engineering, environmental assessment, and policy instruments. The lack of integration between microplastics and polymer circularity systems prevents us from reaching sustainable circularity in polymer systems.

10.9 The Role of AI in Polymer Recycling and Waste Management

The recycling process of polymers and waste management has received a revolutionary boost from AI which solves multiple technical and economic and environmental challenges that have persisted throughout the recycling process. The current conventional recycling methods experience multiple problems which include poor sorting accuracy, mixed waste materials, expensive maintenance requirements, and restricted production flexibility. AI tools which include machine learning (ML), deep learning, computer vision, and optimization algorithms use data to create solutions which improve decision processes, system operations, process management, and sustainability at the system level.

The most sophisticated AI application for polymer recycling uses machines to detect waste which enables machines to sort materials into their correct categories (Längle and Maier, 2025). The system operates through NIR, hyperspectral imaging, X-ray fluorescence, and RGB cameras which produce high-dimensional data streams that convolutional neural networks and support vector machines process. The systems enable users to identify different polymer materials including PET, HDPE, PVC, and PP, and they can detect multiple layers and detect foreign substances like fillers and pigments and additives with higher than 95% precision when operating at peak performance. The deployment of AI-enhanced sorting systems produces better material purity, which determines the quality of recyclates and their eventual market worth. ML models now optimize reagent quantities and airflow parameters and contact duration for flotation-based and density-based separation processes which eliminate the need for experimental testing and decrease chemical usage. The system uses reinforcement learning algorithms to perform real-time adaptive control of sorting lines which handle feedstock variations that rule-based systems cannot manage.

AI serves as a vital component which enhances the sorting process while simultaneously optimizing all three recycling methods which include mechanical, chemical, and thermal recycling technologies. The implementation of supervised ML models enabled successful predictions of three vital process outputs which are melt flow index and tensile strength and degradation degree of recycled polymers by studying processing temperature and shear history and contamination level data.

AI systems help chemical recycling operations through pyrolysis, gasification, and solvolysis by modeling their complex nonlinear reaction systems. The use of data-driven surrogate models helps decrease the computational requirements of standard kinetic simulations, which allows for fast techno-economic optimization processes. These models enable reactor design and catalyst selection and energy integration strategy development which results in better production outcomes and decreased energy requirements and lower environmental impact through reduced GHG emissions.

The system achieves its highest value through its predictive maintenance operations. AI systems which analyze vibration signals and thermal profiles and operational logs enable predictive maintenance of shredders and extruders and reactors which results in reduced equipment downtime and longer operational periods. The implemented methods enhance recycling plant economic performance while minimizing their environmental effects throughout their operational period.

The sustainability assessment of recycling pathways depends on LCA and TEA, but their results depend on the quality of available data and measurement accuracy. AI-based metamodels function as computer software which performs LCA evaluations at increased speeds, while it generates uncertainty and shows vital environmental impact factors in complex recycling systems.

The optimization algorithms for multiobjective problems enable users to optimize four different objectives, which include GWP reduction, energy usage minimization, cost reduction, material recycling, and circularity enhancement. These tools enable policymakers together with industry stakeholders to evaluate different recycling approaches (mechanical versus chemical recycling) based on local restrictions which helps them make circular economy compliant choices.

The combination of AI systems with digital twin technology enables developers to create new waste management infrastructure solutions through their creative methods. Digital twins function as virtual duplicates of physical recycling systems which use real-time sensor information and AI models to create simulated models for system monitoring and performance optimization. The systems enable users to conduct scenario analysis and perform rapid system troubleshooting while they continue operating the system.

AI systems at both municipal and national levels operate smart waste management systems through route optimization, waste prediction, and enhanced logistics operations. The applications help organizations decrease their fuel usage while lowering their operational expenses and waste transportation emissions, which results in better waste collection services and higher material recycling achievements.

The complete realization of AI technology for polymer recycling faces multiple barriers which block its maximum potential from being reached. The availability of data together with its quality stands as major obstacles which block the development of new recycling methods and make it difficult to process different types of waste materials. Model transparency and explainability need to exist for industrial acceptance because they serve as fundamental requirements in safety-critical and regulatory environments. The sustainability assessment needs to stop burden transfer by analyzing both the energy usage and material needs which big AI systems produce.

The deployment of digital infrastructure requires policymakers to create protective systems which defend workers from employment loss while ensuring digital resources reach every community.

AI functions as a system-level enabler that enables polymer recycling and waste management systems to become more efficient while making them more resilient and sustainable. The combination of AI with advanced materials science and process engineering and supportive policy frameworks enables the fast development of high-quality closed-loop recycling systems which will establish a complete circular polymer economy.

10.10 Concluding Remarks Toward Sustainable Polymer Circularity

The path to sustainable polymer circularity needs complete unification between technological solutions, economic factors, governmental regulations, and community participation. The complete environmental impact of polymers requires recycling to work together with material reduction, eco-design, and responsible consumption practices. The recycling process has become more efficient because of technological progress in mechanical and chemical and digital systems, but the recycling industry faces ongoing challenges from market instability, conflicting rules, and deteriorating product quality.

The future circular polymer systems depend on high-quality recycling operations which will succeed through established standards, clear life cycle performance indicators, and partnerships between different industrial sectors. The rubber recycling process shows how circularity encounters barriers which generate innovative solutions because technological advancement needs financial incentives together with suitable governmental oversight. The combination of AI technology with advanced materials engineering and complete sustainability evaluation methods will convert polymer recycling from waste disposal into a core industrial component which enables low-carbon industrial systems to function.

References

Castillo, R., Alva, A., París-Viviana, O., and Bosch, M. A systematic bibliometric review analysis of research on the use of waste rubber tyres in building and construction materials and their applications. *Polymers* 2025, *17*(18), 2480.

Chen, S. and Hu, Y. H. Advancements and future directions in waste plastics recycling: from mechanical methods to innovative chemical processes. *Chem. Eng. J.* 2024, *493*, 152727.

Davidson, M. G., Furlong, R. A., and McManus, M. C. Developments in the life cycle assessment of chemical recycling of plastic waste—A review. *J. Clean. Prod.* 2021, *293*, 126163.

Du, S., Zhu, R., Cai, Y., Xu, N., Yap, P. S., Zhang, Y., He, Y., and Zhang, Y. Environmental fate and impacts of microplastics in aquatic ecosystems: a review. *RSC Adv.* 2021, *11*(26), 15762–15784.

Finkbeiner, M., Tan, R., and Reginald, M. *Life Cycle Assessment (ISO 14040/44) as Basis for Environmental Declarations and Carbon Footprint of Products.* In ISO Technical Committee 207 Workshop. Norway, 2011.

Geyer, R., Jambeck, J. R., and Law, K. L. Production, use, and fate of all plastics ever made. *Sci. Adv.* 2017, *3*(7), e1700782.

Kalauni, K., Vedrtnam, A., Sharma, S. P., Sharma, A., and Chaturvedi, S. A comprehensive review of recycling and reusing methods for plastic waste focusing Indian scenario. *Waste Manag. Res.* 2025, *43*(9), 1378–1399.

Kirk, E. A., Popattanachai, N., Barnes, R. A., and van der Marel, E. R., Eds. *Research Handbook on Plastics Regulation: Law, Policy and the Environment*. Edward Elgar Publishing: Cheltenham, UK, 2024.

Längle, T. and Maier, G. *Sensor-Based Sorting for Waste Plastics Recycling*. In Springer Handbook of Circular Plastics Economy; Springer Nature Switzerland: Cham, Switzerland, 2025; pp. 375–390.

Liu, Q., Martinez-Villarreal, S., Wang, S., Tien, N. N. T., Kammoun, M., De Rcover, Q., Len, C., and Richel, A. The role of plastic chemical recycling processes in a circular economy context. *Chem. Eng. J.* 2024, *498*, 155227.

Ostad Movahed, S., Ansarifar, A., and Estaji, S. Review of the reclaiming of rubber waste and recent work on the recycling of ethylene–propylene–diene rubber waste. *Rubber Chem. Technol.* 2016, *89*(3), 54–78.

Ostad Movahed, S., Ansarifar, A., Karbalaee, S., and Atharifar, S. Devulcanization and recycling of waste automotive EPDM rubber powder using shearing action and chemical additive. *Prog. Rubber Plast. Recycl. Technol.* 2015, *31*, 87–116.

Pakbar, H., Ostad Movahed, S., and Jourabchi, S. Effect of pre-microwave irradiation on the flotation of polystyrene, polyethylene terephthalate, and polyvinyl chloride using traditional depressants. *Prog. Rubber Plast. Recycl. Technol.* 2023, *39*(4), 307–324.

Plastics Europe. *Plastics—The Facts 2024: An Analysis of European Plastics Production, Demand, and Waste Data*: Plastics Europe: Brussels, Belgium, 2024.

Wang, Lin, Peter Rem, Francesco Di Maio, Max van Beek, and Gonçalo Tomás. An innovative magnetic density separation process for enhanced recycling of plastics and electronic waste. *Recycling*. 2024, *9*(3), 48.

Xu, Z., Sanchez-Rivera, K., Granger, C., Zhou, P., Munguia-Lopez, A. C., Ikegwu, U. M., and Avraamidou, S., et al. Solvent-based plastic recycling technologies. *Nat. Chem. Eng.* 2025, *2*(7), 407–423.

Index

Note: Page numbers followed by *italics* and **bold** refers to figures and tables respectively.

S

Printed and bound by CPI Group (UK) Ltd, Croydon, CR0 4YY

24/08/2026

14947830-0003